Raghavendra Rao Althar, Debabrata Samanta, Debanjan Konar, Siddhartha Bhattacharyya
Statistical Modelling of Software Source Code

Also of Interest

Electrochemistry.
A Guide for Newcomers
Helmut Baumgärtel, 2019
ISBN 978-3-11-044340-0, e-ISBN (PDF) 978-3-11-043739-3,
e-ISBN (EPUB) 978-3-11-043554-2

Electrochemical Energy Storage.
Physics and Chemistry of Batteries
Reinhart Job, 2020
ISBN 978-3-11-048437-3, e-ISBN (PDF) 978-3-11-048442-7,
e-ISBN (EPUB) 978-3-11-048454-0

Wearable Energy Storage Devices
Allibai Mohanan Vinu Mohan, 2021
ISBN 978-1-5015-2127-0, e-ISBN (PDF) 978-1-5015-2128-7,
e-ISBN (EPUB) 978-1-5015-1492-0

Raghavendra Rao Althar, Debabrata Samanta,
Debanjan Konar, Siddhartha Bhattacharyya

Statistical Modelling of Software Source Code

—

DE GRUYTER

Authors
Raghavendra Rao Althar
Specialist-QMS
First American India Private Ltd
Bangalore, India
RAlthar@firstam.com

Dr. Debabrata Samanta
CHRIST (Deemed to be University)
Department of Computer Science
Hosur Road
Bhavani Nagar 566003, India
debabrata.samanta369@gmail.com

Dr. Debanjan Konar
Sikkim Manipal Institute of Technology
Department of Computer Science and
Engineering
East Sikkim
Majitar
Sikkim 737136, India
konar.debanjan@gmail.com

Prof. Dr. Siddhartha Bhattacharyya
Principal
Rajnagar Mahavidyala
Birbhum, India
dr.siddhartha.bhattacharyya@gmail.com

ISBN 978-3-11-070330-6
e-ISBN (PDF) 978-3-11-070339-9
e-ISBN (EPUB) 978-3-11-070353-5

Library of Congress Control Number: 2021935544

Bibliographic information published by the Deutsche Nationalbibliothek
The Deutsche Nationalbibliothek lists this publication in the Deutsche Nationalbibliografie;
detailed bibliographic data are available on the Internet at http://dnb.dnb.de.

© 2021 Walter de Gruyter GmbH, Berlin/Boston
Cover image: Gettyimages/Thinkhubstudio
Typesetting: Integra Software Services Pvt. Ltd.
Printing and binding: CPI books GmbH, Leck

www.degruyter.com

Preface

Challenges that are hampering the smooth progression of software development processes are discussed with prospective solutions. Some of the obstacles like fault prediction across functional and security aspects, traceability across software development life cycle, software, and deployment issues are explored. This book focuses on addressing some of the challenges faced by the software development community. With the objective of *Statistical Modelling of Software Source Code*, the book intends to draw inspirations from some of the work happening in the area of data science. Probabilistic modeling for the software domain is a crucial area discussed in the book. The book explores data science approaches in light of the challenges involved in the field. Putting together all the associated efforts helps to get a bird's eye view of the most essential aspects and for further exploration for software development improvement. They investigate the approaches used in different domains like rainfall prediction which attempts to look at the problem at hand from a different lens and add a new perspective. Software systems being the building blocks of the modern-day economy, focusing on optimizing some of the critical software development issues, is helpful. This focus will put power back in the software development community's hand and focus on the creative side of work. With many problems the community faces, currently, significant time is spent on keeping the lights on. Exploration of this book is to fuel the transformation toward the creative side.

The best features of this book are its multitude of focus on data science capability for software engineering. The range of coverage extends from simple approaches like regression to most sophisticated methods like cognitive computing. This book intends to give space for understanding the history of some aspects of software engineering and data science, which helps to get deeper into challenges. Attempt to explore different domains like rainfall prediction and cognitive computing to look at the problem at hand from an extra dimension is a unique attempt. Across the book, a significant effort is put in toward breaking down the domain from data dimension and challenges associated; helps to go deeper wherever needed and stay at the high-level view in other places. Different areas addressed in the book can provide enough information to explore further and create a value-based system for the industry. Expert inputs have been gathered to choose the problem areas from software engineering; this ensures that focus is retained in an area that matters for the industry. Analysis of all data sources, especially the source code, and other natural language data, is an essential part of the book. This study helps build the ecosystem needed to learn from the processes' pattern: probabilistic programming and intelligent integrated development environment are also discussed and is a significant part of the book. This book also provides information on visual depictions used in many of these explorations.

The evolution of the software development processes is dealt with in Chapter 1. Different approaches to software development are explored. Data science's growth is also touched upon to assess the applicability of some of the relevant strategies for

https://doi.org/10.1515/9783110703399-202

solving software development issues. Further, the chapter jumps into the key challenges bothering software development processes. Prospective areas of data science applications to software development are explored. One of the key inspirations is the naturalness of the software source code, which is discussed and sets the context for applying data science for software development challenges. Beyond the source code, all other data points hidden in the software development ecosystem exploration are done. Probabilistic modeling of software source code is another area of investigation. The chapter also focuses on specific problem areas and current approaches, and further scope of improvement.

Chapter 2 deep dives into probabilistic modeling in case of some of the challenges in software development. As the exploration extends toward understanding software engineering ways to handle some operations, artificial neural network approaches are discussed in line with them. Security vulnerabilities being a prominent concern for software development, the same is considered a focus area to solve. Statistical debugging models and program sampling framework exploration conducted – text sources of data in software development taken as critical data sources – discuss the possibility of automating software vulnerability management are done. Since software vulnerabilities are temporal, some of the time series prediction approaches are explored.

Chapter 3 intends to facilitate traceability across the software development ecosystem to understand the system better. Deep learning techniques focus on the software traceability issues in the chapter. Some of the promising approaches discussed to understand some of the best practices that can be adopted. Methods of deep learning across word embeddings, latent Dirichlet allocation, and many other exciting approaches explored. The topic "modeling" is an exciting approach explored in the context of this chapter. In the exploration of traceability, requirements are the essential constituent of the software development processes and are taken as reference.

In Chapter 4, to solve some of the software development issues, facilitating auto code completion will be a significant step. This chapter assesses the possible approach for the same. The formal structure underlying the software program is an inspiration to figure out auto code completion approaches. Data science's recommender system is one associated format to explore in this context – some of the code suggestion tools surveyed to understand the current capabilities and potential future directions. Convolution neural network base attention models explored for code summarization. The concept neural turing machine has been analyzed and is an interesting one to build the chapter's thought process. The capabilities of natural language processing (NLP) also been studied for program comments processing.

Software deployment processes issues tackled in this Chapter 5. Some of the latest approaches in data science such as transfer learning have been explored. The chapter investigates the evolution of software deployment processes to figure out the latest challenges. A deep learning study has been done with mobile devices as a reference in handling the deployment process efficiently. There is a discussion on automation possibilities for the software. The book's central theme is modeling the software source code; deep transfer learning thought process in this area is explored. Various facets of software deployments have been discussed to enable further focus on resolving the associated issues.

Integrated development environment (IDE) being the central focus of software development, making them smarter would enable software development and add value. The IntelliDE system discussed in the chapter helps to get insights into the possibility of more elegant IDEs. Probabilistic models explored to utilize the data associated with IDE. Knowledge graph is another area of interest in data science to aggregate knowledge, explored in this chapter's context. To use all data sources in the software development ecosystem text data looked at from NLP perspective. Probabilistic programming has been discussed to build the context for probabilistic modeling in software development. Applications of the smart IDEs explored to understand the broad implications of this area. Integration of the knowledge done in some of the work on knowledge graph reviewed in Chapter 6.

In Chapter 7, with many natural language–based data produced during software development processes, NLP takes prominence. Deep learning–based NLP explored here to use patterns hidden in these software artifacts and solve challenges associated with software development. Variants of the deep NLP approaches are studied. The knowledge gathered from the natural language data will complement the knowledge gathered from software source code and help build an efficient and effective software development ecosystem.

Chapter 8 takes the exploration to the next level by exploring possibilities of artificial intelligence in the light of cognitive computing. Human thinking simulation is closely targeted in cognitive computing. It will be an excellent exploration to deepen intelligent computing's thought process and provide the right direction for streamlined software engineering. The cognitive computing stack has been explored to understand its architecture. With all the data science approaches explored in previous chapters to enable software engineering, covering different data sources, this chapter helps build the thought process of producing an integrated system. Cognitive computing will also help to create a framework to integrate multiple approaches into one system.

Chapter 9 throws light on the possibility that basic data science can bring in. As the discussion in previous chapters built the complexity of approaches, this chapter attempts to keep the thought process grounded. To make sure that visibility is not lost on the power of simple things. Software systems being temporal will have to consider time series data and its management. This chapter discussing rainfall prediction can give some common themes for this exploration. Also, having a perspective of how approaches are devised in varied fields can throw light on challenges involved in the target field, like in this book, software development optimization. Some of the strategies focusing on data mining capabilities will be a good takeaway for managing data from varied software development sources.

Chapter 10 focuses on an in-depth understanding of the structure of software programming language. After addressing all the facets of software engineering and data science, this chapter focuses on the core of software engineering and the programming language. Understanding the programming language construct will provide deeper insights into challenges associated with the software development processes and the most relevant approaches that are explored. Machine learning algorithms that are carefully investigated in this chapter for understanding some of the critical constructs to examine in the light of specific challenges. Abstract syntax tree approach has been explored to understand the programming language construct. The neural attention model is an exciting approach that the chapter explores.

Raghavendra Rao
Debabrata Samanta
Debanjan Konar
Siddhartha Bhattacharyya

Mr. Raghavendra Rao Althar would like to dedicate this book to his parents Mr. Heriyanna and Mrs. Gulabi and his soul mate Jyothi and son Ramateja.

Dr. Debabrata Samanta would like to dedicate this book to his parents Mr. Dulal Chandra Samanta and Mrs. Ambujini Samanta, his elder sister Mrs. Tanusree Samanta, who have encouraged him all the way, and to his beloved daughter Ms. Aditri Samanta.

Dr. Debanjan Konar would like to dedicate this book to his parents Mr. Debidas Konar and Mrs. Smritikana Konar, his elder sisters Mrs. Barnali Roy and Mrs.Piyali Roy, who have encouraged him all the way, and to his beloved daughter Ms. Sanvi Konar.

Prof. (Dr.) Siddhartha Bhattacharyya would like to dedicate this book to his parents late Ajit Kumar Bhattacharyya and late Hashi Bhattacharyya, his beloved wife Rashni, and his youngest sister's parents-in-law late Anil Banerjee and late Sandhya Banerjee.

Contents

Chapter 5
Transfer learning and one-shot learning to address software deployment issues —— 145

Chapter 1
Software development processes evolution

1.1 Introduction

Advanced extreme programming concepts are an exploration area, which popped out of the need for software development advancements. Following this, it calls for the community to look for advanced approaches built on traditional methods. Software systems development began in the 1940s, with the issues resulting from software; there was a need to organize the software development processes. This field is dynamic and will be up to date with the latest technology advancements. With efficiency as the focus area, multiple models of software development evolve. The waterfall and Agile model of products are well-known ones. The agile model of growth seems to be catching attention recently. Figure 1.1 shows the depiction of waterfall model of software development.

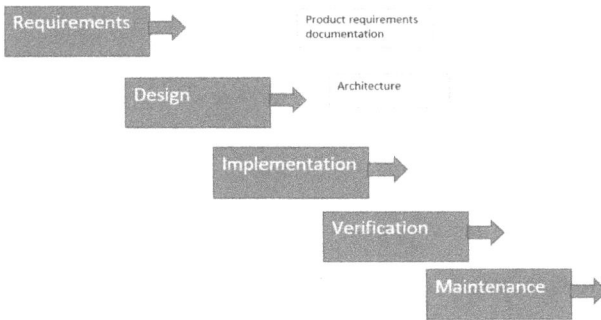

Figure 1.1: Waterfall model of software development.

During 1910, Henry Gantt and Frederick Taylor put together a methodology for effective project management, particularly from handling repetitive tasks. It was a game-changer for the industry to enhance their productivity. Working as a team was another critical factor unearthed by the industry, which is also a backbone of current agile methodology. The waterfall model helped to bring in structure into software development in the late 1980s. Design, development, unit testing, and integration testing were the critical phases involved. Development operations, also called DevOps, was a later advancement that focused on integrating the software modules into production. The waterfall model follows a sequential approach in development with the customer requirements flowing down from a high level to a lower level. The output of one stage is dependent on another location. It helps by having a structured approach, accessible communication with customers, and clarity in delivering the project. But it brings in the challenge of difficulties involved in less flexible structure, and, if there are any

https://doi.org/10.1515/9783110703399-001

issues, they are costly to be fixed. The approach works well if there are unexpected drastic changes in requirements. A V-shaped model develops with the waterfall model's spirit, where the flow bends up after the coding phase. Early testing involved makes it the most reliable approach. A V-shaped model is useful as every stage involves deliverables, with the success rate being high compared to the waterfall model. But V-models do not facilitate change in scope, even though scope changed; it is expensive. Also, the solutions are not exact. Only in case of clearly defined requirements can this model work reasonably even if the technology involved is well understood.

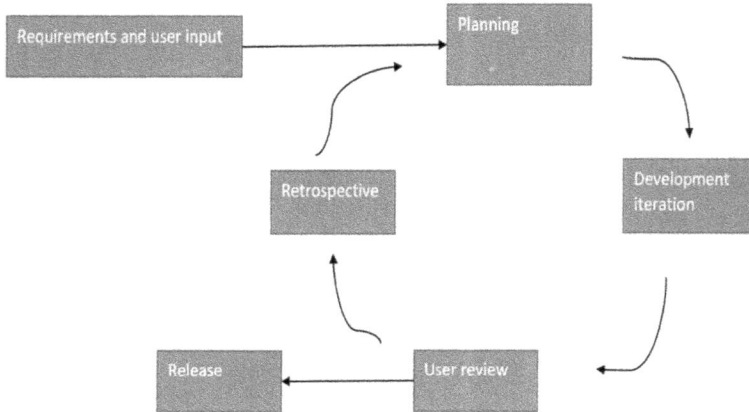

Figure 1.2: Agile model of software development.

The agile model works based on collaboration among all the parties involved. Figure 1.2 shows Agile model of software development. Here the customer is delivered with the expected product on an incremental basis. In this way, the risk involved is minimal. The focus is to create a cumulative effect that customers can visualize and confirm if the development team is in line with their requirements. Throughout the exploration of best methodologies, the focus has been to build efficiency and effectiveness in the software development process. From this point of view, the Agile and DevOps model has been the most preferred one. Scrum, Crystal, and XP (extreme programming) are some versions of the Agile methodology. With a focus on DevOps being to facilitate faster time to market, Agile and DevOps have preferred software development (Wadic 2020). Figure 1.3 depicts V-shaped model of software development.

1.2 Data science evolution

As it is perceived today, data science is mostly influenced by what has evolved after the year 2000. The original form of data science dates to 800 AD, where the Iraq

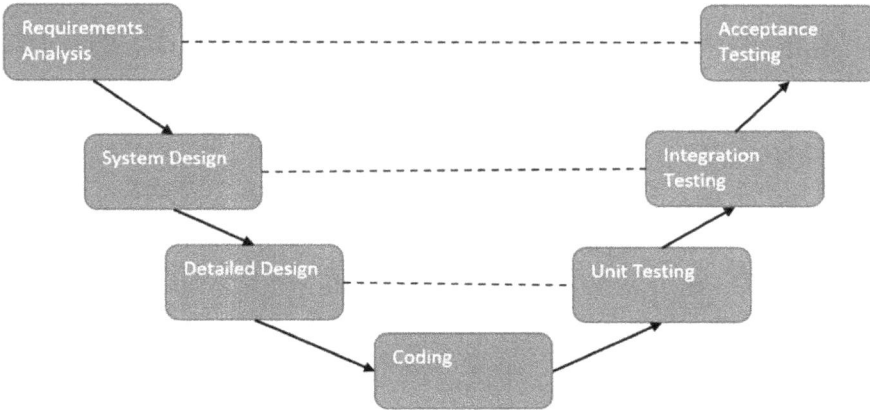

Figure 1.3: V-shaped model.

mathematician Al Kindi used cryptography to break the code. It was one of the initial frequency analysis mechanism that inspired many other thought processes in the area. In the twentieth century, statistics became an important area for the quantification of a variety of community in the society. Though the start of data science was in statistics, it has evolved itself into artificial intelligence, machine learning, and the Internet of Things (IoT). With a better understanding of customer data and an extensive collection of the same, there have been revolutionary ways to derive useful information from the data. As the industries started demonstrating the value created with the data, applications extended to medical and social science fields. Data science practitioners have the upper hand over the traditional statistician regarding the knowledge over software architecture and exposure to various programming languages. Data scientist's involvement is across all the activities of defining the problem, data sources, and data collection and cleaning mechanism. Deriving the deeper insights hidden in the data is the critical area of focus in data science. Operations research was also a lead toward optimizing the problems in the way of mathematical equations. Digital age introduction with supercomputers paved the way for information storing and processing on a large scale. The intelligence associated with computing and mathematical models led the way for advancements in artificial intelligence. A more extraordinary ability to understand the underlying patterns in data is to use it as a base for prediction kept getting better with more data available. There was always a disconnect between the data analysis and software program, which was bridged by the visualizing dashboard solution to make it simple to consume for the core data science team. Business intelligence provided business insight from data without worrying much about how the data transformed into wisdom. Big data was another central area of explosion for handling complex and large data that would be run on extensive computation to understand the underlying data pattern. Data science was mostly about managing this extensive knowledge hidden in the complex data forms. Putting the best of computer science, mathematics, statistics is the primary focus of data science. The massive data flowing in the field helps the decision-makers focus on the critical part of the data landscape for their decision making. Figure 1.4 depicts disciplines of data science.

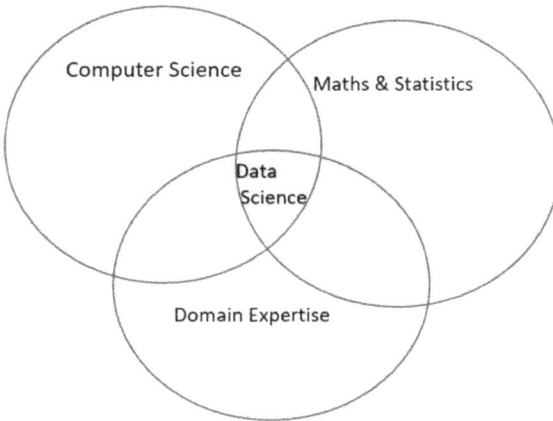

Figure 1.4: Disciplines of data science.

1.3 Areas and applications of data science

Data engineering and data warehousing are some of the areas. Data engineering is the transformation of data into a useful format that will help the critical analysis. Making data usable by an analyst for their required analytics is the objective. Data mining provides an experimental basis for data analysis to provide the required insights. It will help the expert to formulate the statistical problem from the business concerns. Cloud computing is another area that provides a platform across the enterprise for large scale solutions. It takes care of securely connecting with business systems. Database management also finds an important place in the view of extensive data that happens to be part of the ecosystem. Business intelligence improves data accuracy, a dashboard for stakeholders, reporting, and other related activities. Data visualization is another area of focus that strives to convey critical messages in visuals. It also closely associates with the Business intelligence area for providing the required dashboard based on business needs. The data science life cycle also can be viewed as the data discovery phase, data preparation phase, mathematical models, deriving actionable outcomes and communication associated with the process. Data science areas are also spread across machine learning, cluster analysis, deep learning, deep active learning, and cognitive computing.

Drug discovery fields with its complex processes are assisted by the mathematical model that can process how the drugs behave based on biological aspects. It will be the simulation of the experiments conducted in the lab. Virtual assistants have been in the peak of business support, and rapid progress is happening to improve that area's experience. Advance in mobile computing backed up with data helps to take the data knowledge to a large population. Extensive advancements in the search engines are another worth noting area. Digital marketing opened large-scale optimizations like a targeted advertisement, cutting down on the expense of advertising and reducing the possibility of large-scale dissatisfaction that would creep up with large-scale promotions. All this is possible by tracking data of the users based on their online behavior. Recommender systems have a prominent part in the business to effectively utilize the customer data and recommend back most useful things to help the customer experience the best they deserve. Advancements in image recognitions are seen in social collaboration platforms that focus on building social networks to establish a network connection. It further leads to improvements in object detection that have a significant role in various use cases. Speech recognition capabilities are seen in voice support products like Google Voice and others, enhancing customer experience. This capability includes converting voice to text instead of a customer requiring typing the text data. Data science capabilities are leveraged by the airline industry, struggling to cope with the competition. They must balance the spiking up air fuel price, and also provide significant discounts to the customers. Analysis of flight delay, the decision of procurement of air tickets, decision on direct and multi-point flights, and managing other customer experience with the data analytics capability have enhanced the industry performance. Machine learning in the gaming industry had made a significant mark, with the players experiencing the game complexity based on their progression in the earlier level. Developments also include the computer playing against human players, analyzing previous moves, and competing.

Augmented reality seems to be at its exciting point. Data science collaborates with augmented reality with computing knowledge and data being managed by algorithms for a more significant viewing experience. *Pokémon Go* is an excellent example of advancements in this area. But the progress will pick up once the economic viability of the augmented reality is worked out.

1.4 Focus areas in software development

The software development industry focuses on the loss of efficiency due to its unpredictability throughout the value stream of software development. People involved in the processes are not confident about their deliverables at any phase of the life cycle. There is a lot of effort to unearth the code's issues at a much earlier stage of the life cycle to optimize the fixing effort and cost. There is also a lack of focus on an initial analysis of the requirements based on the business needs. Both domain and technical analysis matured enough to demonstrate end-to-end knowledge needed for the customers. Establishing the traceability of the requirements across the life cycle needs to be scientifically based on this. The big picture is missing for the development community, as they do not understand customers' real needs. The focus is more on the structure of code, and the holistic design view is missing. There is a gap in the association of technical details to users' needs in the real world. This concern calls for understanding the domain of the business. management teams are contemplating the need for the system to make development communities experience the customer world. Though business system analysts intended to build this gap, the objective is not satisfied due to a robust system that can facilitate this business knowledge maturity. It points to the need for a meaningful design that makes all the value stream elements of software development, traceable end to end, and intelligent platforms to integrate this knowledge across.

The importance of unit testing as part of building a robust construction phase of the software is not shared with all parties concerned. In the context of the Continuous Integration and Continuous Deployment (CICD) model of software development and integration, quality assurance activities are defocused. Bugs are looked at as a new requirement, resulting in low quality not being realized by team members. On the other end, the CICD model's maturity is not sufficient to ensure good quality to the end users, which points back to people transitioning from traditional software development methods to the CICD model. Also, there is a need to understand the product and software development landscape well. Though there is recognition of the importance of "First Time Right" as a focus, there is a lack of real, practical sense of first time right in the software development value stream, resulting in the suboptimal deliverables in all the phases. Lack of focus in the initial analysis also extends to nonfunctional requirements and leads to security vulnerabilities costing the organization its reputation and revenue.

Lack of knowledge about security vulnerabilities contributes to the issues. Lack of systems to evaluate the best vulnerability assessment tools and framework also contribute. Approaches are adopted to make the system robust against security vulnerabilities by adopting tools that identify security flaws and gated check-in formats. The efficient mechanism to incorporate checks against possible vulnerabilities recognized is lacking. Knowledge is essential to trace down to a specific area, where the openness and defects exist. Examples like tracing back to methods that impacted customers' data will be an essential need. This lack of knowledge also contributes to security vulnerabilities. The multiple parties involved in identifying and fixing the issues create inefficiencies. But all these need a robust framework to make an end-to-end intelligent system, which points to the work happening in intelligent integrated development environments (IDEs). The industry is also looking at the need for robust approaches to adopt new technologies with the least intervention to existing systems. Enabling a new person to become productive quickly and need a smooth transition from one coding language to another are the focus areas that need attention. Companies are also striving to build intelligence to evaluate new business opportunities and leverage current business and technology knowledge to estimate new projects effectively. Making practical usage of the existing data to expand the business is lacking and needs focus. Technical code quality is another area needing direction to optimize the foundational aspects of software development. They are raising the need for the move into the cloud to increase knowledge of the software landscape.

1.5 Objectives

Building a probabilistic model for fault prediction across functional and security related areas of software development based on source code and other artifacts involved is the objective. Debugging is a significant area of the software development life cycle. Debugging also can be extended to build the knowledge repository for future optimization. Security vulnerabilities management is a critical aspect of the software development life cycle due to growing concern about a security threat; this is a priority area among all other optimization efforts taken up in software development. Establishing traceability between software development domain artifacts is another focus area. Advances in this area would assist developers in understanding their code ecosystem better and maintain the same better. It helps developers to manage requirements across the software development value stream effectively. It helps to build an effective collaborative environment among all the players involved. It also contributes to maintaining visibility on customer acceptance criteria for the products delivered. It helps during the validation of the code by the end users. This study area plays a significant role in building the domain knowledge since it connects end to end from customer need to the customer's product. Auto code completion facilitation in software development by a structured prediction-based autocompletion model is another focus area. Automatic code completion is a critical element of a software development process for its efficient operations. It builds the robustness in coding by reducing the possibilities of the error.

Productivity also improves with this feature. This part of the study also helps the effort toward understanding the usage of various components of the product. One of the critical areas of software engineering process optimization is to understand the intent of software engineer. This is explored as one of the focus area. Transfer learning and one-shot learning to address software deployment issues explored. Since there is a large focus on machine learning modeling solutions, the software development community needs to bridge the possible gap between the core Software development ecosystem and the supporting machine learning solutions ecosystem. It aims toward reducing potential complexities involved in maintaining the integrated ecosystems of these entities. With various challenging areas focusing on software development, statistical modeling for software source code tries to derive the knowledge hidden across multiple software artifacts, including code and text, and build a robust and intelligent system. IDE is one area of focus in the domain. Since this provides a platform for entire software development processes, it plays a key role. We then explore natural language processing (NLP) based on deep learning, as this is an exploding area of artificial intelligence. Any efficient approaches from here will provide useful knowledge of software source code modeling. In turn, this will help to facilitate a variety of tasks in software development. We also explore the capabilities of the cognitive computing and how machine learning has influenced this area. It will also provide a strong knowledge to improve the proposed systems discussed in the early part of the book for software source code modeling. To give it a good roundup, we complete the exploration by looking at machine learning and its application in one of the areas closely associated with human life, like rainfall prediction. The objective of picking up a different area from our primary focus area of software development is to get a different perspective and explore all other possible areas that can contribute to our primary objective. For example, there is a good focus on time series data in rainfall prediction.

1.6 Prospective areas for exploration

Statistical debugging modelswould help to facilitate this objective. It was leading toward the facilitation of learning software development artifacts using these statistical debugging models. These models also help build the knowledge repository for the advanced refinement of these debugging models in the future. Software execution traces are a good source of information for this exploration. Possible areas of inquiry include pretrained models and transfer learning for understanding software vulnerabilities. Exploring the ensemble of models with various combinations of machine learning models is a prospective approach. Vulnerabilities injection rate being a time-based factor, time series–based methodology applied. The autoregressive integrated moving average method is explored for time series data. The occurrence and exploitability of the vulnerabilities also play a significant role in the software domain. Probabilistic models are a good fit for the event and exploitability of vulnerabilities.

Establishing traceability between software development domain artifacts like code, specifications, requirements, code fixes, and bug reports is of high importance. These would assist developers in understanding the code ecosystem better and maintain the same better. In this area, since various artifacts are involved, some of the critical approaches of NLP are explored. Since it is about traceability, dependency parsing would be the right approach with co-reference analysis. Traceability benefits inefficient requirements management helps build collaboration between multiple teams involved. Confirmation of source code being reviewed based on established acceptance criteria and many other areas are benefiting. This area also poses the need to efficiently transform the requirements that evolve from the text into fine-grained actionable software artifacts. In other research works, deep learning neural networks are built based on the software development life cycle's text data. The further research scope would be to leverage other source code artifacts for this statistical modeling and study. Auto code completion facilitation by structured prediction-based autocompletion model: Establishing acceptable benchmark, evaluation methodology, and metrics to provide a quantitative comparison, and highlighting strengths and weaknesses of the models built for this objective of auto code completion is an area for research. Generally, these approaches assess what tokens may apply in the current context based on rules configured. These rules are configured based on the syntactic and semantic structure of the data under study. Even with the work done in this area, managing long-range semantic relationships like suggesting a calling of function defined many tokens before in the code is a challenge. One of the research focuses would be to build models that can likely continue a partial permit using character language models for auto code completion. Software deployment optimization using one-shot learning and transfer learning: Deployment is another prominent area of software development that needs attention. There is a need for advanced research in one-shot learning and transfer learning to facilitate some of the critical challenges. This area addresses the problem involved in the concept drift resulting from the drastic changes in the code ecosystem. It results in the need for machine learning models to evolve with these changes, which costs time and money. As frequent training of the machine learning models account for this concept, drift becomes a need and adds complexity.

Enabling intelligent IDEs with probabilistic models enables a computer system to manage noise in the data related to speech and written text, resulting in improving conventional interaction between developer and computer. Providing an ecosystem to aggregate, mine, and analyze software data to offer intelligent assistance in the life cycle of software development. This area will help in building capabilities toward making IDEs smart. All the aforementioned areas can be conceptualized in this architecture. IDEs of IoT are explored, as this is a fast-growing area and would provide good knowledge for this book's primary objective. IDEs are also studied in machine learning to understand how data leveraged to build intelligent IDEs. With various challenging areas focusing on software development, statistical modeling for software source code tries to derive the knowledge hidden across multiple software artifacts, including code and text, and build a robust and intelligent system. IDE is one area of focus in the domain. Since this provides a platform for entire software development processes, it plays a key role.

With various challenging areas focusing on software development, statistical modeling for software source code tries to derive the knowledge hidden across multiple software artifacts, including code and text, and build a robust and intelligent system. Using this knowledge, we can make use of the application of learning in deep learning for NLP effective. Deep learning is topping the list for resolving tasks that are complex like that of text translation. Text data tends to be messy and does not fit into the needs of machine learning algorithms, which expects a fixed set of inputs. The text format to be converted into vectorized format before processing adds another complex layer to the equation. Historically, text data has been dependent on linguistic experts, and the focus area was very narrow. Deep learning has overtaken the statistical methods in this area and providing simple models for the purpose. In exploring the importance of machine learning in cognitive computing, focus areas are to study the same architecture. In exploring the importance of machine learning in cognitive computing, focus areas are, to study the architecture of cognitive computing and challenges that are posed for machine learning. Computing a stack of machine learning for cognitive computing is looked into and is an essential aspect of cognitive computing, and it poses quite a challenge that needs to be studied. Looking into the dimension of supervised and unsupervised learning in the area will be beneficial. Decision making is another core component that must be understood and studied. Supervised classification areas of cognitive radio must be centralized and decentralized learning areas. From the primary time series modeling to deep learning, ensemble methods-based time series analysis will provide needed knowledge. Some of the exciting approaches of self-organized map and support vector machines are prospective areas of exploration. Learnings from this area will provide a solid foundation for software source code statistical modeling that can solve many problems faced by the software development fraternity.

1.7 Related work

Work done in Bennaceur (2016) focuses on learning-based software testing and code synthesis with machine learning. Work also explored, Machine learning technique adoption for automation of software artifacts and software analysis. They combine software analysis frameworks with runtime verification and automated learning to leverage the deployed system's observed behavior. This work also provides an overview of the recent researches around machine learning used in automating key software processes. While the book (Bennaceur 2016) focuses on test case generation area, this book focuses on software development core activities and the possibility of predicting the vulnerabilities much before the testing phase. The book will focus beyond research work review to extend those learnings of research work and explore new applications. Focus is also to optimize software artifacts traceability, code completion automation, and deployments, utilizing the knowledge hidden in all software artifacts in the development ecosystem. Work done in Thouraya et al. (2019) focused on aspects such as reuse and integration of code and other artifacts across the software development life cycle.

Adoption of solution to translate one programming language or a domain to another, is a focus area. Effective representation of software design in contrast to unified modeling language (UML). This book will focus on fault prediction in the software development life cycle. Adoption across programming language and software design are looked at, but approaches related to probabilistic modeling are explored in these areas. Work done in Tim et al. (2016) focused on aspects such as best practices around data mining for software engineering cutting-edge areas of handling and making use of software engineering data are detailed. The focus is to tour all the possible software engineering regions and bring out the data sources' possibilities and not to prescribe a specific approach. This book will focus on particular methods going beyond just discussing the possible data sources. This book explores the key areas that matter for optimizing software development. Work done in Christian et al. (2015) focused on understanding data structure in the software development processes, using various techniques like text analysis and topic analysis. The focus is on studying the data analytics application in industry and open sources. This book will focus on specific software development areas like software artifacts traceability, deployment, and code completion to discover vulnerabilities found early in the life cycle. The focus is on exploring specific approaches revolving around the above areas with probabilistic statistical models' background. Work done in Ron et al. (2018) focused on the following aspects: system and software testing utilizing state-of-the-art tools. Bayesian graphical models were explored for high complexity testing. Automated software testing and test case maturity are key focus areas.

Work done in Frank et al. (2007) focused on aspects such as pattern-oriented software architecture. The study of the patterns is associated with software designs from the literature. This book will focus on pattern recognition but based on the software source code's statistical characteristics. The book also extends to use those patterns to optimize the software development operations by building the designs into probabilistic statistical models. Work done in Mark (2017) focused on the following aspects. This book explores various machine learning models for the application security perspective. The book approaches more conventional machine learning models like Naïve Bayes and Markov models. This book will focus on security vulnerabilities; also, there is a study of broader aspects of software development, which helps solve issues with a more holistic approach. The book explores the latest cutting-edge possibilities of applications in software source code representation and transfer learning possibilities in a software development context. Work done in Pascal et al. (2017) focused on the following aspects. Mathematical models were explored for semantic analysis of logic programs. An extensive study focused on mathematical techniques for a logic program's semantics, breaking down all topics' elements. This book will focus on this on the semantics parsing but from an NLP point of view and a detailed study, and focus on extending the analysis to practical application to solve some of the issues experienced in software development. This book will also focus on the core aspects of making software development a mature process with predictive capabilities to address vulnerabilities. The study focuses on the software development landscape's critical areas that contribute to inefficiency in the life cycle.

1.8 The motivation of the work

Exploration in machine learning, software programming, and software engineering has been focusing on deriving a probabilistic model in software source code. It is possible with a large amount of patterns that are hidden in the code. Here, the critical point is to look at natural language and programming language together, which provides insights on how probabilistic models are derived for software source code. This exploration area has its inspiration from the software becoming a significant part of life, touching various life dimensions. Software has its mathematical base from its programming perspective. There is a well-defined structure associated with the programming language. It extends to building software tools and covers a structured method of defining and deriving the design. The focus area covers verification of the program, refactoring of the code, and bug exploration. Practical applications of the software will be a needed area to understand the structure of software better. An open-source software system body of knowledge has been an additional motivation. In addition to open-source code, these repositories also open ownership of the assets, management of bug, associated metadata, and other related areas. There is a massive repository of the information available publicly. The theme is to use considerable knowledge hidden in the software assets as a base for designing the tool to enhance the software and its ecosystem. It helps to move beyond addressing those that fail to the areas where the proactive contribution made. Analysis of an extensive repository of well-written software assets will provide the required knowledge to understand patterns associated with efficient and useful software. Figure 1.5 depicts association of Software engineering and statistical modeling related areas.

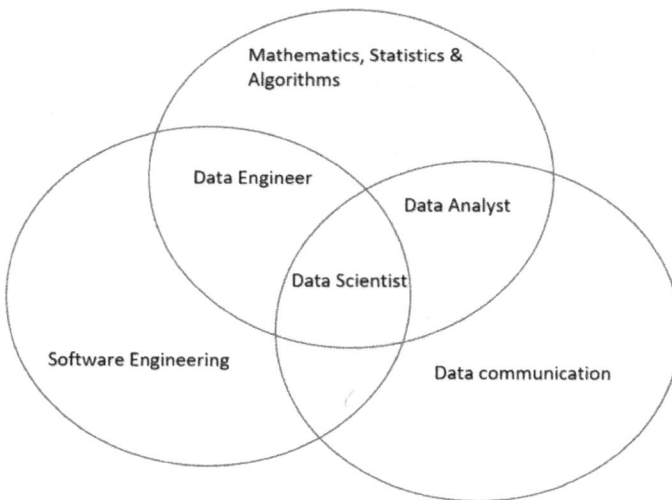

Figure 1.5: Software engineering with statistical modeling.

Greater hope is the machine learning ability to generalize the learning and manage the noise well. In machine learning applications in software engineering practices, the focus has been in a black-box way. The inferences are focused on rather than the algorithms' internal workings and kind of assumptions made in the process. One of the motivations would be to explore this area as part of statistical modeling for software source code. They are taking NLP's example, where there was a shift of focus from rule-based predictions to statistical backend analytics. On similar lines, software engineering should focus on combining traditional structure-based inference from the program with information related to properties of the code that are statistical. Exploration should cover natural alignment of software programs with that of natural language. Exploration should cover natural alignment of software programs with that of natural language and the structured composition of source code. From the source code's probabilistic background, machine learning involvement is understood, like n-gram models and deep learning methods. Other representations of the source code in terms of sequences and continuous expression will explore different areas beyond n-gram models. Possibilities range from debugging, analysis of the program, and recommender systems emerging as new advancement areas. Parsing the source code focus should be on generating general-purpose output rather than programming language–specific one. Characteristics of the source code and natural language are explored to get better design insights for source code prediction systems.

1.9 Software source code and its naturalness characteristics

The program's semantics are influenced by many aspects of source code, such as format used and lexical ordering. It is the reason for abstracting these elements in analysis associated with programs. How the statistical method of exploring source code will help is where natural language similarity of source code comes into the picture. Trigger for the naturalness is associated with the programming points. Building software engineering tools' possibilities evolves with software programs and natural language having similar statistical properties based on this concept. Human communication–based statistics is a developed area for quite many applications. A space like speech recognition and translation uses as its base the studies done in the area of natural language and human communication corpora with statistical models. At its core, programming is about the communication between humans and machines; hence, this opens the possibilities of utilizing the rich patterns associated with natural language from the probabilistic machine learning model's point of view. It merely leads to the application of a machine learning approach in the probabilistic source code modeling context. It will have the intention of understanding the naturalness with which the coders write their code. The software engineering tools combine the probabilistic models. It derives from gathering statistical information to produce improved tools. These approaches provide the capability to build a hypothesis and the probability of the decision.

Probabilistic modeling also associates the coverage of other sources of information related to source codes such as requirements, blogs, and comments that are more ambiguous like natural language. This helps to improve the confidence about the probabilities projected by these models by keeping them close to reality. Not all areas of software engineering would provide a promising path for the exploration of its statistical properties. The source code's naturalness comes from the fact that the developers would prefer writing and reading code that is familiar and structured with routine maintenance. The code based on the known structure will be more favorable and predictable from an experienced consumer perspective. That makes code predictability an outcome of the naturalness of the code, including a similar pattern hidden in verification statements and token sequences of the code which can also be leveraged. Exploring all possible source code files for probabilistic models, to estimate the data distributions is the approach Artificial intelligence applications like computer vision and NLP, which has its base in probabilistic machine learning models, have a crucial feature of learning during uncertain situations and noisy data.

How will the uncertainty come into the picture in the source code when the code is deterministic, as the code is a known one? A good fit for the machine learning approach in the software model is its ability to handle multiple sources of information and unambiguously combine them. A natural framework is also available to connect various knowledge sources, putting together statistical properties associated with source code. Interest would be to integrate the patterns from source code and all its associated assets of information involved. Due to the ambiguity involved in natural language text, it is essential to figure out the associated uncertainty between code- and text-related information of the source code. Probabilities help cut down on the stricter requirements of having useful data related to the program's properties. Figure 1.6 depicts probabilistic modeling outline.

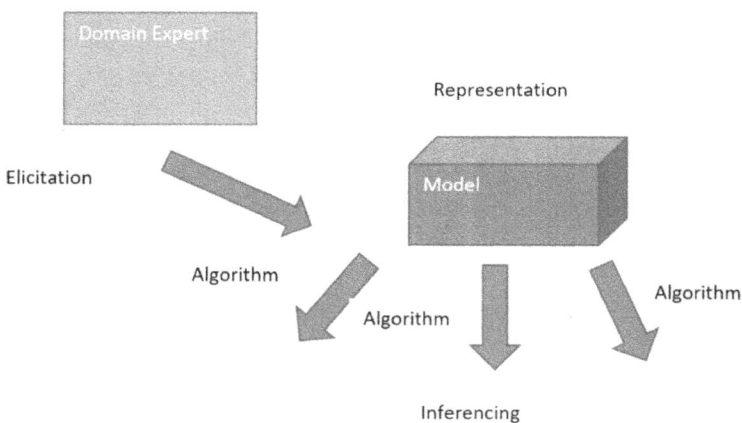

Figure 1.6: Probabilistic modeling.

1.10 Machine learning in the context of text and code

Programming languages are intended to reduce the distance between humans and computers in terms of communication.

Source code is a two-way communicator on one side, a machine and a human on the other side. Figure 1.7 depicts interaction between human language and programming language. Code understanding is critical for a human to present it to the computer where the computer needs to understand the same for execution. This aspect makes the code demonstrate its commonalities and differences with the text. Coding for a general-purpose language would be a new exploration area that focuses on the similarities involved with code and text. Explorers have demonstrated a similarity between text and regulation, with the code also having less ambiguity in comparison. Out of this conversation, though, it is evident that the code and text have considerable differences; understanding these differences will provide deeper insights into NLP's modification for applying in code context.

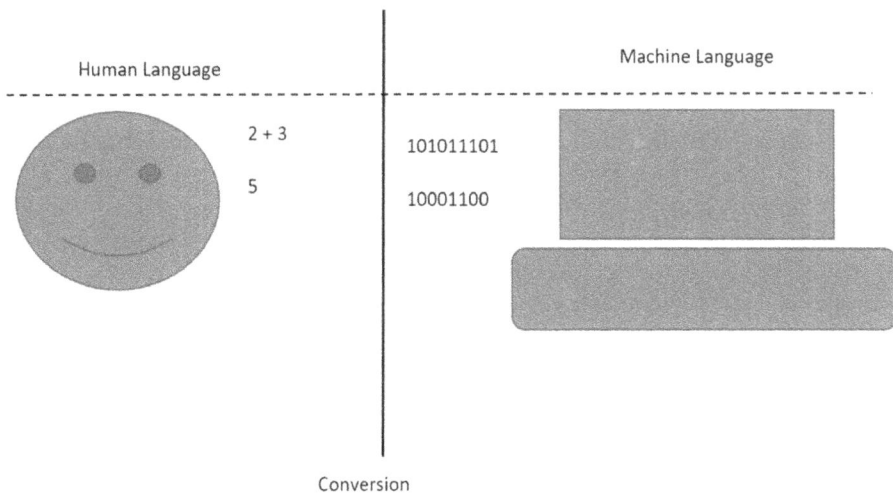

Figure 1.7: Human language and programming language.

Though the code is human-oriented, and code is sensitive to noise, it needs to combine probabilistic and formal methods. Since there is a proper and probabilistic model working together, there is a need to bridging them to work together. Possibilities of changing over between natural language without losing the context are not completely practical. Comparatively, programming language porting from one to another is more accessible, though it has its challenging aspects. These challenges are not explored from machine learning perspectives except for some common base languages like Java and C#. Extensive usage across various domains makes natural language take the upper hand compared to the programming language used for communication across a wide variety of disciplines.x Control and data flow within the code is part of its executable characteristics that cannot be easily related to the text. This coded character also provides its static and dynamic nature, like execution traces missing in the text. Looking at these traces or flows would be the right direction to explore. Programming languages provide formality to the language; also, they are mathematical representations of the natural language. Many users consume the programming language that is designed as top-down by designers. Social dynamics make natural language to evolve bottom-up. Programming language evolution is structuredwhen compared to natural language which graduates slowly. Programming language, also having the characteristics of not being backward compatible in many of the circumstances.

Formatting is essential in code; the text is robust in its environmental dependencies; in regulation, they need an exact specification of its ecological dependencies. This environment of code execution will evolve at a higher pace when compared to natural language. Reusability becomes possible due to the source code's format structure. It helps the program to have semantic similarity within the program. These similarities may not be expected in the competitions of the coders. Code completion possibilities fuel from the demonstration of the rich pattern that is existent within the code.

Syntactic and semantic consistency becomes possible due to programming language naturally translating into code. NLP models will have to handle ambiguities in the text on the same lines; probabilistic models can use rich code structures. Code ambiguity is still prevalent since aliasing and other practices contribute to code analysis. In the case of dynamically programmed languages, when viewed statistically, co-reference ambiguities creep in. Nonstandard compilers contribute syntactic problems, and the undefined behavior of the programming languages can introduce semantic ambiguity. Since there are few compilers for many languages, this will handle this problem. Though we discussed the bimodal nature of code from algorithmic and explanatory nature, the challenge of code components' association with text units is still challenging. Identifiers, blocks, and functions are the semantic components of the code. These do not associate with the text semantics component. Readers' knowledge level needed to understand code and associated text is significantly different when seen in isolation. Length and semantic information in blocks are significantly additional and generally cannot be restricted with boundaries for readers to understand. Functions are differentiable and have semantic rich information but run longer. Text, on the other end, has limited tokens or words in its semantic construct. But functions and text are not easy to equate. Parts can be named and called in other areas for references, whereas the text in the form of sentences or paragraphs is not callable in a similar way, nor do they have identifiers. One function can be a source of action with a long-range of tokens encompassed with multiple activities, but rarely paragraphs can be complete action-based content. Processes exhibit abstract syntax trees deeper and more repetitive than parse tree structure shown by sentences in the text. Unlike text-based corpus, programs are diverse in the languages involved and they facilitate several tasks that influence its semantics' frequency. Most of the code is identifiers that need developers to name it. There is no need for names in the case of text, whereas the existing word is utilized. Figuring out the relevance of semantic code units for the type of tasks is an exportable area in machine translation, where the code needs to be explained in the text. Corpus is used as a reference for learning in the case of statistical machine translation (SMT).

For cases of code, search engines need to align queries as code units of semantic structure. The answers are derived from mapping the code and text with their granularity influenced by context. Figure 1.8 depicts Machine translation outline and Figure 1.9 depicts a machine learning model's architecture.

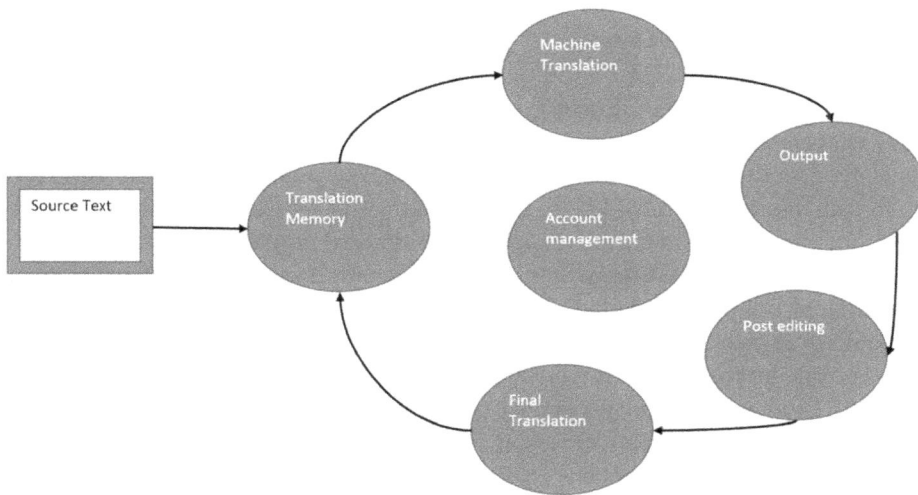

Figure 1.8: Machine translation.

1.11 Probabilistic modeling of source code and its inspiration

Probabilistic modeling of source code is about the understanding of the probability distribution associated with code artifacts. Like in the case of any other modeling approach, probability models make their assumptions about the domain. Though these assumptions help to create a learning situation, they also introduce errors. Since the type of model influences these assumptions, each of these models will have their strengths and weakness and different capability in their applications.

To group these models, we look at the form of equation used to represent these models, inputs, and outputs involved, but, still, some models may appear in multiple categories. We will have to explore these models in line with the NLP models.

Hyperparameters tunning

Machine Learning Model

Parameter

Control System
Pipeline:

X Train
Y Train
X Test
Y test

Paramter

Parameter

Scoring System

F1 score
Adjusted R score
R Squred score
Confusion Matrix

Parameter

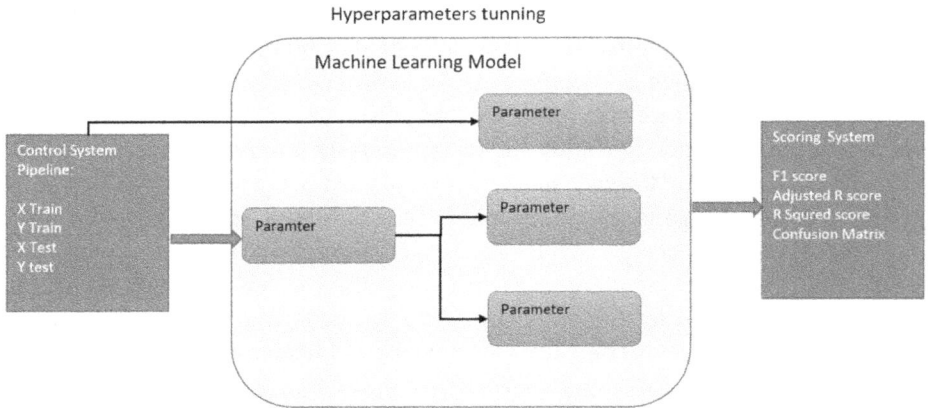

Figure 1.9: A machine learning model.

Code-generating model's category group models can be defined probabilistically on code with stochastic nature to generate small and straightforward tokens. The symbolic code model will consider the abstract format of representing code. The symbolic code model will consider the abstract format of representing code with data flow and token context taken as examples. Conditional probabilities over code components concerning their properties like the type of the variables used are base for predictions. The pattern exploration model depends on the unsupervised structure recognition in code. These are the context of clustering but in the context of regulation. Looking for the human interpretable pattern is a base here. In the case of generative models of text, the code-generating model is used in machine translation models and language models. Code representative models are like text classification, named entity recognition, and sentiment analysis in NLP. Code pattern mining models are also in line with probabilistic topic models and machine learning translation to mining the structures in information-based machine learning approaches.

1.12 Applications of the source code statistically modeling

Probabilistic modeling has extensive applications in software engineering and programming language research. Handling uncertainty with probabilistic modeling is the key feature. Concerns in these models are contributed by ambiguous data involved and anonymous information. Another key benefit is the computational ease introduced by the probabilistic model in the analysis task context and helps to accelerate the same. In software engineering tasks, recommender systems are involved in code autocompletion and highlight the code changes to benefit the code reviewers. A key focus in these contexts is to mine data and apply machine learning approaches. Developer intent modeling is another area; though the expectation is to separate the developer's intent from code itself, it rarely happens. Code completion is the most prominently used aspect in the recommender system feature of the IDE. Most of the widely used IDEs have some form of code completion aspects, and it is one of the most used parts. But the code suggestions are made in alphabetical order rather than the relevance of context as predicted. Statistical-based code suggestion would focus on providing recommendations with probabilities of learning the recommendations in a ranked list. There is automatic completion concentrate on the request of constructs like method calls. In all these probabilistic systems existing codebase will be the training scope for learning.

Earlier code completion methods were improvised with Bayesian graphical models for better accuracy. This model that banks on context uses the objects to model the next call's probability distribution in sequence. Language models of source code are used inherently for code completion purposes. Initial works have focused on the token-level language model that uses the previous set of tokens for its prediction. Semantic context addition to n-gram modeling improved the code completion accuracy. The formal code properties are explored. Figure 1.10 depicts a basic recommender system.

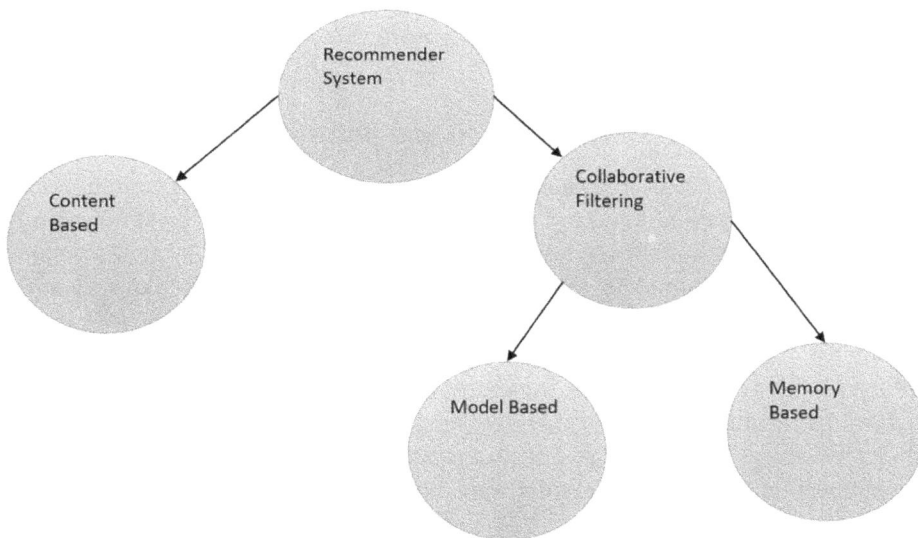

Figure 1.10: Recommender system.

1.13 Code convention inferencing

Beyond grammar involved in the code conventions associated with the code are the constraints for syntactic nature. These are needed for a governance point of view, like newline placement use of braces or camel casing. These assist in easy comprehension of code, easy navigation, maintenance, and prevention of specific categories of bugs. In online courses and online competition, conventions play a crucial role in helping the tutors rectify common issues made by the participants. Enforcement of these conventions in code is a challenging process. In the rule-based system, it becomes the worst scenario to finalize the format of traditions. Machine learning can contribute by learning the coding conventions from the codebase. This supports understanding the ways followed by codebase and prevents the need to define the convention and enforce it or utilize pattern implementing tools. Source code models of machine learning that look at the basic code structure will suffice for these applications. Source code learning data can be used to figure out conventions associated with quantifications of the uncertainties associated. A key challenge is the code constructs' sparse nature due to a less repeatable form of source code within various projects. The code surface's statistical similarities are used as a base for learning methods, class, and variables. Software engineering tool design can be facilitated with conventional syntax and semantics that are mined from software source code.

1.14 Defects associated with code and patterns recognition

Figure 1.11 depicts defect injection rate across various stages of software development and importance of initiating actions to make sure that the defect injection is shifter towards left, so they are less costly to fix. Code appearing at high frequency will get a higher probability in the case of probabilistic models. Defect identification is a prominent focus in software engineering and researches associated with a programming language. Maintaining high precision and recall rate for the correct detection of defects by categorizing the source code is a key challenge. Diverse source code and rare nature of defects add on to complexity. There are hopes of using probability assignments of language models as a base for identifying weaknesses. There are pointers on language models assessing the source code's complexity to figure out code that has bugs and demonstrate that the code that has bugs tend to be less natural (Baishakhi et al. 2016, Miltiadis et al. 2013). Deep belief networks are used for learning features from source code tokens for defect prediction (Song et al. 2016). Simple statistics on the code assessed by language models are based on many of these models. The sequence of API calls is modeled with probabilistic models with topic models with recurrent neural networks. These models identify a high improbable API sequence of calls to bring out bugs in the Android code (Vijayaraghavan et al. 2017).

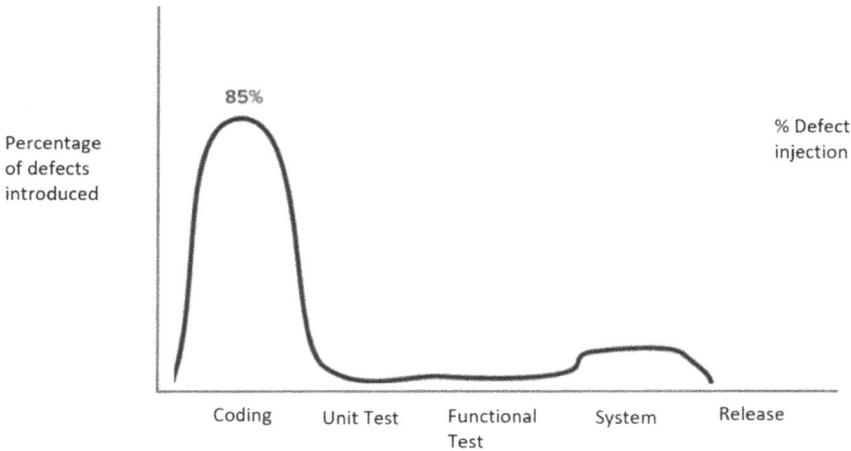

Jones, Capers, Applied Software Measurement: Goal analysis of Productivity and Quality

Figure 1.11: Left shift for defect injection.

Source code abstraction at different levels needs to be used for detecting code due to the sparsity of the code. Based on how the designers have chosen the conception, another class of defects is exposed. Syntax error detection seems to be relatively more straightforward. Consider that not all strange behavior would be defect, but the executed code's peculiar behavior mostly points at bugs. With this mix of nature, probabilistic models are the best fit for defect identification in code. Lack of precision has resulted in the industry not leveraging on these much. The sparsity of code construct and its diversity contributes to the suffering of anomaly detection methods.

1.15 Motivation in the area of code translation and copying

With the inspiration of natural language, success in SMT has motivated code being translated from one source language to others. Rule-based system writing has been in practice, but that makes rules maintenance tough as the programming language evolves. In the light of invalid code being also produced, SMT has prominence in this activity. Semantics were added to this translation process to improve performance (Ann et al. 2015, Svetoslav et al. 2014). SMT models are also applied to texts in existing explorations. These models learn mappings between languages like API constructs; their applications have been focused only on languages with a standard structure, like C# and Java, as they both are object-oriented languages that have managed memory. This is a significant limitation as there is a need to extend these beyond common construct language and across languages with different memory management. The conceptual difference between these other structured languages needs to be understood to translate object-based code to functional languages while retaining the semantics like learning to translate loop into the functionality of map-reduce. Evaluation of the translation model investigates the exact match of semantic and syntactic correctness after cracking the code using a suitable measurement mechanism. Developers will copy code during coding. This must be facilitated with variables renaming and handling the collision between names. This adds to the advantage of creating clones of code and putting together common code snippets from different codebase locations. Cloning also presents the opportunity of refactoring of the code and reusing the cloned code. Autoencoders and recurrent neural networks assist in clones finding in code snippets with similarity in distribution representation. A continuous similarity metric is learned among code locations using distributed vector representation instead of edit distance. Figure 1.12 depicts Recurrent Neural Network architecture. Figure 1.13 represents the outline of Autoencoders.

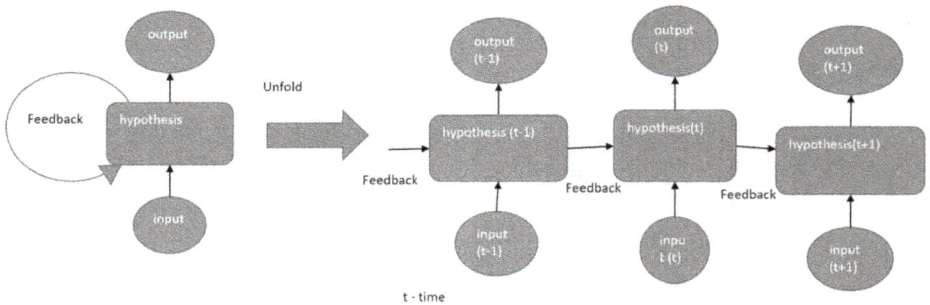

Figure 1.12: Recurrent neural network.

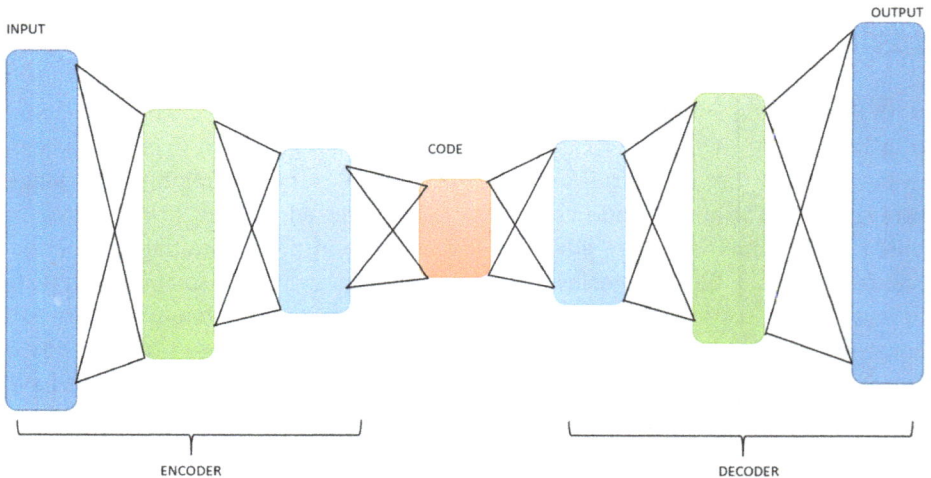

Figure 1.13: Autoencoders.

1.16 Inspiration in area of code to text and text to code

Natural language inspiration for text to source code can assist in multiple applications like documentation, search, and traceability. But the diversity of text and regulation, the ambiguity of the text, and code compositional nature, and software's layered abstraction make connecting text and code hard. Probabilistic modeling provides a moral base for modeling the code to text, and vice versa also helps resolve ambiguities. Making code readable and assisting code documentation is facilitated by regulation to a text with natural language generation from code. Code to text summarization with the neural attention model is explored (Srinivasan 2016). N-gram models and topic models are used for comment generation from code (Dana et al. 2013). Text to code conversion effort supports the coders and end users to program efficiently. NLP's semantic parsing area is closely related to this. Natural language utterance conversion to its meaning representation is the objective of semantic parsing. Presentation of the meanings is in the form of logical queries that are often used in question answering.

1.17 Inspiration in the area of documentation, traceability

Software engineering has a central question in the area of code search and improvements of documentation. The obvious choice is probabilistic modeling naturally as they provide integration between the text, which is natural language and code. Code search can be possible with natural language queries, as it is a standard task developer. Information retrieval techniques have been explored for code search problems. Sequence-to-sequence (Seq2Seq) neural models are used for statistical source code modeling for API sequence mapping to natural language (Xiaodong 2016). Figure 1.14 depicts Sequence-to-sequence modeling format.

Documentation puts together specifications, descriptions of codes, and requirements for code. This helps engineers to prioritize the code and efficiently maintain the code. Documentation includes deriving information from unstructured documents.

1.18 Program synthesis opportunities

Using specification generating full and limited programs is program synthesis. A formal statement with relevant logic is the specification. In recent times the partial specification and combination of input and output pairs are the focus for researchers. In cases where the specifications are a natural language, that will be a semantic parsing task. Programming language research focuses on program synthesis based on examples or specifications. The more considerable challenge is to search among the large space of possible programs to figure out the plan that fits perfectly. Probabilistic programming helps to keep the search scope limited to the most probable programs. Machine

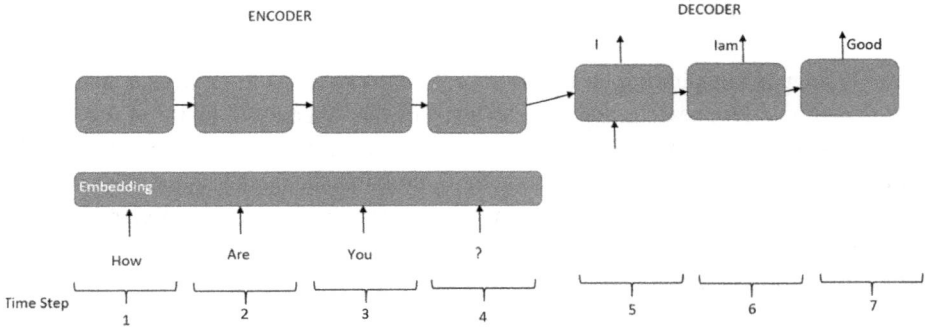

Figure 1.14: Sequence-to-sequence modeling.

learning methods are utilized to synthesize code in case of research on programming by example. Graphical methods are used to learn commonalities of the program across tasks similar to program synthesis search improvement. The general context of using program synthesis is in the area of generating a program that is in line with specifications. Probabilistic models are utilized in the area of synthesizing the random functioning of the program to benchmark (Miltiadis et al. 2018).

1.19 Conclusion

Probabilistic models exhibit a greater chance of supporting new tools in all the possible areas of software engineering. Here the inspiration was derived after understanding the taxonomy of machine learning models of probabilistic nature for source code and its applications. The potential area that is looked at is the result of recent advancements in the research, which indicates the industry's interest and its potential. Probabilistic models in the case of source code open opportunity for learning involved in the existing code base; reasoning the new source code–related artifacts in a probabilistic manner also assists in knowledge transfer between developers and other project team members.

Chapter 2
A probabilistic model for fault prediction across functional and security aspects

2.1 Introduction

Probabilistic models that are in use to ease recognition of the software program's highly improbable state will be focused here. Statistical debugging models would help achieve this objective, leading to learning using these statistical debugging models. These models also help build the knowledge repository to refine these debugging models in the future. Software execution traces are a good source of information for this exploration. Possible areas of investigation include pretrained models and transfer learning for understanding software vulnerabilities (SVs). Exploring the ensemble of models with various combinations of machine learning (ML) models is a prospective approach. Vulnerabilities injection rate being a time-based factor, time series–based methodology applies. The autoregressive integrated moving average method is an area of exploration for time series data. The occurrence and exploitability of the vulnerabilities also play a significant role in the software domain. Probabilistic models like the Hidden Markov model, support vector machine (SVM), and artificial neural net versions can be a good fit for the occurrence and exploitability of vulnerabilities.

This chapter's organization is as follows. Discussion extends to various statistical debugging models background, program logging, and different statistical debugging methods. Chapter also cover the program sampling framework exploration to understand feature selection and utility function. Identification of multiple bugs and probabilistic programming are also reviewed. The next section looks at SV identification with deep representation learning. We discuss the use of artificial neural network (ANN) for SV assessment like web service. The review involves the concept of text analysis for SVs assessment. We also look at the possibilities of automating SV assessment. The chapter further extends on the exploitability of SVs using stochastic models. Exposures under study as a time series component get explored (Eric et al. 2016).

2.2 Statistical debugging models

Tracing software bugs' location in the software program is a critical component in the software program delivery's success. Arriving at the approaches that facilitate this is a key focus area, complicated, vital, and expensive. An increase in complexity and scale of software has added to the criticality of the situation. In the case of life-critical systems, these faults in the system can cost a life. Studies also have indicated the rolling-up effect of these faulty systems causing economic losses leading up to

https://doi.org/10.1515/9783110703399-002

impact on GDP (NIST report). A significant part of the expense involved goes to the users of the software program. Conventionally the fault locating has been a manual process and depends heavily on the experience of the developer. A large corpus of research is done to explore various approaches with varying possibilities of advantages and disadvantages of the methods.

2.2.1 Program logging

Program logging is one of the traditional mechanisms that facilitated the fault location where the log gets printed as part of the program state's status. Breakpoints in the program enabled the programmer to observe the program behavior and specified point at which the action happens. Adding an assertion to evaluate the condition for running a program was another aspect. Metrics profiling for memory usage and execution speed also helps. For an advanced approach for fault logging, the failure's causality plays a role in defining the probabilistic and statistical models.

Among the advanced techniques of fault logging, program slicing is the approach where the program gets sliced, such that the sliced version still represents the overall program and scopes down the focus area of fault location (Weiser 1979). Static and dynamic program slicing approaches get applied. Execution slicing is an improvised approach where this focusses on data flow tests; here, execution slices are targeted based on the set of statements executed by the test case. This is an improved approach compared to static slicing, as this one considers the input values as a base for its analysis. Test case focused slicing also is optimal compared to dynamic slicing, which does not have a specific scope of slicing and makes it complicated. Statistical and probabilistic models inspire program spectrum–based models. Program spectrum methods use a program's execution information with a perspective like conditional branching. Focus is on program behavior; while the execution fails, information related to such failures get analyzed to arrive at the cause of failure. It uses information about part of programs that are considered as part of testing and then refines down to that part of the program where the loss happens. Though there was an attempt to try this with failed test cases, later, it was needed to shift focus on both failed and successful test cases. Approaches based on the nearest neighbor that relates failed circumstances with success cases to bring out the contrasts and locate the faulty areas based on the cases' distances are also experimented with (Renieris et al. 2013). The time spectrum–based approach computes the execution time of all methods associated with successful and failed executions. Observations are devised as the time spectrum model reference. Failed cases are then validated against these models to identify the failure prospects (MITRE).

Some of the spectrum-based fault localization methods are that they do not calibrate the contribution of successful and failure instances of the test. Just with one test case failure, the suspicious and unsuspicious grouping is arrived at with risk being computed for questionable category only and lowest risk assigned for the unsuspicious type. The

approach proposed was to appoint a high contribution for the first failed test case followed by the next failed issue being the next highest, likewise for subsequent failures.
Similar format applied for successful test cases and failed test cases together, will have a
larger magnitude compared to application on successful test cases alone.

2.2.2 Statistical debugging methods

For each of the selected bases or segments, the probability of failure in the case of this
base being true is computed. Likewise, importance scores are arrived at for various com-
mands that are examined. The importance score here provides information about the
association of the bugs with these bases. Higher importance scored bases are the priority ones for examination. In place of the program base, the test cases' path profiles are
later proposed as a better approach (Ben et al. 2005). Path profiles of multiple test cases
are rolled up, importance scores are computed for every path, and top outcomes are
targeted for potential failure causes. In the improvised approach, a ranking of the suspicious scoped program base is applied. Here for every test case, the scoped base is validated for being true or false. Distribution is derived out of this; based on the
distribution, assessment is made to compare the significance of the difference between
successful execution and failed execution to associate the base with the fault. Nonparametric hypothesis tests are further used to enhance confidence in the importance of the
difference. Fault localization has been scoped as a causal and inference-based problem.
The linear model is built for the control flow of the program, which is represented as
graphs. Graphs, in turn, identify the effect of the given statement being considered on
failure possibility. A later model was enhanced by including the data flow dependencies as well into the linear model. Further, this list was enhanced by including program
execution phase data like memory and CPU usages to debug the program statistically.

2.3 Program sampling framework

The program sampling framework collects the program behavior information during
the run time of the program. This framework samples the execution time responses.
Checks are planned for the program but are executed on a sampled basis across
multiple program code runs to make it efficient. Outcomes of these runs are rolled
up without focusing on the information sequence, which also helps to optimize the
data space. Critical aspects like values returned by function calls and memory cor-
ruption are acceptable references. Guessed assertions are introduced in the program
in anticipation of capturing the specific behavior of the program. These assertions
are randomly executed by Bernoulli random variable generation and geometric random variables. The frequency of argument being true or false will be the object of
analysis for the program sampling framework.

2.3.1 Feature selection

Feature selection is the next logical step after the data points were created through assertions in the previous actions. Here the identification of the most important features is the target. Based on program exit status, program labeling fails, and success in to, 1 and 0 respectively are done, and is a classification problem. The overall objective here is to identify the assertions that are influencing the program's crash and identify the scoped base of the program or predicates that have a positive influence on program crashing. Cases of the program crashing at an early instance create a situation where these instances may miss out in sampling of this approach as there is less chance of them being covered in the sample. This may create a situation where this scenario may resemble the successful run chase. In a classifications problem, these are false negatives. Other systems are the cases where the failure is evident after the appearance of a buggy feature every time. If there are any false positives during the training phase of the model, they must be penalized heavily.

2.3.2 Utility function

This approach's utility function is a classification model where "x" factors are the features identified as specified in the previous section. "Y" is the class label of 1 or 0 for failure or success. Prediction outcome would be true positive (TP), true negative (TN), false positive (FP), or false negative (FN). The utility function will have the objective of maximization of the classification correctness. The actual distribution of input factor "x" is determined by software under examination, making it complicated and a non-Gaussian distribution. This situation calls for using a discriminative classifier. The logistic function is used to model the class probability, which has Sigmoid function $[1/ (1 + e^{-z})]$ in action underneath. The decision boundary is at 50% for the possibility to be classified as 0 or 1. One option of probability being more significant than 0.5, equal to zero or otherwise, would be the result. L1 norm regularization is the parameter chosen, which retains only the most significant factors in the model. The model attempts to optimize each of the features involved as part of the objective function. Extra penalties are added in the case of false positives. Log loss metrics of classification is leveraged to evaluate the objective function.

Log loss function can be expressed as:

$$\text{log loss} = -1/n \sum\nolimits_{i=1}^{n} yi . \log(p(yi)) + (1 - yi) . \log(1 - p(yi)) \qquad (2.1)$$

where n is the number of samples, yi is the label, and $p(yi)$ is the predicted probability.

L1 normalization or Lasso regression regularizer weighs every class prediction to optimize the objective. Stochastic gradient ascent is used for learning the parameters. Weights closer to zero are not considered; those with the most positive consequences are shortlisted finally. Parameter for balancing the imbalanced data set where each of the classes is disproportionate is also included. This imbalance in the levels is attributable to the failures being of low probability compared to success. There are parameters to manage the deterministic and nondeterministic bugs. In the case of deterministic bugs, parameters penalize any FP cases. In contrast, there is relaxation in the case of nondeterministic ones to accommodate the patients where the program should have failed but did not due to their nondeterministic nature. There is an approach to eliminating the assertion, which is universally zero, which means they have never failed earlier.

The approach attempts to pin down the location of bugs in software. Different types of bugs are handled by tuning the parameters configured in the model, which makes the model custom designed. The real world brings the challenge of a multiclass of bugs that would be difficult to be labeled and got into the scope of training the model. In that case, the scenario changes from microscopic changes between failure modes to the macroscopic usage pattern (King 1976).

2.4 Identifying multiple bugs

This section will discuss multi bug challenges that make the situation complex in bringing a clustering dimension to the problem. But since the overall outcome is to assess successful and failing runs, it is a classification problem situation. The objective here would be to identify features that would facilitate easy identification of the clusters and help us classify succeeding and failing instances. Testbed used in this approach is the MOSS plagiarism determination program operated by a large community. It was clustering the failed run occurrences based on the different bug results in representing the program's usage mode rather than the failure mode, which defeats the primary purpose.

Figure 2.1 depicts bug histogram visualization format, in this experiment. Here a plot is drawn for each of the group which has different cluster number. This helps to identify the best cluster scenario that can highlight a specific bug id which fails the most. It was also noticed as a challenge in the experiment of a single bug scenario that the predicates [scoped base of the program considered for debugging] ended up being redundant, which calls for clustering the predicates. While pressing mutual redundant predicates would be challenging to be gauged as the data is sparse in this case. The reason would be the lack of enough runs to unearth the redundancy of predicates. The spectral clustering methods are employed, which will need a similarity metric like the correlation coefficient for the predicate pairs at different moments. With this setup, the predicates' clustering is run for clusters in a range of 5 to 20. The experiment is run on various cluster combinations of the predicates. The plot of the histogram of the frequency of bug on multiple runs, in each of these clusters, is observed. In these histograms, it is seen that most of the cases have various types of bugs that

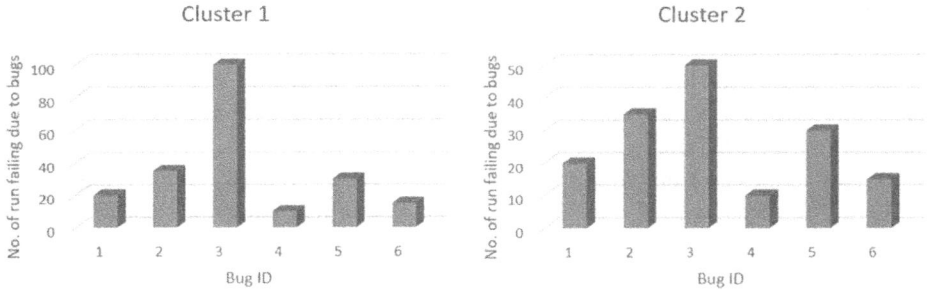

Figure 2.1: Bug histogram: Every plot is drawn for a different number of clusters. To identify which is the best cluster scenario that provides a specific bug id which fails mostly.

have high occurrences. This situation is due to a primary superbug predicator; these are generally a precondition of failure. Since they influence multiple failed runs, they tend to affect various types of bugs. So, this is an issue in this approach.

The proposed approach to handling this issue represents the predicate samples in a graphical way to assess the probabilities from observed data. Chances are then used as input to run the bi-clustering model, which will identify features based on combined clustering of runs and predicates.

With the observed frequency of the failures and knowledge of the sampling process adopted, we can arrive at each of the predicates' probability. This helps manage the data sparsity issue and sampling bias so that clustering and feature selection can be focused upon. This approach depends on the symmetric principle of predicates that should be grouped up by the runs in which they predict, in turn, to group by the predicates that indicate them. The standard bi-clustering technique and standard distance metrics will not be enough. The bi-clustering algorithm is enhanced with the iterative collective voting mechanism. Each failed run votes the predicate that was optimal for it. Further predicates are evaluated based on the votes they get. Every run will have one option, and initially, this vote is distributed across predicates based on their quality. Predicates are competing for a vote of the run. So higher the predicates, the more distributed the votes get. Since predicates face off for voting, the redundancy issues of predicates are managed. Iteration runs until a convergence point where the runs will have a firm option on a predicate. Finally, predicates are shortlisted based on rank achieved, considering the options. Overall this approach would cluster bugs with failed runs and also help to identify the predicate associated with the bug. The simple one-variable model will not be able to capture the exact bug predictor without run clusters being configured (Alice et al. 2006). Figure 2.2 depicts Truth probability model.

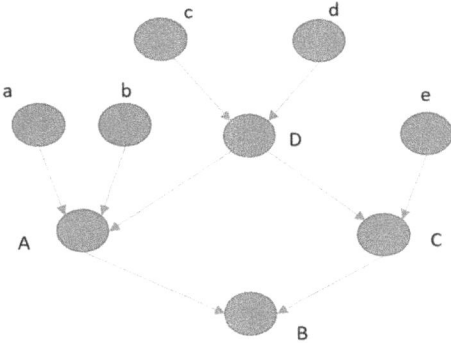

Figure 2.2: Truth probability model: C is the random variable number of times predicate is observed in a run. B is many times it is observed as true. A is an actual number of times it was true. D is the number of times observations are taken. a, b, c, d, e are the parameters associated with the algorithm.

2.5 Probabilistic programming

In this section, the focus is to assess the probabilistic programming's applicability to software engineering's programming language. Unlike other programming languages, the probabilistic programming language explains the probability distribution of the study process. It has two components: the possibility of drawing random values from distribution and modifying values of variables using observations into distribution.

Probabilistic programming intends to automate Bayesian inference (Debian). Probabilistic programming uses computer science as a tool for performing statistics. The general programming pipeline is creating a program with the specification of arguments and evaluating output. Modeling in statistics starts with production, then specifies abstract mathematical models relating output with input to arrive at posterior distribution. In the case of the programming language model, learning is in the form of program induction. Graphical modeling works based on the model built using the data structures; later, observations are fed to run graph model inferencing. In the case of the probabilistic programming model, they act as data structures (MITRE).

Bayesian networks are the graphical probability model based on the Bayesian inference. It focuses on conditional dependency so that causation can be explained; the same is represented as edges in a directed graph. As the network is built based on these conditions, it is possible to derive inferences on the network variables. Probability theory is at the heart of the Bayesian network. In the Bayesian network, each edge is a conditional dependency, and each node is a random variable. If edge X and Y exists in a graph connecting random variable X and Y, that infers $P(Y|X)$ being the factor of a joint probability distribution, and we need to have values of $P(Y|X)$ for all values of Y and value of X for any inferences (Shin et al. 2011).

The discrete-time Markov chain is related to the stochastic process at a discrete-time. This targets to evaluate the progress of a real event over the period, like the stock price of a stock after "*n*" number of days, the bandwidth of telecommunication after "*n*" hours of usage and so on. This banks on the complexity of the model to capture the phenomena that are studied and enough simplicity to compute the event of interest in the critical point of time.

As an example of the application of probabilistic models, let us consider online games' skill ratings. Here player's skill is rated related to other players, based on the outcome of games played until the point in time. The Bayesian model can be built for this scenario and further be represented as a probabilistic program. The skill of the player is initialized based on randomly generated Gaussian distribution. Later few sample games are initiated, and results are presented as random variables of players' performance. Additional Gaussian noise is introduced to accommodate the uncontrollable factors like the amount of sleep the player had last night. The joint conditional probability is generated based on the multiple game results, which are used to evaluate the players' relative skill. This evaluation will help to score the players and decide on matching players for them to play with.

Calculating distribution based on a probabilistic program is a probabilistic inference that can be represented by a posterior probability distribution. It is part of inferencing a probabilistic program, a status inferencing approach where the probabilistic program is described as a probabilistic network like Bayesian network. Inferencing algorithms like belief propagations are used. In dynamic inferencing, the program is run multiple times to arrive at probabilistic states, then observe the values of expected variables of valid runs to compute the metrics on observed values. Approaches explored are mostly based on Bayesian in ML. Figure 2.3 depicts Simple Bayesian network representation. In contrast, optimization techniques that use gradient descent would be a good exploration area compared to the search-based Bayesian approach (NIST 2017).

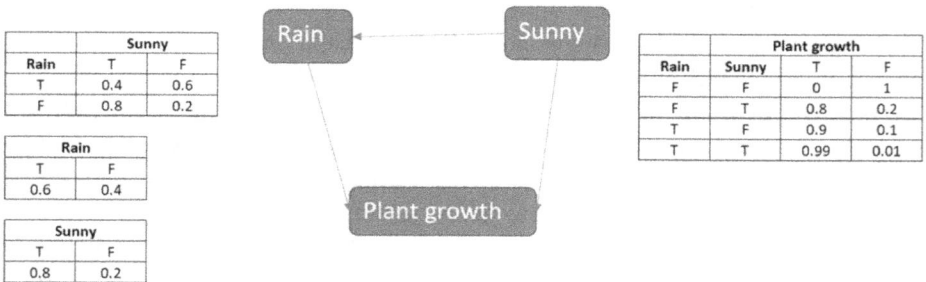

	Sunny	
Rain	T	F
T	0.4	0.6
F	0.8	0.2

Rain	
T	F
0.6	0.4

Sunny	
T	F
0.8	0.2

		Plant growth	
Rain	Sunny	T	F
F	F	0	1
F	T	0.8	0.2
T	F	0.9	0.1
T	T	0.99	0.01

Figure 2.3: Simple Bayesian network representation with a conditional probability table.

2.6 Deep representation learning of vulnerabilities

A large corpus of C and C++ open-source code is used to develop a functional level vulnerability identifier in this approach. Hidden flaws due to the developers' subtle mistakes move in the system due to the open-source nature of the program and the code being reused. Many of the program analysis tools are constrained by only unearth the bugs based on the predefined rules. Since the widespread movement of open-source repositories, data-based approaches for identifying patterns of vulnerabilities have come to the forefront. Dynamic analyzers execute the program with various test inputs to identify real or virtual processors' weakness, but both static and dynamic analyzers are rule based. Symbolic execution is another approach that uses symbols and analyzes the utilization of those symbolic values over the control flow graph of the program (King 1976). Though it covers various program paths, it is expensive and does not scale appropriately for big plans. Vulnerabilities assessment has been carried out with SVM using a bag of word (BOW) representation of tokenized Java source code. Later it was expanded by using n-gram for feature representation using SVM classifier (Pang et al. 2015). Deep learning has been explored to perform the program's analysis, where the source code is represented as an abstract syntax tree further to train a convolutional neural network (CNN) on the tree-based structure. This was a simple supervised problem. Recurrent neural network (RNN) has been prepared using code snippet related to function calls to identify calls' improper use. An approach that we are discussing here stands unique in terms of leveraging deep learning capabilities. In contrast, other techniques have not been able to use deep learning capabilities due to limited data.

The focus is to study the software packages at a granular level, so the study is done at functions. That is the granular component that provides insight into the overall flow of the subroutines. SATE IV Juliet test suite used in the study has code examples from synthetic code and covers about 118 common weaknesses and was meant to be used for code analyzers (NIST 2017). Since this was a synthetic code it differs from natural code and may not suffice for training. Debian packages and git repositories also have been used to balance the naturality. Debian and GitHub repository were not labeled, so static analysis tools were used for labeling.

To generate meaningful features out of source code, a custom-designed tool was used to summarize the key tokens' critical meaning to generalize the representation and reduce the size of the vocabulary involved. Using a standard model for the key tokens of the source code from across the open-source code repositories creates space for transfer learning. A delicate balance was needed for the details identified by the feature identifiers so that the possibility of the overfitting would be avoided. One hundred fifty-six tokens were finally representing the entire code base. Part of the code that does not influence the compilation is removed. Data types from the standard libraries are grouped into generic tokens, and these affect the vulnerabilities.

As part of data preparation, redundant functions were removed to make sure that they do not hamper the model's performance, avoiding overfitting. Challenging were the functions that were near-duplicate and were difficult to differentiate. This was handled by assessing the tasks for lexes representation, which was duplicated for source code or duplication of the feature vector. The control flow graph for the function was used to assess the compile level feature. Also, the process operating at basic blocks was looked at for identifying feature duplication.

Labeling of the vulnerabilities related to code is a complex task since the target training data majorly was open source, where the ground reality is not fully comprehensible. For tagging purposes, the dynamic and static analysis was used, also making use of bug reports. Since dynamic analysis is resource-intensive and commits level message based labeling resulted in low quality, the simple approach of looking for keywords from commit around the pair. of functions in the beginning and end were used which resulted in an optimal result and relevant one. The simple approach of looking for keywords from commit around the pair of functions in the beginning and end resulted in an optimal result and relevant one. This reduced the possible labeling required for the operation, but the manual inspection was substantially needed. Multiple static analyzers were employed to make use of the analyzers' varying features. The output of these was refined to remove those findings that are not associated with security vulnerabilities. Security experts analyzed the outcomes of the static analyzer and mapped them to the Common Weakness Enumeration (CWE) database (MITRE). Based on this mapping, the labeling of vulnerable and non-vulnerable would be generated.

Lexes function from source code was used as neural features in ensemble classifier and random forest. Leveraging upon the commonalities involved in writing the code, natural language processing (NLP) was leveraged. Feature extraction associated in the case of the sentiment analysis technique of NLP was adopted. RNN and CNN were the frameworks used. Embedding representation was done using the tokens that made up the lexes function of the code. These lexes presentation was embedded using "k" dimensional representation for the range of -1 to $+1$, which was learned in the classification algorithm training using the back-propagation approach. This is targeted to generate a linear transformation of one-hot embedding. Other unsupervised techniques like word2vec, which used a larger unlabeled dataset, were used but did not give good results compared to earlier methods. Fixed one hot embedding also did not provide satisfactory results. The overall size needed for vocabulary in this experiment is minimal compared to the usual context of NLP. Since the embedding was limited to avoid overfitting, a small amount of Gaussian noise with specific mean and standard deviation values was introduced. This resulted in substantial control over the overfitting and resulted in better working compared to other regularization techniques that focus on decaying the weights.

RNN and CNN were used for attempting feature extraction with representation from embedded sources. CNN "n" number of filers with the size of $m \times k$ is used. All the filters will scan through an entire space of feature embedding. Filter size, "m" denotes the number of tokens that will be scanned in a moment. An optimal number of filters with batch normalization can be employed. Rectified linear unit (ReLU) layer on top of these was the architecture in CNN's case. RNNs were intending at capturing the longer token dependencies as a feature. These dependencies were fed into RNNs of multiple layers. At every step of time, output was concatenated for a specific length of the sequence. Two-layer gated recurrent unit–based RNNs were employed with a specified number of hidden state size. Long short-term memory (LSTM) versions of RNNs also work well. Since the length of C, C++ functions can be large, a max-pooling layer was utilized with a specific sequence length to generate the required size representation. The architecture's objective would be to extract features that can identify the vulnerabilities' pattern across a component sequence. A fully connected classifier followed the quality removing layers, where dropouts were configured on the max-pooling layer to control overfitting. Optimal hidden layer numbers before the final softmax output layer would have to be configured. Softmax function plays the role of aggregator at the end of the architecture.

For training purposes, optimal batch size needs to be configured with a specified token length range with padding employed to make the consistent size of batching across input data. Adam optimizer with appropriate learning rate needs to be configured with cross-entropy loss as the metrics to assess the model's performance. As highlighted, since it is an unbalanced data, the loss function is configured to have larger weights in vulnerable positions. The consequences employed here are the hyperparameter in the model, which is tuned to attain optimal range. The validation set is the hold out set to validate the model's final setup, once training and testing iteration is run on train and test set. The validation set helps to prevent data overfitting to train and test settings.

Ensembling-based learning on the representation derived from the neural network was a customized approach. A neural network is automatically characterized at feature recognition in a black box mode, but in cases like this, such a black box approach results in suboptimal. The best combination was to use the convolution layer with the sequence being max-pooled and similarly sequence max pooled output of RNNs being fed into a robust random forest classifier, or utterly random state trees worked well. The format of optimizing classifier and feature generators separately helped to optimize overfitting. This set up also facilitates accessible retraining of the model with new features. The random forest classifier run on BOW depiction of software source code was tried. In the BOW representation sequence of tokenized elements are ignored. In this setup, the model uses the correlation of the labels with source length indicators and complexity. Also, this arrangement brought out the group of function calls, which led to vulnerabilities usually. Now, this setup is based on any further restructuring of the model, and getting better performance on top of this was a clear indication of the model getting better at recognizing more complex vulnerability patterns and identifying specific patterns. CNN models have the upper hand over RNN in this setup for feature generation. CNN is also featured as fast training options with lesser features expected. Random forest classifier, which was trained on neural feature representation, was performing better than cases of standalone CNNs and RNNs. This also worked better than the benchmarked BOW based classification model. Figure 2.4 depicts architecture of solution implemented in the work discussed here.

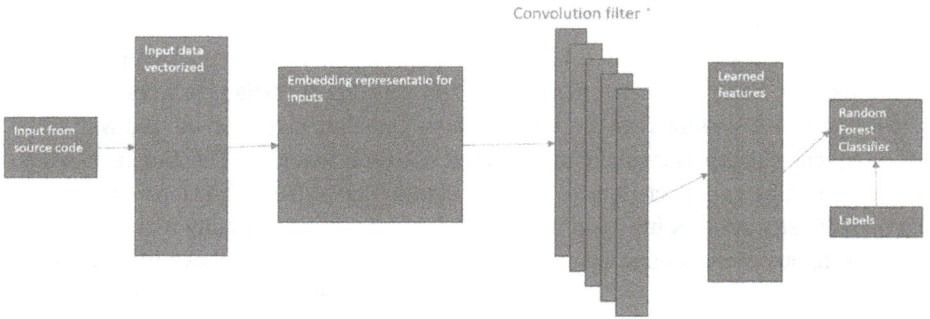

Figure 2.4: Illustration of the approach used in the work reviewed here.

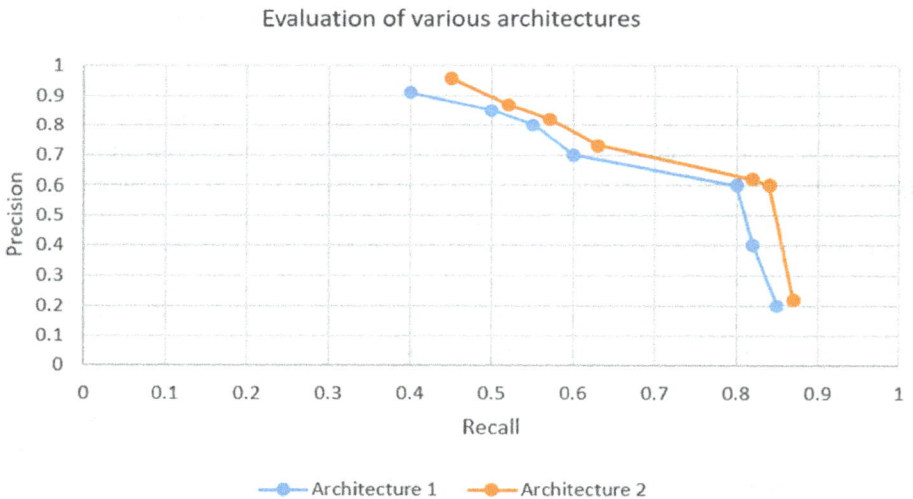

Figure 2.5: Graphical representation of evaluating various architectures based on precision and recall metrics.

Precision versus Recall metrics plotting helps to visualize the performance of the various combination of architecture experimented with. Figure 2.5 depicts graphical representation of precision and recall being plotted.

Receiver operating characteristic (ROC), which is the plot of TP rate vs FP rate, is another visual representation of the model performance; this is also called area under curve (AUC). Figure 2.6 depicts Receiver Operating Characteristics curve. The higher the

Receiver Operating Characteristics Curve

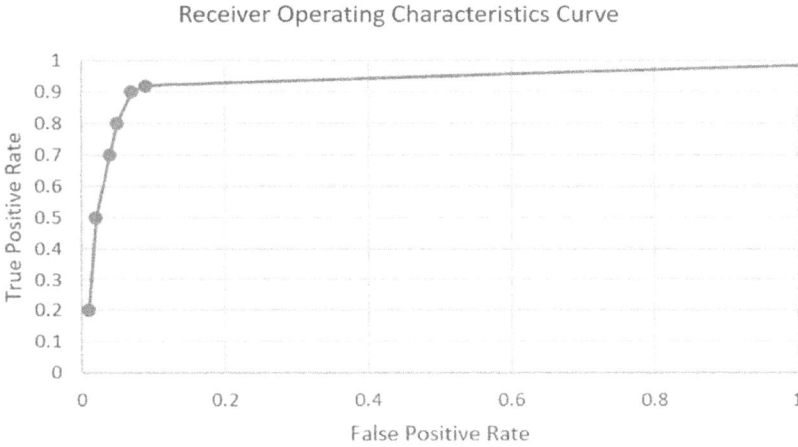

Figure 2.6: Receiver operating characteristics curve.

area under the curve better is the performance. F1 score, which is the harmonic mean of precision and recall, provides an overall indication of the model's performance. The static code analyzers' version has also been assessed against the ML models in the approach. ML models had the upper hand on the static analyzers by not needing to compile the source code but can assess and score on an extensive source code repository. ML models also specialize in outputting probabilities, which provide the option to tune the modeling and balance the precision and recall as needed. Static analyzers may come up with many findings for large codebases or end up figuring out a few of them in the case of critical applications. ML models help to visualize the outputs by using an approach like feature activation maps for convolutions. This feature activation map is laid over the original code to visualize the vulnerabilities.

This approach demonstrated the ML models' role in assessing vulnerabilities from source code. Further enhancement of the system can be done by improving the labeling of functions using derived information from security patches or dynamic code analysis; this will assist ML model scores in acting as an add on top of static analysis tools. If the data set labeling can be advanced, deep learning for the source code can get extended into a wide variety of options (Rebecca et al. 2018).

2.7 ANN for software vulnerability assessment

ANN for SV assessment as a web service: This section focuses on web application vulnerabilities being unearthed to facilitate better verification mechanisms. The approach aims to build a vulnerability prediction model on the Azure cloud computing platform. The experiment on various models available in the forum showed that multilayer perceptron dominated with performance. Set up with provision to pass on the specification of the metric provided by users, from a web form, into vulnerability assessment model is devised. In turn, predictions were sent back to a web form for users to refer. Data sets had vulnerability from Drupal, Moodle, and PHPMyAdmin projects. ANN stood out as the best format for web service provision to model the vulnerabilities assessment.

Security is a large part of focus during the development of software. Though security tools play a large role, vulnerabilities continue to be part of the system. This becomes more critical in the case of the software-intensive system used in health care. Hackers use these vulnerabilities and create damages, and these go unnoticed for a long time in many cases. Open Web Application Security Project (OWASP) provides top threats for web applications cross-site scripting, broken authentication and session management, insecure direct object references, sensitive data exposure, missing functional level access control, and others. AUC metric will assess the performance of the models. The best model was hosted as a web service to generate the metrics performance of the model. Some of the metrics associated with source code that were considered features were total external calls, lines of code, cyclomatic complexity, number of functions, nesting complexity and so on. The label or dependent variable was vulnerable and non-vulnerable modules. So, this is a case of two-class classification. Vulnerabilities in the data sets used were classified into various standard categories. Azure Machine Learning Studio experiment was built for the model. Figure 2.7 depicts Design Of Experiment in Azure Machine Learning Studio.

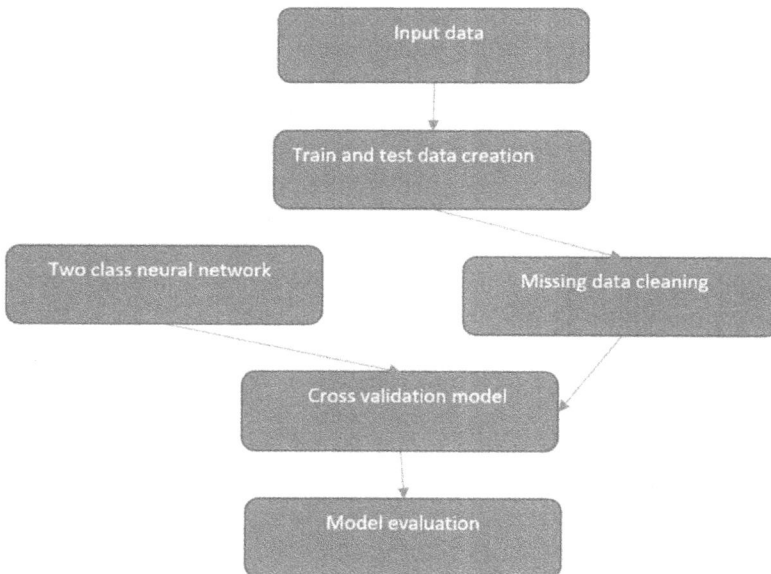

Figure 2.7: Representation of design of experiment in Azure Machine Learning Studio.

Two-class neural networks were employed after experimenting with several other algorithms like multiclass decision tree, multiclass decision jungle, multiclass neural network, multiclass logistic regression, one vs all multiclass, and two-class boosted decision tree, two-class locally deep SVM, and others. Three-fold cross-validation was employed, and the AUC score was considered as a reference for scoring. Acceptable AUC scores to be configured, this value will depend on the domain in which the experiment is conducted. Generally, the AUC value of 0.7 is a reference for the software engineering discipline. The best combination of count of neurons in the input layer, hidden layer, and one output layer to be worked out. Output neurons will classify the module as vulnerable or not. Input neuron represents the software source code metrics for web applications. The basic form of the neural network is an averaged perceptron that works well with linear patterns. The Bayesian approximation approach(approximate Bayesian computation) is employed in the Bayes point machine case. The ensemble of the tree is configured in a boosted decision tree. Decision jungle is an extension of a decision tree specializing with a combination of acyclic graphs that are decision-directed–nonlinear SVM like locally resonant SVM. Logistic regression, a basic supervised learning algorithm, and multilayer perceptron neural network were the models experimented. A neural network that had more than 0.7 AUC was compared to all other approaches. Bayes point machine and logistic regression have good performances as well. Web service was built on the Azure ML platform, with a web form for users to plug in the metrics related to software source code, which is passed to the prediction model, and results are published on the web form. Random forest has been reported to have been performing well in some studies where a combination of Azure ML and Weka platform has been used, which would also be due to parameter tuning being experimented with.

In contrast, in this study, parameter tuning has not been accounted (Neuhaus et al. 2007). Metrics calculation systems can be integrated with this modeling system. Multilayer perceptron model excels in the study, and further study on ANN like radial basis function network and Boltzmann machine can improve the performance (Cagatay et al. 2017).

2.8 Text analysis techniques for software vulnerabilities assessment

Most of the studies have banked on feature creation from the software source code, but this approach utilizes software source code as text. This approach is promising on 18 large-scale mobile application versions. Textual analysis of source code is the key feature of the source's system, and monograms will be treated as features. The email client application of Android is used as a base for the prediction model. Eighteen subsequent versions of this application have been utilized, with good precision but low recall reported. Comparatively, vulnerabilities prediction has lesser approaches being tried compared to one on defect prediction. Vulnerabilities are also lesser than defect, and work done on defect prediction cannot be directly transferred to vulnerabilities prediction. Vulnerabilities are bugs associated with the security flaws in the software. There is work on establishing a relation between import statements and exposures. Though the correlations have been found between code metrics and vulnerabilities, such correlation's statistical significance is low. Works on establishing a relation between developer activity, complexity, and code churn on exposures using linear discriminant analysis and Bayesian network classification methods have resulted in good relations among the parameters (Shin et al. 2011). The approach discussed here does not bank on the derived feature from code metrics and do away with the assumption of the importance of the feature's influence on vulnerabilities. The disadvantage would be in failing to establish useful features out of the study.

The approach taken is that Java files are considered as text. Each file of the program is a feature vector, and every word in the file referred to as "monogram" will be considered a feature. Before feature extraction, data must be preprocessed to remove the unwanted entities such as comments, which do not have added value to identify vulnerability patterns. Strings and numerical values are other entities that can be removed. Preprocessed source code to be converted to a feature vector; before that, all the content of the textual representation in the code must be tokenized to monograms. White spaces, Java punctuations, logical operators, and mathematical operators are also counted as monograms. In a feature vector, the count of the authorizations is considered as the identification of the feature. During training, each feature vector that is the file will have a label of vulnerable or non-vulnerable. Later, once the model is built, it must be fed in as feature vector-like in the training phase to predict the new case.

SVM concept is utilized for the training phase where the prediction model is built and the prediction phase where the feature vector is passed into the model for inferencing based on a model built in the training phase. Radial basis function with appropriate other parameters are used in SVM; these are experimented with using a grid search algorithm. The trend shows mobile sales going up than compared to computer sales. A large base of users is on Android. So that explains the selection of the Android platform for this study. Open-source repositories of the Android applications can provide an open platform for study. The prediction model is built on a version and being inferences on the subsequent performance. Labeling is done by leveraging on Fortify static tool analysis that provides the severity of the issue on top of labeling vulnerable or not. Fortify analysis is based on the common vulnerabilities reported. There are claims of correlation between vulnerabilities reported by static code analysis tools and vulnerabilities reported in real. But there is quite a problem of FPs out of prediction done by these tools.

Accuracy, provides overall correctness related metrics, precision is, the count of correct predictions out of all that is predicted as vulnerable and Recall is the count of correct predictions among all that are labeled as vulnerable. Naïve classification of the Fortify static code analysis stands as the base for the study. And accuracy demonstrated by the model studies stands better than the base performance. The study also featured new files being tested in the testing phase with high accuracy this is a good trend of model performance. The study can be further built on exploring various possibilities of the vector representation of the text features. Since Fortify also provides the severity of the vulnerabilities, it opens the opportunity to construct a multiclass classifier to classify the rigors. This study can also be complementary on top of models that work on code metrics as their feature (Aram et al. 2012).

2.9 Software vulnerability assessment automation

This section throws light on the problems associated with NLP-based models that fail to account for the models' drift due to new SVs getting introduced into the ecosystem. This issue is mainly attributable to a lack of work on handling these SVs' dynamic entries over time. In this approach, it is targeted to predict few standard SVs. Best models are built for each of the SVs' distinct categories with cross-validations methods across the timeline, utilizing various NLP representation and ML models. National Vulnerability Database (NVD) is used as a reference to study thousands of SVs based on their descriptions. The model features the possibility of performing well without even the need for retraining the model. The approach also explores the possibility of building the aware drift models with lesser features and best classifiers and appropriate representations in NLP.

Different SVs have an extra level of impact, and it is essential to understand the effect to make sure that the right prioritization is done for action initiation. The change influences vulnerability databases in the characteristics of the descriptions over time, which is to do with, the new products and technologies. Thousands of unique SVs' descriptions get updated into NVD every year. This nature makes the model building complex. Approaches have not accounted for the possible drift in the SVs' descriptions. Even the exploration of machine translation to handle out of vocabulary words has not been sensitive to feeling future changes for predicting the vulnerabilities. Concepts such as using embedding vectors that are randomly generated have failed to represent vocabulary terms. This approach discussed focusing on increasing the SVs' assessment framework's robustness to handle this drift experience due to a change in vulnerabilities.

The study of the descriptions existing in the SVs at character, and word-level features,are experimented with machine learning. A customized cross-validation method based on time opts for a selection of the model. The cross-validation methods intend to split the data based on time, which helps account for the time-based changes happening in SVs. The study was focused on the SVs' key characteristics, like authenticity, integrity, confidentiality, and severity. The study characterizes by bringing out the influence of the time-based drift on SVs as per NVD. Since the time-based split of the data was done, it reduces the possibility of future characteristics of the current models' data being learned. Character word–based model is proposed to address this drift over time.

NVD is a reliable source of SVs as government entities maintain it. This database is also well validated using a standard validation framework such as the Common Vulnerability Scoring System (CVSS). Severity, impact, and exploitability are the key factors accounted for in the framework for validation. Confidentiality, availability, and integrity would be the critical building blocks for the factors that are mentioned. Each of these characteristics has three levels of outcomes. This will be a multi-class classification problem from an ML perspective. Some of the attributes like authentication, access vector, and access complexity have class imbalance due to lack of data compared to other classes. Figure 2.8 depicts format for representing frequencies of each class of vulnerability characteristics.

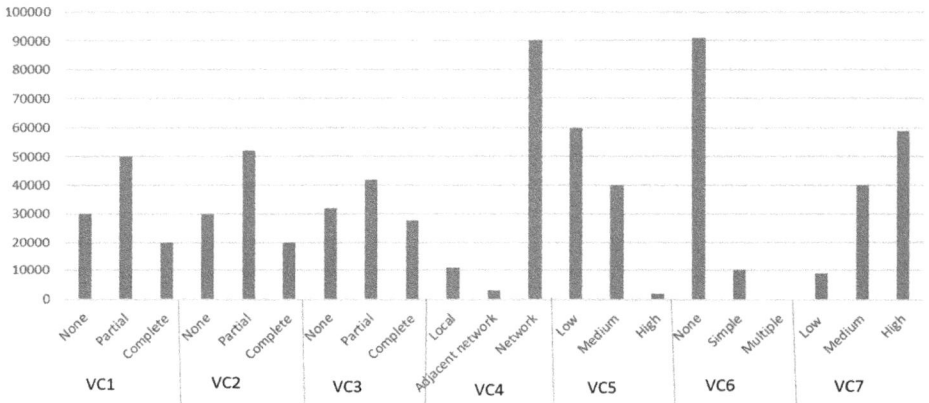

Figure 2.8: Frequencies of each class of vulnerability characteristics are visualized as shown in figure.

Model selection and building phase work on training data. The prediction phase is run on a separate set for testing. The model selection part has to preprocess the text and time-based k-fold cross-validation. The next step is time-based k-fold cross-validation is to derive appropriate representation for the text. In this step, the focus is on the word-level model and not to do wholistic concentrate on the word- and character-based model to avoid massive data space complexity. As the character level representation will get complicated due to the need for different NLP representations. Model building part encapsulates word and character n-gram generation, rolling up the features and model building for the components. Preprocessed data is fed into these characters and word n-grams to arrive at the best NLP representations. These representations are focused on each of the vulnerability characteristics (VC).

Character n-grams facilitate the coverage of the out of vocabulary terms to facilitate the management of concept drift. Feature aggregator rolls up both versions of n-grams. Optimal NLP representation is derived from this rolled up an aggregator for each VC,

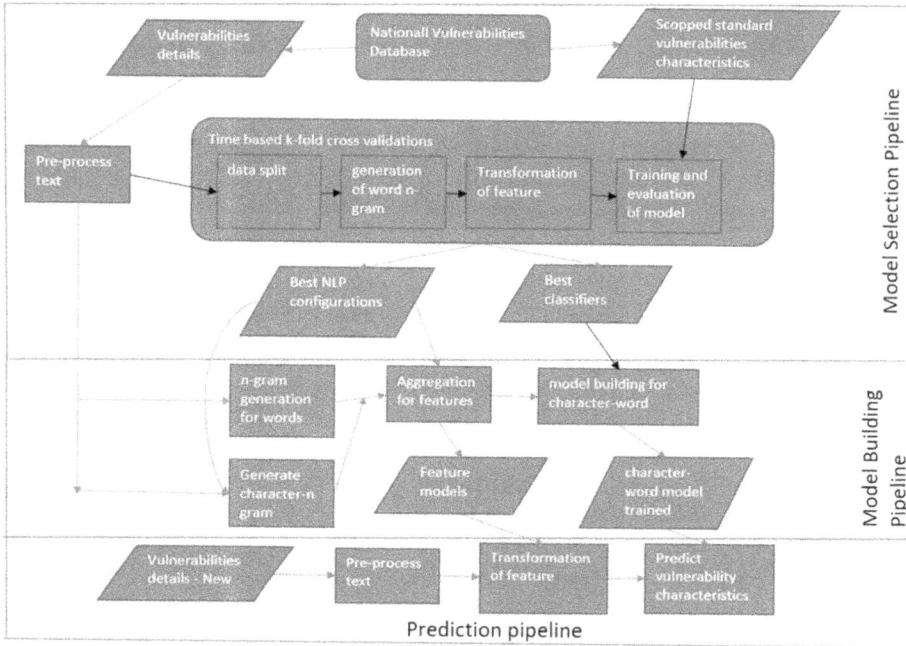

Figure 2.9: Workflow of the model proposed in this work that is reviewed.

Sentence: Hello All

n-gram	Word	Character
1	Hello, All	H,e,l,l,o,A,l,l
2	Hello All	He,el,ll,lo,o_,_,A,Al,ll

Figure 2.10: "Hello All" sentence represented as word and character n-grams. "_" represents space.

which will become a base reference for future prediction. Aggregated features are trained with optimal classifiers based on the model selection pipeline of the framework. Ultimately the character word–based models are built for VCs to facilitate automated vulnerability assessment with accounting for drift over time. Figure 2.9 depicts workflow of model proposed in the work that is reviewed here and Figure 2.10 word and character n-grams representation. Later in the prediction phase, new descriptions of the VC are processed using the same processing pipeline; then the processed content is run into the saved model weights from the model building pipeline. The final character word model is used to classify the VC class for the test case.

Text preprocessing involves removing the stop words included from NLTK and Sci-kit to learn libraries of the ML. Punctuations removal, a lower casing of the content, stemming down to a needed version for NLP context, and space removal are some of the other processing steps. Punctuation removal is customized to retain the punctuation essential to maintain the relevant phrases to software and security context like "cross-site." Porter stemmer is used, and stemming helps cut multiple versions of the word into its root forms, like fishing and fish transformed to "fish." As the algorithm just needs the root word from the content to gather the pattern from the content. Lemmatization is another approach that can also be experimented with; this retains the context around the word.

2.9.1 K-fold cross-validation-based model selection on a time scale

In time-based k-fold cross-validation to identify best NLP representations and classifiers, the data pipeline is organized and timeline-based to ensure that the time-sequenced information is not leaked into a training set that facilitates the handling of concept drift. Each split of the data is organized year-wise, so every next set has features of the SVs' new characteristics. Different combinations of word n-grams and the term frequency or tf-idf (Term frequency, inverse document frequency) are employed to create features. Term frequency applied on unigram is nothing but BOW model. Various combinations of the NLP representation are validated with a combination of classifiers and then validated with the metrics' variety. Average cross-validated performance of the model is assessed, and one with the best performance is selected as the final model for VC. The process is done on each of the VCs, and the final model is stored for each of it.

2.9.2 Aggregation of the feature space

An aggregator works on word and character n-gram model combination to build the model. With word n-gram, character n-gram, descriptions of SVs, the maximum and a minimum number of n-grams, and the best variety of NLP representations are all utilized. The final output is the weights that transform the files into the word character model. These models will also serve for the prediction of future inputs. In feature building, most care will be taken to make sure that the model does not overfit or fail to generalize the learning. Strike a delicate balance such that the model would still handle vocabulary words in the future. Once the model configuration is done, the input document is converted to the word feature matrix and character feature matrix. Both feature matrixes are concatenated together to form the final feature matrix. We have a word model, character model, feature matrix of the word, and the sentence's feature matrix. NLP representations n-grams, tf-idf, and term frequency are run on nltk and scikit learn libraries of python and classifier are built using it.

The first focus area is to demonstrate that this time-based data validation approach is better than other conventional methods that do not account for time influence. Thus, the study reviewed the traditional process and its lack of handling information leakage into the past configuration. For this review, each of the terms was mapped to their release year and the relative study was done in the vulnerability assessment scenario. To ensure optimal model selection for multi-class classification of vulnerability characteristics, it was another key focus area for the classifier and NLP representation. For each of the VC word feature experiments with five-fold, timeline associated cross-validation methods. Different combinations of single models and ensemble of models can be experimented with. The character model's influence was demonstrated by comparing the word's performance only model in terms of their ability to handle out of the vocabulary terms and, in turn, managing concept drift. Since the feature space is a large and sparse, a dimension reduction approach to validate the lower dimension feature space concept was looked at. Sub-word embeddings and Latent Semantic Analysis were the technique explored; in this review expectation was to assess how much of the pattern can be retained as we move down to lower dimensions. This attempt helps to optimize the model and make it more efficient.

Top ML models were chosen to build a multi-class classifier. Naïve Bayes (NB) that operates on Bayes theorem (Bayes' theorem) is a model that assumes each of the features are independent of each other. Parameter tuning was not done for this study. Logistic regression (LR) is the linear classifier where the logit function classifies linear output in terms of the probability. Format of one vs rest is used in this classifier; here, multiple binary classifiers are run to conduct multi-class classification. For parameter tuning, regularization is done using the learning rate parameter with an optimal range of values.

SVM classifier works on identifying the maximum margin between the classes. The linear kernel is the parameter that works better for NLP contexts as it handles sparsity and is also scalable. The learning rate will be the parameter tuned in this case. Random forest algorithm utilizes the concept of bagging, which controls the noise and optimizes the variances. A maximum number of leaves, depth of trees, and trees configured are the parameter to play with. Center can be left unlimited to make sure it can handle extensive data and flexibility to assess new data, which is needed for our use case.

Extreme gradient boosting (XGB) is an approach in which a combination of weak learners results in the model's optimal performance and robustness. Like the Random forest, the hyperparameter is tuned. Light gradient boosting is a lightweight version of XGB, which is characterized by the option of subtree being grown in leaf wise manner rather than burdening the model with the depth. Hyperparameter to be tuned are the same as XGB. Naïve Bayes, SVM, and LR are run as standalone models, whereas LGBM (Light Gradient Boosting Machine), RF, and XGB are run as an ensemble of models. As discussed earlier, experiment is run as multiple binary classification problem. Evaluation metrics are chosen based on the combination of TP, TN, FP, and FN combination.

The first part of the representation denotes how the right a prediction is true or false, and the second part indicates an actual label. For example, in the case of TP, classification is correctly done to suggest SV's specific characteristics. In the case of FN, the sort is wrongly done as particular attributes of SV is missing. Accuracy, recall, precision, and F score are a combination of the factors mentioned above to provide different model prediction perspectives. F score is devised to assist the computation of performance in unbalanced data, such as with the metrics like authentication and severity. To get a more progressive outlook of the model, F scores can be computed in a macro and weighted F-score. The Macro F score does not consider the count of elements in each class, whereas the weighted F score counts in multi-class classification. Weighted F score also helps to play the decider's role in case the models have the equal performance on accuracy and macro F score. Occam's razor principle (Occam's razor) recommends using a less complicated model and a smaller number of parameters. Model training time can act as the ultimate parameter to decide.

Qualitative and quantitative analyses of the models selected and their influence on the concept drift have been evaluated. Variations in the SVs are influenced by the release of new products, cyberattacks, and new software. This results in NVD being updated with thousands of SVs every year. There has been a strong correlation between new terms appearing in the NVD and the period in which the major change happened in the industry in terms of influencing factors mentioned above. Attacks such as code red, technologies like Node.js, operating systems like Android, and products like iPhones bring in the new wave and further lead to new vulnerabilities. The degree of overfitting also studied between time-based models and models that did not account for time. Each method was weighed based on the difference between the weighted F score of cross-validated samples and the testing sample; this helps to bring out overfitting possibility. Access complexity, access vector, and availability are the SVs that have a higher degree of overfitting when time-based and non-time-based models are experimented with.

For each VCs to arrive at the optimal model, five-fold cross-validation was used across different classifiers that we discussed earlier and various combinations of the NLP representations discussed. Also, words appearing at least in 0.1% of all descriptions have been used as a study threshold. A local optimization strategy was adopted to reduce the data search space. Logistic regression would require a relatively large learning rate value, as its boundaries are sensitive to change in hyperparameters. Also, tf-idf being topped up with L2 regularization will help faster optimization but it needs more regularization for avoiding overfitting. In the case of an ensemble of models, since multiple models are involved, many parameters are to be tuned. In the case of LGBM and XGB, a drastic increase in several leaves did not help but increase computational time. In the ensemble methods, the tree's max depth had a significant influence than other parameters. Five-fold time-based cross-validation was applied to the combination of various NLP representations, classifiers, and SVs categories. Combinations of NLP representations were n-grams, tf-idf, and term frequencies. Tf-idf did not have better performance compared to term frequencies. Cases of n-gram, greater than one, influenced improving the results. To validate the statistical significance of this improvement, right-tailed two-sample *t*-test unpaired was used. *P*-value resulted in being greater than the confidence level of 0.05. This meant there was no evidence of performance improvement due to an increase in *n*-grams. This inferred that the BOW model works well rather than complex NLP representations. Another primary validation area was among the single models of LR, SVM, and NB against the ensemble model of LGBM, RF, and XGB. Ensemble models performed better compared to an available model; ensemble models also demonstrated consistent performance mostly. Figure 2.11 depicts format for comparing average cross validated weighted F-score, for various vulnerabilities characteristics, for their single and ensemble model.

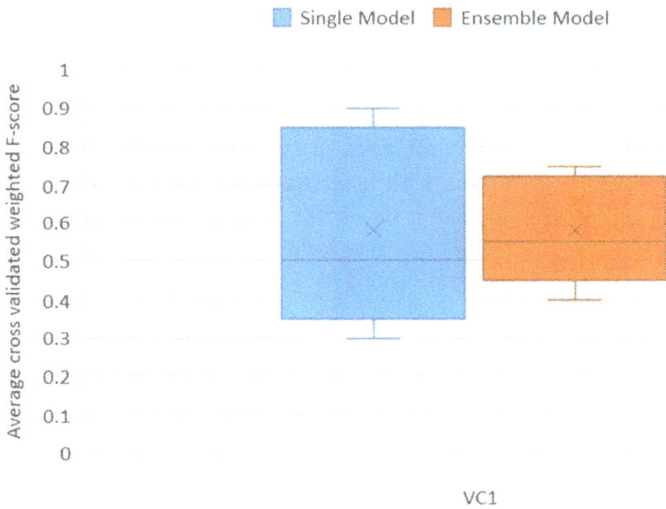

Figure 2.11: Plot like this is generated for all vulnerability characteristics for both single and ensemble of models to visualize the variation in "average cross-validated weighted F-score," to assess the best performing model for each of the VC.

Above mentioned hypothesis tests confirmed that the ensemble methods are superior. Compared to LGBM, XGB models take more time to train, which is significant when tf-idf representation is used. LGBM, XGB with BOW, would be the right combination.

2.9.3 To assess the influence of the character-based model

Out of vocabulary (OOV) significantly influenced the word model, not sufficing the situation and need for a character-based model. Study of the cases where the words were not occurring in the text of vulnerabilities description is an area to focus on. Data was split into train and test set based on the period, past data for a train set, and post data

as test data. With this check, all the words which occurred at least in 0.1% of all descriptions were shortlisted. All zero cases of words can be reduced by increasing the
vocabularysize but with a trade-off on the computation time. Some of the patterns that
came out of the study of vulnerabilities description are that vulnerabilities have a
shorter length than non-vulnerable reports. This indicates the lack of information related to vulnerabilities in words. This lack of knowledge is attributable to the fact that
the vulnerabilities' description is restricted to attack type and assets affected. For
building a character-level model, the condition of words to be available in at least
10% of the descriptions was considered. Considering 0.1% availability as cutoff, in
the word model case, led to large dimension space, and performance was also not
that significantly high. For character model, two n-gram models were taken as a
minimum, with a max of three n-gram models. With this setting, data is broken into
training and test set on the timeline, as discussed earlier. It showed that at least
there was one nonzero feature for all descriptions just for the first year of data understudy for the training set. This gives confidence in the approach used and its
capability to handle the concept drift, even with fewer data. To build in the generalizations into model n-gram, a range of n-gram values were taken and mapped
against the vocabulary size. Figure 2.12 depicts elbow method to arrive at an optimal n-gram size for overall size of the vocabulary.

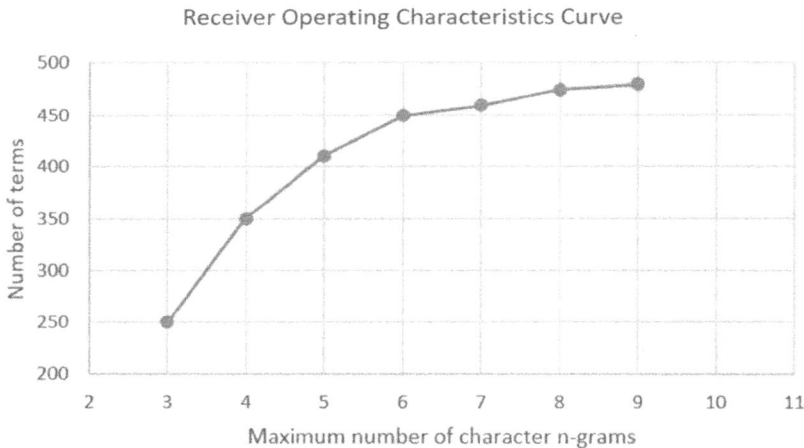

Figure 2.12: Elbow method to arrive at an optimal n-gram size for the size of the vocabulary.

n-Gram range versus the vocabulary size to arrive at the optimal n-gram size is plotted. This is the elbow method for cluster analysis. The minimum and average length
of words in NVD was in line with the max and min n-words that were chosen as in
the study of the n-gram. Feature aggregation for character n-gram that are chosen

for selected range and the word n-grams, were used to build the model. This model was compared against the baseline model of the word-only and character-only model (COM). Word-only model did not have concept drift being handled. With a combination of BOW and RF with specified hyperparameters, character word model (CWM) performed slightly better than word-only model (WOM). CWM also resulted in five times more nonzero compared to WOM. Overall COM was poor enough, understandably due to less information. CWM does not outperform WOM significantly, but the key selling point is CWM's ability to handle OOV terms.

Another critical part of the study was how well the pattern be retained if the dimensionality of the data space is reduced. Generally, NLP models of n-grams are impacted due to the high dimensional nature of the data and its sparsity. Subword embeddings, like the fast text method, which is an extension of the word2vec embeddings framework for word, were utilized. Latent semantic analysis (LSA) was also experimented with for dimensionality reduction. The fast text considers the character-level features. The fast text also exhibits the capability of using the context for words based on its surrounding words, which is unique compared to traditional methods. In this study, a sentence was represented based on the fast text's average embeddings based on characters and words. Gensim library in python was utilized to run fast text. In LSA's case based on the elbow method and considering the amount of variance explained by the principal components range, 300 dimensions or principal components were shortlisted. Combinations experimented with were CWM, LSA-300 (LSA with 300 measurements), fast text 300 (fast text trained on the same 300 dimensions as of LSA), and fast text 300 W (fast text trained on 300 dimensions from English Wikipedia page).

Fast text 300 performed better than LSA-300; the fast text model got better than CMW in some VCs. Fast text 300 was performing better compared to fast text 300 W. This infers that the vulnerabilities database has a different vocabulary range compared to regular domain vocabulary. Overall, LSA and fast text demonstrate the capability to build efficient models at reduced dimensions. As a validation tool, nltk and sci-kit learn were used, which are popular. Still, since many hyperparameters are involved in the configuration, performance may not replicate across all types of SVs. But it provides an excellent base to build on. Overall there is a possibility of improved performance with even necessary parameter tuning, reason being time-based cross-validation is incorporated, and generalized trend of SVs is accounted. Since the source of data is NVD, which is a comprehensively maintained public data repository, the data set chosen for the study is also significantly large and representative of the latest vulnerabilities at the point of the study. Also, CWM has been demonstrating the ability to handle OOV words efficiently with limited data involved. Though there are scenarios where rare terms will fail, the model can be re-trained to build new entities' knowledge. This helps to build the capabilities for more SVs beyond those targeted for the study. The randomness of the study was accounted for by taking the average performance of the five-fold cross-validation method. The confidence level of the NLP representations was validated with a hypothesis test by looking for statistical significance based on the *test's p values* compared with a confidence level of 5%.

CVSS method of SVs' evaluation was used as a reference among many other possible approaches. A recent Bayesian analysis study on CVSS has proved the effectiveness of the CVSS as an evaluation technique (Johnson et al. 2018). So, the approach used in this study can be generalized to other vulnerability rating systems with the common theme of multi-class classification. There is work that has focused on the possibility of vulnerabilities getting exploited and the time at which the exploitation happens, using the SVM model (Bozorgi et al. 2010). There is work on analyzing the VCs in SVs based on multiple sources of the vulnerabilities data such as dark web and social networks, whereas this study discussed here has focused on NVD (Almukaynizi et al. 2019, Edkrantz et al. 2015, Huang et al. 2013, Murtaza et al. 2016, Nunes et al. 2016, Sabottke et al. 2015). Studies have been done on the VCs of SVs' annual effect in NVD, but no success in accounting for OOV terms. Vulnerabilities description can be studied to assess the severity of the vulnerabilities, look for most used terms in VC, and map the types to SVs using the concept of topic modeling. Deep learning techniques have been explored for assessing the severity of the vulnerabilities, and this has resulted in useful information (Han et al. 2017). Still, this study has not considered the concept of drift issues.

The time series approach has been explored to handle the time dimension of SVs. The autoregressive integrated moving average is one such area with exponential smoothing (ES) methods being explored to predict vulnerabilities' prospects. Time series analysis has also been used to expose the trend in SVs (Roumani et al. 2015, Tang et al. 2016). Occurrence and exploitability have been subjected to study with Markov models, ANN, focusing on identifying the possibilities of the vulnerabilities' event across time. Further study on utilizing deep learning to integrate the dependency of character and word model in lower dimensions can be focused. With the drift of the concept, handling the imbalanced data scenario also is key (Triet et al. 2019).

2.10 Vulnerability assessment with stochastic models

Exploitability of the vulnerability's assessment with stochastic models: Though gathering all the needed information for the vulnerabilities plays a significant role in making the prediction model effective, there are challenges. In the earlier studies of these authors, before this one, there has been an effort to use the Markov approach to explore the probability of the vulnerabilities to be in a certain state across the life cycle, which brings out the associated risk factors. The study here builds on exploring more data to align with probabilistic exploration to make the model reliable and build with new modeling strategies to enhance the estimation of the risk probabilities. The focus is to explore the nonlinear model possibilities, which provide the probability of vulnerabilities. The study targets the generic security system and the possible vulnerabilities, which can be replicated by the users in their own vulnerabilities system settings with time as a factor. This aspect plays a major role in having a stronger security strategy for the companies so that the systems are strong enough and not exploitable.

From essential personnel apps to enterprise-wide apps, they are exposed to the risk of cyberattacks and losses. The hacking fraternity is growing more robust, and system security teams have been up to it, but the battle has been fierce, and advancement on both sides has been fast paced. Understanding the strategy of hackers and counteracting has been a critical need for security experts. To make a foolproof security system, the essential element is to understand the risk of system exploitation, so that reasonable measures can be put in place and required resources provisioned. For this, there is a need for building effective models that can highlight risks involved as a function of time. An earlier study of these researchers that is covered in this section was focused on arriving at a transition probability matrix for all the states of the vulnerability with time as a factor utilizing the Markov process. Markov processes (Markov decision process) were analyzed to examine the state of vulnerability, reaching exploited state and patched state. Based on the exploiting state, a risk factor needs to be arrived at for vulnerability being exploited, ultimately to model the risk without going into Markov processes. In a previous study logical approach was used to set initial probabilities of the vulnerabilities. The current study proposal is to consider more refined techniques to arrive at the same. CVSS score will be continued to be used as a reference like earlier study; also, Common Vulnerability Exposure (CVE) database will be used upon to arrive at initial probabilities. Overall, three different models will be built based on a different vulnerability range in

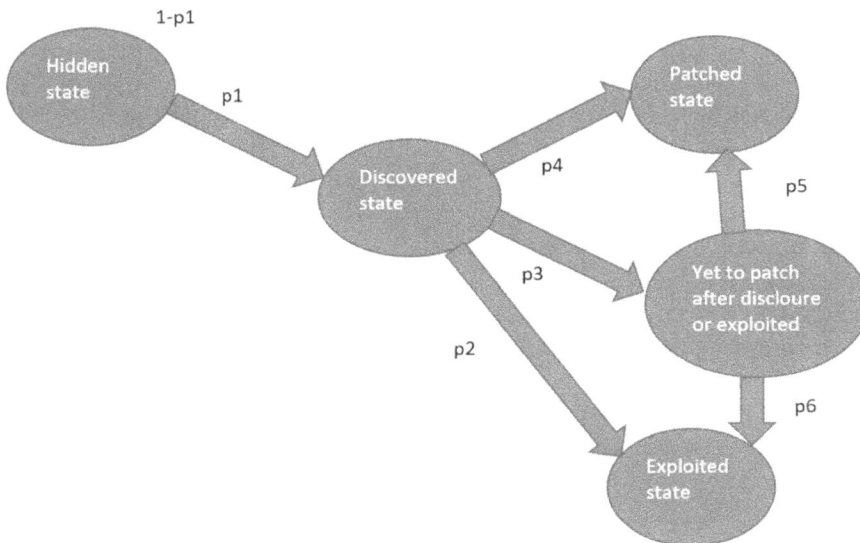

Figure 2.13: Vulnerability lifecycle states represented as Markov model with state transition probabilities.

the class of low, medium, and high risk. These models provide a base for identifying the risk of vulnerability getting exploited at other points of time in its life cycle. Figure 2.13 depicts vulnerability lifecycle states represented as Markov model with state transition probabilities.

In the life cycle graph of the vulnerability, each node represents the state of exposure in the life cycle. The probability of the hacking happening on each node was arrived at based on the property demonstrated at every node for vulnerability. Patched and exploited are the end states of the cycle. From every node, there is a probability of the state changing from one to another or remaining in the same state. Since the probability of discovering the vulnerability is small, it starts with a small value, and accordingly, the probabilities are distributed across other states. When random analysis of the probabilities was done, it was noticed that the probabilities, were the function of time. In the study, probabilistically, more relevant methods are used to arrive at the initial values. Though it is challenging to reach complete information on the vulnerabilities, reliable data sources are explored to build this.

CVSS is the standard mechanism that scores the intensity of the vulnerabilities. Scoring is done on a scale of ten, accounting for various parameters. NVD provides the CVSS score on the base aspect. CVSS are computed on base, environment, and time temporal aspects. The reason for NVD providing only base aspect is due to other two factors being influenced by the workplace using the computer systems. CVSS provides the relative importance of the vulnerabilities so that the focus of resolution can be planned accordingly. Models intended to be built are based on three classes of CVSS score with high, medium, and low. CVE is another reference database for vulnerabilities having CVSS scores.

To arrive at the initial probabilities, a large amount of information is needed, such as details of each category's vulnerabilities across the high, medium, and low categories. Also, we need to know the information about the identification of the vulnerabilities across a time scale. On the same lines, for other states in the life cycle, information is needed to count vulnerabilities for different CVSS score classes. CVSS and CVE are the primary references for getting the required information as stated; if the information is not available, other valuable sources are explored. To arrive at the initial probability of a vulnerability being discovered, we may have to look at the total population of the known and unknown vulnerabilities across all CVSS categories so that we can get a proportion of known ones. Since it is not practical to compute the system's total vulnerabilities due to the ecosystem's dynamic nature, to logically gauge the situation, the cumulative count of vulnerabilities in the year is looked at. Then the ratio of newly discovered exposure for next year is compared against the cumulative vulnerabilities, of previous year. One of the key considerations is that the vulnerabilities that are not found in a year would be found in the subsequent year, and the cumulation of the vulnerabilities in a year will be population size of vulnerabilities for the previous year. CVE database was base for computing the exploitation possibility before being found or patched. In this case, the calculation was done for all ten ranges of CVSS scores. One of the literature's key learning was the possibility of exploitation being higher than patching once the vulnerability is found. Literature provided information that there is a 60% possibility for the vulnerability to be exploited after being found, so this value was considered for model building. Based on the transition probabilities of the state of vulnerabilities, the transition probability matrix is derived that is based on the Markov chain transform method for probability.

2.10.1 Transition matrix for the life cycle of the vulnerability

General form of the probability matrix in transition need to be derived across five states; not found, found, exploited, patched, found but not patched or exploited. The initial representation can take the form of [1 0 0 0 0], which represents that the initial state of not discovered in 1 and other state is at 0. The time dimension is accounted as we look at other state probabilities, which are the Markov process's key role. The Markov process's complex computation is used to account for various categories of vulnerabilities across time until the stop point of a steady-state is reached. Probability of the vulnerability at every state and different point of time is used to build the vulnerability transition matrix for every category, which plays as input for deriving the model. For every category, we transition to steady-state with patching or exploited as the ultimate state. At the steady-state, we would get a probability of being patched or being exploited, with all other states at zero. Several steps taken to reach the steady-state are the same as the number of steps involved for a hacker to exploit the vulnerability, and the probability arrived in the process provides the possibility of being exploited or patched. On top of probabilities arrived above, more available information can be used to refine these. One of the key findings would be to look for the probabilities of exploited and patched state, summing up to one, which means the state end up in one of these possibilities. Final probabilities at steady state are the steady-state vectors.

2.10.2 Risk factor computation

A risk factor is computed based on steady-state vectors; it will be the product of the probability of vulnerability at time t for which risk is calculated, at exploited state, and exploitability score. An exploitability score can be derived from the CVSS score. The risk factor was more refined as information from more reliable sources is plugged in for calculation, compared to an earlier study. At any point in time, we need to get the state probabilities for vulnerability, making use of the Markovian process, but this is computationally intensive. So, to optimize, we must arrive at the models that are nonlinear. With risks found with date and risk factor, we

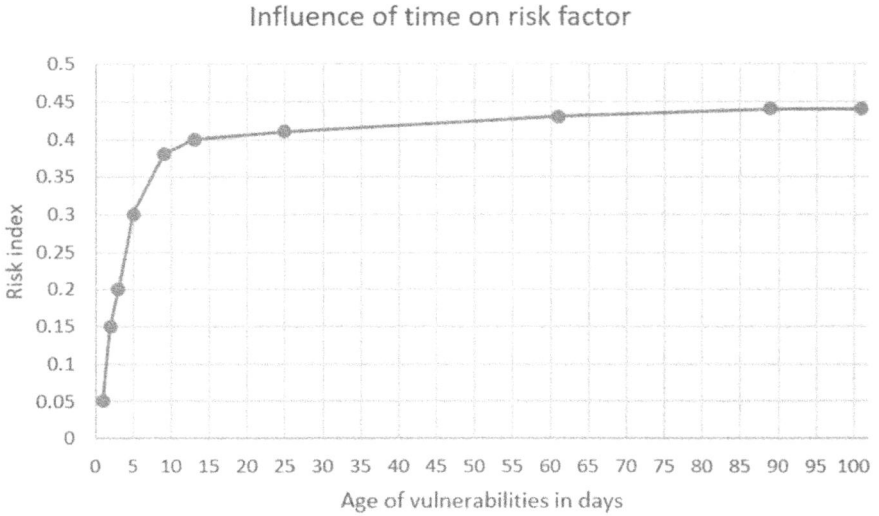

Figure 2.14: Risk factor plotted as a factor of time.

would find out the probabilities of exploitation at any given time. Overall vulnerability detail, published date, CVSS score, exploitability score, age of vulnerability, and risk factor information are available. Figure 2.14 depicts Risk factor plotted as a factor of time.

Risk factor when plotted as a function of time, it is noted that the initial increase of risk factor over time is high from the point of time when the vulnerability is being published. Later, there is no decreasing trend that is due to the way the scenario is configured in the study. Exploitability is considered as one of the final states with patching state. So, there is no further state beyond these, so there is a continued increase in risk though short, over time. This trend of the risk need not be considered the reference, to decide that the risk never comes down. The current Markovian model does not account for some of the key events from the real world; this study intends to bring out the risk until patched. Another variant to be considered, if the holistic view is looked at, is to consider patching attempts and exploiting attempts for both before and after exposure of the vulnerability.

2.10.3 Exploitability exploration model

The previous attempt was too analytical to build an algorithm to explore several steps involved before reaching a steady state based on the transition state's probability matrix. Since the Markovian process is time-consuming, we use the information about steps involved for each of the categories of vulnerabilities and the state probabilities at each state for exploitability to build the model as a factor of time. Since the data behaves nonlinearly with time, multilinear regression is out of consideration. The statistical model built for nonlinear consideration had two weight factors and time as an independent factor and exploitability probability as a dependent factor.

$$\text{Model 1} \rightarrow Y = A0 + A1^*(1/t) + A2^*\ln(t) + B \tag{2.2}$$

$$\text{Model 2} \rightarrow Y = B0 + B1^*(1/t) + B2^*\ln(\ln(t)) + B \tag{2.3}$$

Y is the output, A and B are the weights, and t is the time with B, as error terms. Maximum likelihood estimation is adopted to optimize the model weights. R-square and adjusted R-square values are the metrics observed for optimization. The R-square score indicates the goodness of the fit of the model or explainability of the model. Based on the learning of model 1, model two was refined by introducing another layer of the log, which resulted in improved performance. With R-square value, which is the coefficient of regression and adjusted R-square value, the residual of the original data was also conducted, though the same was not used in model building. R-square values are the ratio of sum of square regression (SSR) to sum of square total (SST). R-square adjusted value balances the unusual addition of the parameter and accounts for possible bias in the model on top of R-square value. Closer the R-square and adjusted R-square value to one, it indicates goodness of the model. Residual analysis of the model is crucial to make sure that the error pattern has not contributed to the model. Testing conducted on the set of data that was not included in model building demonstrated the useful model.

2.11 Vulnerabilities modeled based on time series methods

This study's key aspect is to account for trend, seasonality, and level elements of the vulnerabilities. Top browsers were the focus area to build a vulnerability trend for vulnerabilities reported across. Web browsers is taken up for study as it is an essential element for all information system access. On the flip side, browsers have been the target for the hackers. Though there is a more significant attempt to make software development processes vulnerability proof, naturally, the focus shifts on timeline and quality with security getting hit. That demands a greater need for good vulnerabilities prediction system. A significant exploration area has been source code mining, ML models, source code text analytics, historical vulnerabilities analysis, and software components analysis. Most of it needs the exposure of source code, so that limits the study to only open source, except for the vulnerabilities data that is publicly available. A time series study considers the inherent parameter of periodic, autocorrelation, and seasonal components of the time dimension. ES and autoregressive integrated moving average (ARIMA) models of time series are explored. Unlike other modeling techniques, time series techniques vary the weights based on providing importance to more recent data. This study facilitates the planning for the patching process in the future, based on the prospective trend of the vulnerabilities. This also plays a prime role in planning software security testing.

Earlier works have focused on code-related metrics and the application of mathematical models for prediction purposes (Shin et al. 2011). Developer metrics, complexity, and code churn have been proved to have a good predictive capability with low FPs. But the studies have been limited for a browser and are said to be not generalizable. Works have explored using the component dependency graph for vulnerability prediction (Nguyen et al. 2010). Vulnerability discovery model (VDM) has explored the probabilistic modeling based on the dependency built between the process used for vulnerability discovery and the factors affecting them. VDM has accounted for vulnerability type and its severity to enhance the predictivity. The major drawback of VDM is not accounting for seasonality, level, and trend of vulnerabilities. This information helps to bring out the regularities involved in data across time. VDM needs historical data to manage this vulnerability prediction method based on code properties that have been proposed. A stochastic model based on the compiler static analysis of the code was proposed for this. To avoid dependency on source code, proposals of supervised and unsupervised models built based on open source code have been done. Semi-supervised models have been reported to perform better over a supervised model for scarce vulnerabilities (Shar et al. 2014).

Time series modeling has found its place in forecasting. Range of utilization of the time series modeling can be seen across stock market, inflation rate assessment, software reliability, and security aspects. Time series methods do not intend to bring out the root cause but tend to learn the pattern to facilitate forecast or future prediction. ARIMA and ES work well in the context of the count of vulnerabilities being the focus of prophecy.

2.11.1 ARIMA

These are the univariate model for linear categories. Making sure the vulnerability data is stationary is the first need for ARIMA modeling. Static characteristics are about the mean of data; in this case, several vulnerabilities show no trends over a month on month. There are tests under ARIMA to ensure this characteristic. The next one is to figure out the structure of the model needed and its order. ARIMA model has three components: count of autoregressive terms (AR), count of moving averages (MA), and several differences. In the case of AR terms, it is computed if the number of vulnerabilities at t moment is autocorrelated with n number of previous terms. Several previous terms can be a variable. Different component considers the type of adjustment needed to achieve stationarity. MA accounts for several previous predictions to predict future steps. The best ARIMA model is derived based on partial autocorrelation (PACF) and autocorrelation (ACF) to derive three factors discussed above. Later, the model parameters are assessed based on the least square method for nonlinear or maximum likelihood methods for vulnerabilities. Complex iterations as per standard statistical techniques are to be conducted to derive the parameters. Next, the estimated parameters' statistical significance is to be derived and ensure that the residuals follow a random pattern. The simplest model selection is the criteria to choose the model. The variance of residuals and mean over time are plotted to check if there are any missed specifications.

2.11.2 Exponential smoothing

Exponential smoothing (ES) specializes in separating the seasonality component from the rest of the variations. It decreases the weights on the data points as they go down the timeline into the past. Holt winter additive (HWA) and simple seasonal ES models work best for setting in this study. HWA model focuses on seasonal adjustment, level, and trend of the forecast. The series' level is the relative measure of the magnitude of the vulnerabilities count over time, which may change or not. Trend focuses on movement up or down of the data over time. Seasonality refers to variation in data for fixed intervals in the short term. Level (relative magnitude of vulnerabilities), the forecast, and seasonal adjustment trend to the estimates are the three smoothing parameters. Other parameters involved are several vulnerabilities for a specific period as output, time is measured in months, a future period of the forecast being 12 months, and length of seasonality generally 12 months. In the case simple seasonal model, data is fitted with only constant seasonality factors. It will have a level and seasonality as the smoothing parameters.

Data collection was done from NVD, a publicly available database for vulnerabilities. Data collected included published date, vulnerability identifier, severity, and affected products. One of the sets of data was all vulnerabilities for a period. Another dataset was for the top browsers selected for the study. Top browsers were selected based on the net market share study of 2013. Vulnerabilities of browsers were based on browser published date or as per the data availability for the browsers. Descriptive statistics of data are compiled based on different datasets such as all reported vulnerabilities and browser-wise reported vulnerabilities, month on month and average per month. The study showed a high trend of vulnerabilities for Internet Explorer, Firefox, and Chrome, compared to Opera and Safari. SPSS statistical software was used to analyze the best time series model identification for a time as an independent factor and count of vulnerabilities as a dependent factor across months. The stationarity of the data was assessed with autocorrelation and other statistical tests. Model parameter appropriateness and seasonality were validated with partial autocorrelation and autocorrelation test for a month as periodicity. Residuals of data were plotted against time to make out any possible systemic pattern in data; the acceptable case is for residuals to be randomly spread over time. After analysis of all these factors, the simple seasonal model turned out to best fit mostly (Rajasooriya et al. 2017).

2.12 Conclusion and future work

The probabilistic model and its ability to predict faults in software covering functional and security was explored. Software fault localization needs greater emphasis as the software is growing more complex. This makes the process of fault localization not so easy to automate. Through various methods of fault localization is explored, expert opinions must be considered. Techniques explored here should act as back up for the expert inputs.

The real-world system presents the complexity of multiple bugs and not just one. Context is not simple to just figure out these categories of bugs through simple clustering; it also heavily relies on the software's usage pattern. As future work, in the case of more extensive programs, there is a need to customize the program sampling mechanism to focus sampling on more critical areas of the program, reducing the data's sparsity. In the program to address multiple bugs situation, bi-clustering algorithms are used which are based on iterative collective voting framework.

Probabilistic programming is explored for programming language in software engineering to help build probabilistic inferencing to solve some of the domain's challenges. Future work on the exploration of the optimization algorithms like gradient descent will be better compared to search-based algorithms.

Automated vulnerabilities detection systems explore the detection of vulnerabilities directly from source code using ML techniques. Classification-based algorithms from natural language are refined to the context of software engineering. Features are extracted based on a CNN that was effective. Further work can be focused on improving the labels by using the information mined from security patches.

Software vulnerability prediction is offered as a web service for a user to plug in software metrics from their domain and predict the defects. A variety of algorithms are experimented on Azure Machine Learning Studio to arrive at the best model. The multilayered perceptron performed well in the experiment setup. Further work on ANNs like RBF network, Boltzmann machine, and Hopfield network is prospective. Software vulnerabilities prediction based on text involved in the software source code beyond the software metrics provides good possibilities. Further work can focus on exploring multiple alternatives for building feature vectors. The ultimate target would be at integrating the features obtained from code metrics and those derived from treating complete source code as text.

It is essential to get detailed information about the vulnerabilities discovered. To build statistical models, it is essential to get as much information as possible as well. So, the risk factor of vulnerability is studied as a factor of time, using a Markovian approach. The Markovian method was looking at the states of vulnerabilities and their probabilities at those states. All the data sources coming from the probabilistic background are accounted for to arrive at modeling strategies.

Time series modeling is explored for predicting vulnerabilities trend prediction. Among the time series smoothing parameters, "level" is the only significant parameter for prediction models. Though, the study was focused on the browser development, which is critical, as the vulnerabilities are mostly not in control of the development team. The further focus would be to refine the proposed model by including more parameters based on the vulnerabilities' frequency and severity.

In this chapter, fault prediction was the focus using probabilistic modeling format; we now look at software traceability. As part of making software development more efficient and effective, requirements management is one of the key areas that need to be looked at. Requirements management is a key area of software management; traceability of the requirements from its form in the text to fine-grained actionable is a critical need.

Chapter 3
Establishing traceability between software development domain artifacts

3.1 Introduction

Its important to establish traceability between software development domain artifacts like code, specifications, requirements, code fixes, and bug reports. These would assist developers in understanding the code ecosystem better and maintain the same better. In this module, since various artifacts are involved, some of natural language processing (NLP) approaches are explored. Since it is about traceability, dependency parsing would be the right approach with co-reference analysis. Traceability benefits to tackle inefficient requirements management which in turn helps to build collaboration between multiple teams involved. Confirmation of source code being reviewed is based on acceptance criteria and other areas are benefitted. This area also poses the need to efficiently transform the requirements that evolve from the text into fine-grained actionable software artifacts. In other research works, deep learning neural networks are built based on the software development life cycle's text data; further research scope would be to leverage other source code artifacts.

This chapter is organized as follows: we start with understanding the requirement traceability problem and then explore on addressing software traceability with deep learning technique, deep NLP, word embeddings, recurrent neural network (RNN), long short-term memory (LSTM), and gated recurrent unit. The concept of tracing networks is discussed. The cluster hypothesis for requirement tracing is reviewed in this section, further exploration is to understand, utilizing issue reports and commits to establishing the traceability. Topic modeling for traceability is the next exploration, where we look at retrospect-based tracing, latent Dirichlet allocation (LDA) method, and explore the tool for implementing the same. Then there is an exclusive focus on the classification technique of machine learning used for traceability. Nonfunctional requirement traceability is also discussed.

3.2 Overview of requirement traceability problem

Requirements traceability (RT) has two perspectives: prerequirements specification traceability and postrequirements specification traceability. Most of the problems with requirements are attributable to prerequirements specification traceability. Though the associated research in improving the requirements process is on, this is a challenging aspect due to a lack of problem analysis. The prerequirements specification is the point before the requirements come into specifications documented.

https://doi.org/10.1515/9783110703399-003

Solutions developed for the problem have not assessed the core of the issue proposed. Hence it has not been comprehensive. Lack of a universal definition of the problem is the contributing factor; the purpose is biased. The solution-focused description, purpose-driven, information-driven, and direct driven, has been defined as requirement traceability. In purpose-driven, the focus is on what is needed. In case of a solution, it is about how it should be. In information-driven focus is on information that helps trace back and forward; direction-driven is around the end of the trace flow. So, this has impacted the comprehensiveness of the solutions.

Practitioners' viewpoint of what is the root cause of the RT problem also plays important role. Attribution by practitioners is on the granularity of the traceable entities, long-term nature of the project, lack of integration technology, hidden information, and so on. Understanding how RT improvement will help was around various areas like audit, safety, and better understanding of systems. This understanding shows that RT is used as an overarching subject to address multiple areas, which may be areas that are conflicting with each other. The definition becomes essential to arrive at a framework to address the understanding of the RT. Literature defines RT as requirements specified are traceable if the origin is exact, and there is a way to reference the conditions in the future development course. This RT points to backtrack to all old artifacts and forward trace to new artifacts. The importance is given to prerequirements specification traceability and post one as this distinction makes it easy to tackle the problem in a more refined way. Each of these phases has its influence on the traceability problem. The critical difference between these phases is the information they handle and how they can assist the situation. Post traceability depends on the possibility of tracing the requirements back to baselined requirement specifications (RS). This tracing back happens through a series of documents across which conditions are spread out.

Any changes to baselined requirements must be sequenced through these series of artifacts. In the pre-phase case, the focus is on the ability to trace the requirements back to their source document. This trace would have been set with the process of requirement refinement through a series of sources from which the conditions were gathered. Any changes in RS must be validated through this process across the line of origin. This calls for knowledge about more delicate aspects of requirements hidden in multiple artifacts in the beginning phases of requirements evolution.

Supports provided for the improvement usually are focused on the post phase. The usual problem area is information artifacts associated with the development. This can be controlled by formalizing the development settings so that the pipeline of creating executable deliverables from RS can be ensured and any changes happening can be simulated with the same transformation. Issues associated with the pre-phase are not understood appropriately. Post-phase traceability is considered a black box with no clarity on how requirements transform into the product.

Acknowledging these issues has been happening only lately. Most of the problems are attributed to the pre-phase as there are not enough approaches to trace and record the information. Pre-phase traceability improvement has its direct impact on how well the quality can be managed to understand issues that are addressed earlier and use the learning. A key thing to look at is the conflicting interest of the end users and the people for whom this matters, as part of their work. The problem associated with the requirements provider is that the time needed to establish traceability is optional and extra. So mostly, it ends up getting less or no time and resources. There is no focus on roles that are expected to be played to make sure that the requirements are organized and maintained well. Lack of focus on the work needed for organizing requirements contributes to this situation. Though there are efforts to manage traceability, it is very localized, where focused involvement is needed from all. There is no objective of ensuring end user requirements are met, making the focus remain on the immediate purpose. As the RS gets documented, concerns over pre-phase are reduced, and the post phase picks up. This is mainly fueled by the immature change management processes and activities involved that are unpredictable. Deliverable being the prime focus leads to less attention to required information gathering. Lack of strong feedback process also does not help the situation.

End users' inconsistency in the expectation, quantification, diversity, and depth missing in the specification are contributors from end users. Reliance on personal contact for some form of information from users also does not help.

Challenge is in the satisfaction of both end users and the people involved in work. Problems get escalated when users and providers are from the same role. Technology alone cannot take care of the issues. To look at a solution for this, it is essential to understand that attention is needed in diverse areas.

It is clear that the information needed for a different phase of the lifecycle is additional and not possible to generalize. So, it is crucial to develop a plan of information required for different scenarios. Models on RT have their top focus on awareness of practitioners regarding trace links to be maintained around information gathered. Dedicated roles such as documentation specialists can smoothen this challenge. There is quite a progress in improvising on the ability to collect required diverse information. Tracking the history of requirements evolution and following collaborative activities for requirements management are important. Increased focus has been in automating many of these elements and getting these traceability focused activities being done as an outcome of preliminary work. To manage an interactive process, information needs to be flexible. Frameworks to model the RS to ensure gradual progress, visual representation of the structure, relationship, and change models are some attempts to address the issue. More focus is needed to manage unstable and unstructured data. Data librarians' clear roles in collecting data, repository guardians to focus on data, and facilitator's integrity ensure traceability would help.

Requirement traceability is already encapsulated in terms of what can be traced and presented. The focus should be on developing artificial intelligence around information retrieval (IR). Graph-based traceability system and interactive visualization capabilities are needed to assist impact analysis. Presentation of the requirements in an animated way would help. To make the system robust, dynamically strengthening the traces as the queries of trace comes into the system would be a need.

More focus needs to be given to increasing the amount and different types of information needed. When this required information is gathered via models, methods, and guidelines, they introduce subjectivity. Information unavailability or the customization of information to the audience or inappropriate information all contributes to the challenges at hand. When these situations are encountered, the plan is to interact with the personnel concerned and gather the needed information. It is statistically significant that a large part of the information about prerequirements specification plays a significant role. Also, the critical element is the people involved in the work on the requirements. Though the techniques to solve these problems have focused on providing needed information to people, it has always been required to supplement this with face-to-face interaction to gather the missing part. That this is not being taken care of has played a significant role in problems associated with RT.

Out of date Requirements Specification, due to lack of focus on evolving the specification as they change adds to the concern. Lack of collaboration to manage responsibilities create more vagueness to requirements management expectation. People moving in and out of the project have an impact on the problem (Gotel et al. 1994).

3.3 Software traceability establishing via deep learning techniques

Traceability plays a prominent role in a safety-critical domain where the regulatory and certifying body looks for explainability. Maintaining such systems manually would be a complicated task. Optimizing this process is possible through machine learning and information extraction techniques through creating the links that can be traced back. Still, the approaches cannot understand the semantics of the artifacts associated with the software. Domain knowledge consideration is also weak in this process; this leads to inefficient and inaccurate results. Here the attempt is to account for the domain knowledge and semantics as part of tracing the solutions. The architecture will include the RNN and word embeddings for building the tracing framework. Word embeddings target to represent the domain knowledge as word vectors. RNN produces on top of word vectors to assess sentence-level semantics of artifacts related to requirements. After experimentation with the multiple configurations, bidirectional gated recurrent unit outperformed other combinations of latent semantic indexing (LSI) and vector space model (VSM).

Requirement traceability has its root in establishing the forward and backward trace within the software development life cycle, particularly for requirements when it is undergoing better understanding resulting in refinements and getting converted to reality. Traceability plays a significant role in cost estimation, change impact analysis, regression test, and compliance verification. In the case of critical systems such as aviation, there is a need to establish a trace from the possible hazards across to all phases of the software development life cycle to demonstrate the effort of addressing those hazards. The overall process of establishing this traceability is cumbersome and does not ensure the robustness even for safety-critical systems. The primary concern of automation is the difference in the words used across artifacts. Some of the best algorithms like VSM, LSI, and LDA fail to get the domain understanding to relate the items based on their groupings as per domain. These algorithms focus on bagging the words and do not account for the semantic associations; even if the phrasing is done, concepts relation may be missed.

An earlier framework developed by authors had challenges around setting up cost, for a domain and manual tracing needed between domain and artifact semantics. Conventions syntactic parser involved parsing the semantic of terms involved induced the errors due to the parser's inefficiencies. Deep learning has proved NLP tasks around machine translation, sentiment analysis, question answering systems, and parsing tasks as effective ones. NLP's key feature is to represent the focus area into nonlinear processing nodes, with the unsupervised or supervised learning format, and use these representations for the required size of implementation like parsing and others mentioned. The first part of the approach focuses on learning word embeddings for domain knowledge based on unsupervised learning with many domain artifacts. Representation here is of higher dimensions for word vectors to track the semantics based on the distributions and information about the words occurring together. Another part is to use the existing traceability information of the domain and train the network to identify the trace link. The trace system uses RNN, where for each artifact like source code and requirements, every word is represented as the vectors based on word embeddings training. This representation is passed on to RNN, which outputs the vector space representation of the artifacts. Based on this representation, the tracing network compares the artifacts and generates the probability of artifacts being linked. Since there is manual training of the artifacts' trace, the solution cannot be generalized for other domains. But the usage would be done with various approaches firstly, build initial small manual trace for projects, and then utilize the same for automation of production as project proceeds. Or secondly, attempt to build a manual trace for the full project and predict if there any other linked traces that have not been accounted for in the manual process. The last approach would be to manually build a trace for a specific set of artifacts of the project and use it for the same project itself and then extend the same to artifacts or projects in the same domain. This study focuses on the first approach mentioned. The method is validated against the industrial dataset of the positive train control (PTC) system. The focus would be to configure effective RNN to improve the accuracy of the links and compare the RNN performance on other baselined techniques.

3.3.1 Deep NLP

Deep learning techniques are influenced by the human brain neurons and are the byproducts of ANN. Inspired by neuroscience, the ANN intended to build computation units of the network that are integrated into the multilayer architecture, which was an attempt to mimic the human brain's complex functioning. Complexity in the architecture is attributable to the need for multilayer abstraction for the data space to account for in-depth pattern recognition. This is well suited for conventional learning cases where the focus is on the hand-building of the features. Backpropagation is the key function where the network adapts to the data space incrementally by learning the data pattern. Network adaptation is via configuration of the network parameters to get well structured to capture representations across the layers.

3.3.2 Word embeddings

Traditionally vector representation is focused at the center of the word, with each word accounted on its own. But the word embeddings feature the vector space representation of words, where the words are placed in space based on their commonalities; expression is in the form of high dimensional vector. The exciting aspect is capturing the syntactic and semantic elements of the words to establish a linear relationship between words. The syntactic aspect is about the language construct structure, whereas the semantic is associated with the meaning associated with the domain. Global Vectors for Word Representation (GloVe) and skip-gram with the negative sample are improvised versions of word analogy representation compared to conventional methods. Word embedding operates on the joint occurrence of the words from the raw natural language text, which are not labeled. In the skip-gram case, the corpus is scanned with a focus window to assess the context across training data. Every window's focus is on the middle word and increasing the probability of the expected word around this and decreasing the probability of a less possible word appearing around the center word.

3.3.3 Architecture

Deep learning architecture for NLP is built on a neural network that has feedforward processing across multilayer perceptron. Several layers and the neurons in the layer can grow significantly, but various variants have been proposed to accommodate this complexity. Convolutional neural network (CNN) is specialized for image processing and video analysis. RNNs have been the base for NLP tasks, and mainly there has been a more significant advancement for NLP tasks with RNN around, including machine translation capabilities.

3.3.4 Recurrent neural network

RNN specializes in sequential data like audio and text. Network unit and taking next word embedding as input, hidden state of the unit takes previous time step input. This helps build memory based on the current state and previous state. Figure 3.1 depicts Standard recurrent neural network and its expanded form across time.

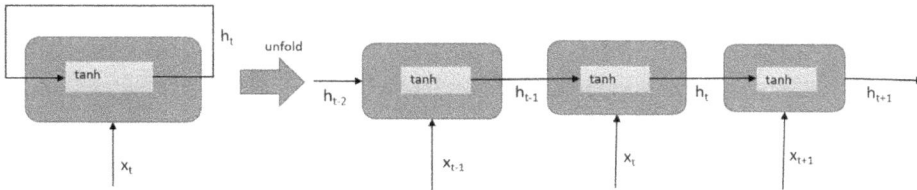

Figure 3.1: Standard recurrent neural network and its expanded form across time.

One of the network's key features is the consistent parameters used by all units of the network across, which helps process the input sequenced data of random length. The standard output of the RNN can be represented as Output = tanh $(Ax_t + Bh_{t-1} + C)$, where A, B, and C are transforming parameters, *tanh* is the hyperbolic tangent function, x_t is input vector at time t, and h_{t-1} is output hidden vector of the previous time step. A major concern of the RNN has been the vanishing and exploding of the gradient in the process of backward propagation. To tackle this suggested mechanism is to reduce the gradient based on the context, which is called gradient clipping. The gradient is the network's learning as it forward propagates and learns; this gradient is passed back to refine the structure for the next run. There have been proposals for handling long-term dependencies with gated recurrent unit (GRU) and LSTM to address the vanishing gradient.

3.3.5 LSTM

Network cell specializes in handling long-term dependencies. There is a gated check to decide how the input must be captured and what input to capture before it is written into memory. Sigmoid function (sigma(z) = $1/(1 + e^{-z})$) is used at the gate to control the inputs. Sigmoid value of zero and one represents if the values from previous stage can pass through or not respectively. Overall LSTM unit has input, forget and output gate; states of these gates are influenced by the input at that point of time t and hidden state identified previous time step output vector. The input gate decides on the amount of information gathered from current input whereas the forget gate focuses on how much information to be retained from the previous memory state. Based on these gates, the current memory state is updated. Figure 3.2 depicts Long Short-Term Memory representation.

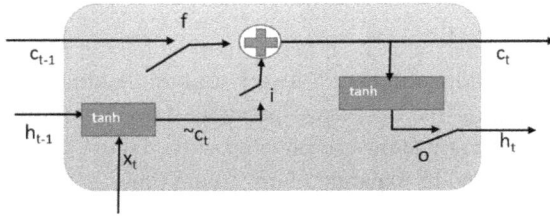

Figure 3.2: Long short-term memory representation; "*c*" is memory cell vector; "*i*," "*o*," and "*f*" are input, output, and forget gates, respectively.

3.3.6 Gated recurrent unit

GRU offers a simplified structure with no dedicated memory in the network but a streamlined structure with an update and reset. The update gate controls the amount of previous state input and current state input to be combined as the final output for the current state.

In addition to these other variations of RNN that are tried are multilayered RNNs and bidirectional RNNs. In multilayered, there are multiple layers of RNN trying to get a much deeper abstract of representations. In the case of bidirectional feedforward and feedback are managed so that the network learns from the past as well as future data.

3.3.7 Network for tracing

As a tracing network setup, source artifact must be identified, and then similarity must be established between the source and other potential artifacts. Then the associations are ranked based on the relations. Later, manual validation must be done to confirm the established connections. This activity is to be done for all artifacts as the source. Multiple deep learning methods have been explored for this task to compare source artifacts and other artifacts. Generally, these are text comparison based, with knowledge of the domain accounted for in the study. The scenario exposes the challenge of establishing the semantic relation even if there are no common words across artifacts. Based on the assessment, RNN and word embeddings were chosen as an approach for this process. Extraction of the word embeddings from a corpus of words would be the first stage and then use those embeddings to establish artifacts.

Source and target artifacts are words embedded for all the words in the artifacts into a vector representation. The word embedding layer is then trained on a domain-based corpus using a skip-gram model. Word embeddings converted words are then passed into RNNs for creating semantic-based common vector representation for an artifact. A notable feature is the word vectors are sent in sequential order. In the case of bidirectional RNN for the backward pass, the reverse order of word is passed into

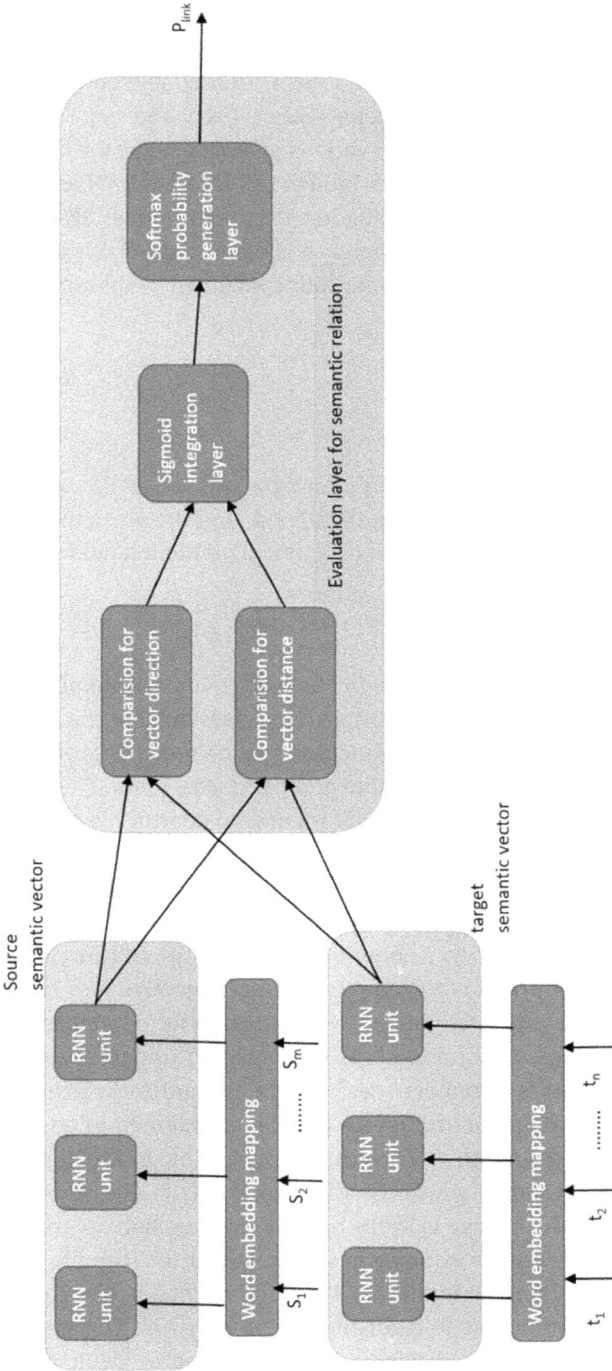

Figure 3.3: Tracing network configured in work reviewed here. The embedded word vectors get software artifacts mapped, which further generates semantic vectors bypassing into RNN layers. The probability of this being linked is then evaluated at the evaluation layer for semantic relation.

RNN in addition to forwarding sequential pass. The second part of the network is a semantic comparison of the vectors. Sigmoid functions combine the values and represent the final probability of a trace link exists between them. The sigmoid output will be of n dimensions, where n is based on the number of sizes chosen, and some of the values will be one. Architecture is refined based on various parameters like the choice of GRU or LSTM units in RNN, number of layers for hidden cell for both RNN and semantic evaluator part, number of layers for RNN, bidirectional and normal RNN. All these configurations are experimented with. The approach is built on the Torch framework (http://torch.ch/). Figure 3.3 depicts Tracing network configured in the work reviewed here.

3.3.8 Tracing network training

Basics of the useful model are to learn well on training data and generalize on the new data. The loss function used was negative log-likelihood, and the objective was to reduce this. This loss function usually is used in categorical prediction models.

$$J(\theta) = -(1/N) \ \Sigma logP \ (Y = yi \,|\, xi, \theta) + \lambda/2^*(||\theta||_2{}^2) \tag{3.1}$$

Theta is a parameter associated with a network for training, N is the number of training data, xi is input for i^{th} training data, yi is the actual class label of the input data that is, linked or not linked. So $P(Y = yi \,|\, xi, \theta)$ represents the network probability of predicting the correct values given input value and parameter theta. The later part of the equation represents L2 normalization that intends to control overfitting. $||\theta||_2$ is the Euclidean distance for the theta parameter and manages the amount of regularization. With loss function as the reference, stochastic gradient descent method optimization is employed to optimize the parameters. In this process, multiple epochs are run, where each period is a complete run of all the training data once. Epochs are continued until there is the convergence of objective loss function to its optimal level. Periods also are features with multiple mini-batches for training purposes. An end of each run gradients is computed to update the network parameters. The learning rate parameter decides the amount by which the parameters are modified. The optimal learning rate is to be maintained to ensure that the learning process does not go into loops without converging, or the quality is so small that it takes ever for convergence. Gradient direction is also another essential aspect of ensuring that the change is in the optimal loss function approach. Learning rate change was adaptive in this experiment to ensure that the learning rate is changed based on the runs' training performance. A specific method is adopted to characterize the choice of the learning rate to regulate learning behavior. The study's hyperparameters are the learning rate at the beginning, gradient clipping value, mini-batch size, learning rate regulator parameter, and total epochs planned.

Here we talk about data preparation, network setup and hyperparameter tuning, and network performance compared with standard ones. Word embeddings were trained with corpus from the PTC domain for software artifacts and domain artifacts. Wikipedia dump of text corpus was also used as a variant for the embedding. Text cleaning included converting to lower case, removing the non-alphanumeric content except for those that are needed as per domain need. The dataset included software requirements and design-related artifacts. Manual trace links were built between requirements and design artifacts for the experiment purpose. The dataset was broken into training, development, and testing sets. Training set helped set up parameters for the network, development set to validate the trained parameter and ensure no overfitting, and final test set to validate the final configuration. This is needed as there is a chance that the learning would have adopted to develop a dataset.

Software data had a unique aspect of the low probability of linked artifacts; this creates unbalanced data. However, there are techniques to handle this by weighing more for minority sets, which is traced to artifacts. But during gradient descent, these weighing may influence the cost function amplification for minority cases, resulting in inefficiencies in training. Another way is to downgrade the majority class on non-linked documents, but this approach led to the loss of keynote linked pairs which had a prominent role in differentiating between linked and non-linked pairs. Also, the increasing sample method was not efficient as they led to a big set of training data and increased training time. The approach finally designed was to dynamically build the balanced dataset at each epoch. Selection of valid links from the training set and randomly choosing an equal number of the non-link set at the beginning of each epoch. This helped in sampled non-linked items to be a representative set and balance the contribution from both the classes.

A crucial aspect of the experiment was to arrive at an optimal network setting and hyperparameters for effective deep learning. Large data space made it complex for trying various iterations and high training time as well. So, the attempt was to shortlist the configurations that are optimal in terms of monitoring the training loss rate during epochs. Grid search technique was run on these shortlisted combinations to arrive at the best ones. Word embedding learning was done with the skip-gram method of word2vec. Two settings experimented were PTC with 50 dimensions and PTC plus Wikipedia corpus with 300 dimensions. The difference in dimension is due to the difference in the size of the vocabulary of PTC alone compared to PTC and Wikipedia combination. A large dimension is needed for later combinations to make sure that the semantic of the entire corpus is accounted for. Various variations of RNN tried are GRU, LSTM and bidirectional version of each of this. These were tried out for 1 and 2 layers each, hidden dimensions of the layers in RNN was tried for 30 and 60 combinations, and similarly 10 and 20 dimensions were tried for semantic evaluation network. To set a baseline bag of words model was run where the semantic representation was just the average of the word vectors in artifacts.

Tracing methods return a combination of links between source artifacts and target artifacts. All valid ones will be on top of the returned list. Mean average precision (MAP) is used to compute the effectiveness of the algorithm. In general IR, MAP is computed for top n returned links, but in this experiment, the computation was done for all returned links. MAP was computed on test data to compare the performance with another well-known method. LSI and VSM are the popular methods. An attempt was also made to consider the optimal configuration of VSM and LSI by adopting a genetic algorithm approach to explore the data space of various parameters. VSM shortlisted configuration used inverse document frequency (IDF) which is localized for weighing during computation of cosine similarity. On top of basic text cleaning for both algorithms' porter stemming was used for steaming purposes. Precision vs recall graphical analysis helped to assess the performance.

Overall, it was noticed that for all combinations of RNNs such as GRU, LSTM, and bidirectional versions of each of it, best configurations were similar. Figure 3.4 depicts format of how learning curves are compared for different variations of configurations.

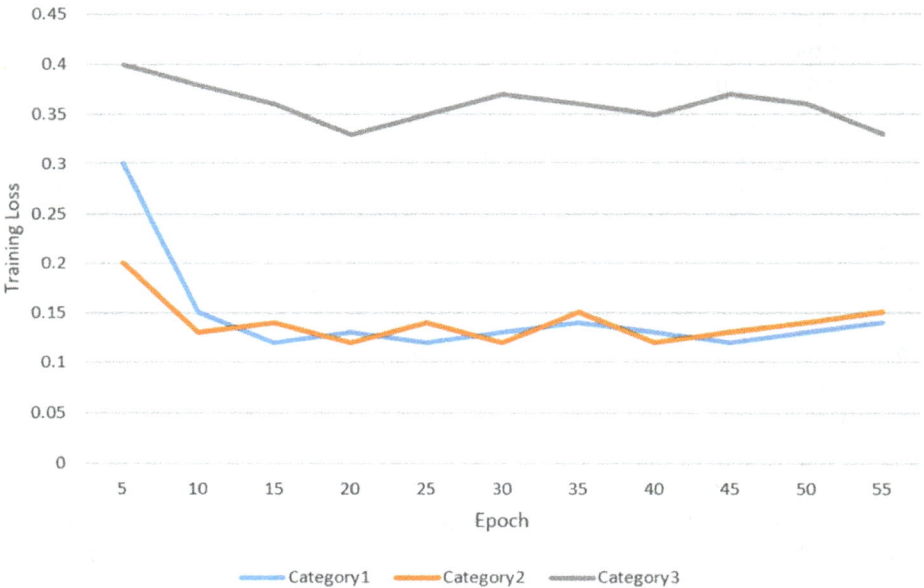

Figure 3.4: Figure shows the format of how learning curves are compared for different variations of configurations. This visualization helps to assess which structure converges to lower training loss faster.

For comparison of the performance, several epochs are to be plotted against training loss. All four combinations of RNN will have to be compared with the average vector method (AV), where the order was not accounted for, and the average word occurrence rate was only considered for artifacts. All RNN versions' performance was better than AV, which signifies the importance of the order of words. Both versions of GRU demonstrate faster convergence compared to LSTM versions. In general, bidirectional models of GRU and LSTMs are slightly better performing compared to their regular versions. Overall bidirectional GRU was the best choice. It was also noticed that the PTC corpus alone performed better than the combination of PTC and Wikipedia; this may be because Wikipedia would have a corpus that significantly is of the generic domain and not specific to PTC. Even though there are common words, the context in Wikipedia would be generic, which will impact semantic reasoning, which is domain-specific. Three hundred dimensions were experimented with PTC but did not yield significantly different results. In the test set with links between source and target pair, the source being ranked based on decreasing probabilities scores, the average precision (AP) score arrived as discussed earlier. This score compared with LSI and VSM showed significantly better performance of bidirectional GRU. Friedman's test of statistical significance was used to provide that the difference in performance was statistically significant. The tracing network outperformed that other versions in precision vs recall plotting as well. At a higher level of recall chosen network was performing better than LSI and VSM. With a 100% recall focusing on tracing tasks, results look favorable on the context of network looking for deriving semantic information build association between artifacts. Figure 3.5 depicts output behavior for GRU at a point of time for a dimension of sentence semantic vector.

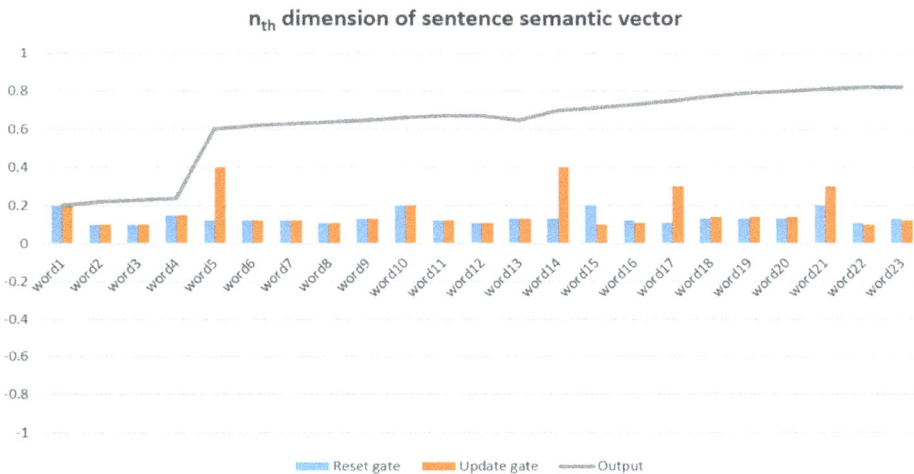

Figure 3.5: Output behavior for GRU at a point of time for a dimension of sentence semantic vector. Y-axis depicts the value of update gate, reset gate, and output. In comparison, X-axis depicts the input of words in sequence.

Interpretability of GRU is tough to assess how the network figures out the semantic relation. But we can derive the sense of how the network behaves at the update and reset gate by taking an example phrase and monitoring the variation in the gates' values and their final value at the respective dimensions of the vector space, as shown in Figure 3.5. Though the study focuses on recall rate, the trends show that the precision values do not go up beyond certain recall values, indicating that there are false positives that are getting into space. This also has to do with the method chosen for managing an unbalanced dataset. Though all the linked cases are considered for non-link random selection, sampled cases were taken for linked cases due to the scarcity of cases. As the project evolves, training data increases with more linked scenarios available, performance improved with the ability to differentiate between linked and non-linked cases. So, there was a good improvement in precision and recall. Particularly improvements were better with lower levels of recall. There are attempts to utilize phrase detection for tracking the impact of changes on the requirements, but the semantics are not accounted for (Arora et al. 2013).

Since the setting up process is time consuming, and this study is localized to the PTC domain only, it cannot be generalized to another domain. But PTC includes the text from various sources also different individuals involved in drafting, like system engineers and requirement engineers. This would account for diversity, and the nature of the artifacts and their characteristics also would help. Though the tracing network has focused on text involved in the domain, the future focus can be the core source code. Since gathering the corpora is difficult, combining human knowledge with the neural capability can also be looked at (Guo et al. 2017).

3.4 Cluster hypothesis for requirements tracking

IR methods have been dominating RT, but there is a lack of accuracy due to possible false positives in the context of balance between precision and recall. This discussion's focus would be to explore the effective methods for improving the candidate links for requirements tracing using the cluster hypothesis that clusters the subject area into two prominent clusters. Since the clusters are based on low and high quality, low-quality ones can be sampled out to improve the data space. The study is conducted on open datasets and later validated with industrial datasets. Since the dominance of manual methods for traceability makes it ineffective. All the applied approaches for IR have been focusing on recall against precision. The reason here is due to false positives are easy to handle compared to false negatives. However, with a high focus on recall retrieving many links may be practical but not useful. Therefore, the focus has been to target top-ranked traces to save time for the manual reviewer. To shortlist top ones, there is a threshold score set above which the links are chosen. Threshold must be optimally selected so that there is a balance between many correct links chosen and incorrect links.

Approaches have not been able to increase correct links at a high threshold or reduce the false positives at lower values. The focus of this study is to enhance the

IR approach by using the cluster hypothesis. The idea is the similarity between relevant documents will be high. The approach also helps cluster out the low-quality ones to make the threshold selection process much effective. Earlier works have taken clustering on top of the ranked list of links (Duan et al. 2007). The effect has been no change in precision and recall. In this approach, clustering has been used to reorder the traceability ranks of links improving precision. There is work on combining clustering with other techniques (Cleland et al. 2005, Chen et al. 2011, Wang et al. 2009). In this study, a detailed analysis of the clusters is done; appropriate metrics are used to validate the clustering parameters and their influence on each other. The study provides detailed guidelines for using the cluster hypothesis in IR-related RT and throws light on all aspects of clustering in a traceability context, enhancing the study of automating traceability tools.

Clustering is an unsupervised method of learning where the groups are divided based on their pattern. The document, query raised, and search outputs are views of exploration. From a document perspective, retrieval efficiency improvement is the focus. As the count of documents grows, matching up the required query with the target documents will impact the system. An optimal way would be to develop clustering patterns for the entire corpus and then run the required query on top of it. Here the query is compared with the centroid of the cluster, which represents the cluster characteristics. This optimizes the search time involved and is a statistic clustering method. Dynamic clustering approaches create the cluster based on the query and make users easily distinguish between low-quality and high-quality clusters. The query information is not enough; then, the historical question can be gathered and used for search.

Approaches have also worked on forming query classes to be used in case of insufficient information. If the query cannot be mapped to the clusters, then the reverse map of the query to the class of the query and in turn identifying the cluster of that query class will help identify the cluster. From result purview, clustering helps as it provides results to in an interactive format for users. In the case of the ranked list, the type of results preview context was not maintained. Clustering the results in terms of related topics will help in better comprehension of the results. But still, to build full context in terms of labels for the clusters and ordering of the clusters, human intervention is still involved. Work on this shows that the clustering of the results is good enough to make the result presentation interactive for users (Hearst et al. 1996, Leuski et al. 2001, Yi et al. 2009). So, the clustering provides control on search based on their relevance intended by users.

Clustering has been widely used in IR, but recent trends have been in the area of traceability. One of the initial works on traceability is around identifying automated trace-related characteristics that are unique compared to IR. Since this is related to document and artifacts involved in tracing and are limited to test cases, classes, and requirements, while compared to other web search domain, there is no focus in terms of reducing the search space. From the angle of the query, it is derived from requirements or other artifacts in the software domain. Generally, the requirements are very specific and clear. There are challenges of them being longer. This makes it difficult to maintain the precision of results but tackles the short query problem, which is prevalent in IR.

It is also important for traceability to maintain high recall, compared to general IR. The expectation is to at least return a small set of relevant artifacts, as there are challenges associated with scanning into all documents set and the chance of expanding the query in IR. That is where the studies focused on result clustering, to increase the interpretability and reduce the need for human evaluation. There is a mechanism to evaluate the cluster granularity and measures to rank the traces. This led to less decision need than the ordered ranked list, so it saves time for evaluation. Though result-based clustering helps improve comprehensiveness since the outcome is like those produced by other non-clustering methods, precision and recall are not improving. Though integrated approaches have been tried, results have not been promising, which makes a need for studying clustering methods more in detail, which is the focus of this study.

3.4.1 Methodology

The clustering hypothesis has been validated to separate clusters of similar links from dissimilar ones. So, the study's focus is to establish granularity in these clustering of the links to make sure we arrive at high-quality and low-quality clusters. Which will enable us to remove out the low-quality links? Figure 3.6 depicts candidate link generation of requirement tracing being enhanced with a clustering approach.

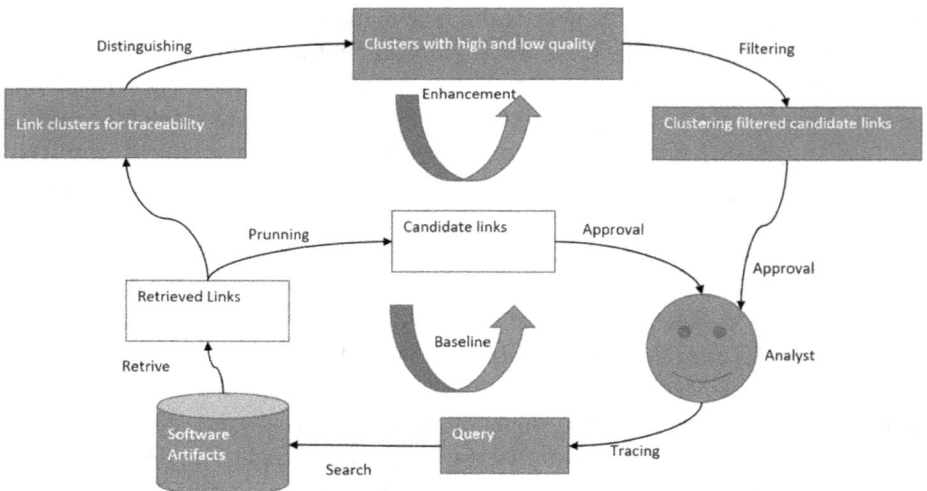

Figure 3.6: Candidate link generation of requirement tracing being enhanced with a clustering approach. The baseline part represents the current process, and the enhancement part represents the improvisation.

The figure shows the traditional IR-based tracing process on top of which the clustering-based enhancement process discussed in this part is built-in. In the baseline process that is used in basic tracking tools, humans send some queries to trace from requirements; then the algorithm will trace back all possible links based on the similarity between software artifacts and queries. There is a refining stage that tries to cut down on the resulted traces that has a lower density of correctness. With the outcomes presented to the user, the possibility of rectifying the false positives will take the lead over the validation of correct links. Thus, the focus is to improvise this pruning of the refining stage. Some of the filtering formats are setting the threshold on the similarity index or setting a threshold on the number of retrieved top results.

The approach discussed here includes the first step of running the clustering after the initial search is done. This will result in a dynamic search, but there are no assumptions such as if two artifacts end up being the right link or maybe wrong link, then they are related to a particular requirement. So, the clusters of links are query based, so it is more likely to be specifically associated with the requirement that is being traced out. After the separation of incorrect and correct links, it is important to explore the automated techniques to validate the groups in terms of high- and low-quality clusters explored in this study. In this approach, reliability is completely not on the relation of link with the query and extended to the association of link to the cluster it belongs to. That means the neighboring links also have a role in deciding. The selection is clustering based rather than depending on the link's rank or similarity score alone. Approach targets to determine the best clustering algorithm that can separate the links that are correct and incorrect, and also to assess the optimal granularities or number of clusters that are needed to classify effectively. The mechanism to ensure the quality of the link cluster needs to be figured out. We also need to figure out after the bad quality links are removed and how to organize the remaining links.

Many studies focus on making the clustering mechanism part of working applications to make it an end-to-end optimization problem (Lawler et al. 1985). This study focuses on looking at general clustering features that provide acceptable optimization with mutually exclusive clustering where the identified link will be associated with one specific cluster. They are all from different languages, with good variations among requirements to design linking and requirements to source code linking.

3.4.2 Experiment on clustering traceability

Three algorithms explored for comparing a cluster of traceability links are hierarchical agglomerative clustering (HAC), bisecting divisive clustering (BDC), and K-means clustering. These methods cover a larger arena of clustering methods and work well for document clustering. Three variants of HAC that are explored are the single, complete, and average link. These variants operate upon the distance of the data points under study, linkage criteria between clusters under comparison. All the five algorithms will depend on the terms of the software artifacts understudy for recognizing the similarity. Tf-idf cosine is used for computing the similarity in the experiment. This also helps to set tf-idf as a baseline tracking mechanism as it is the reference for some of the top methods like the vector space model (VSM). Source code, requirements, and design documents are included in the study. Tokenizing, stemming, and filtering are involved. Several clusters and cluster size influence clustering granularity. Several clusters are a common element among all five algorithms. In the case of K-means and BDC generalrule of thumb used for clustering, is, to use uniform size clusters.

Clustering quality can be assessed externally or internally. The external evaluation would be based on the human expert's knowledge. The internal evaluation considers the closeness of clusters to each other. Pairwise correlation analysis of all clusters generated by the algorithms under study showed that clustering granularity was reasonable enough and consistent across algorithms. This kind of clustering granularity analysis works well for exploratory analysis set up where the pattern in input data is discovered with clustering. With external evaluation, the comparison is made with the established labels. These labels developed by experts are based on the composition of the data into a cluster in the natural process. This provides a good base to explore the clustering algorithm to see how much of the pattern can be recognized. Since the study considered here is about exploring the goodness of fit and the data's granularity, this method suits. There should be two clusters for the correct traceability links and others for all false positives or wrong links in an ideal situation. To validate the clustering granularity externally, MoJo (move and join operations based) distance is considered. This is the base metric considered by the clustering community for the software domain. MoJo computes the distance between two entities based on several of the moves and joins needed to make two entities alike. Basically, the smaller the MoJo distance closer the clusters are.

3.4.3 Outcome

Before assessing the outcome, one factor to note is the constraints associated with clustering to understand the analysis better. The process of separation of multiple entities to specific groups such that the entities with groups are bounded by a constant is called NP-complete. NP here stands for nondeterministic polynomial. In this context, most of the clustering techniques are NP-hard as all the datasets' constraints cannot be satisfied. So, the objective is to choose from multiple options that are optimal for selected measures. MoJo distance is the parameter chosen for optimization. MoJo distance is averaged across queries of trace across the dataset. If the local minimum is reached for the average MoJo distance for an algorithm, then the clusters' level is taken for the minimum of average MoJo distance. This process is repeated for all the clustering algorithms under study. Several clusters are initiated at two, as that is the level from the reference standard used for this study. Local minimum for lower bound values of several clusters is selected as the optimal clusters as part of optimization. The range is plotted on X-axis and MoJo distance plotted on Y-axis for visualization purpose to figure out number of clusters. Trend was drawn for each of the algorithms and the same visualized on three different datasets. Figure 3.7 depicts Visualization to figure out the right number of clusters across various clustering approach on different datasets.

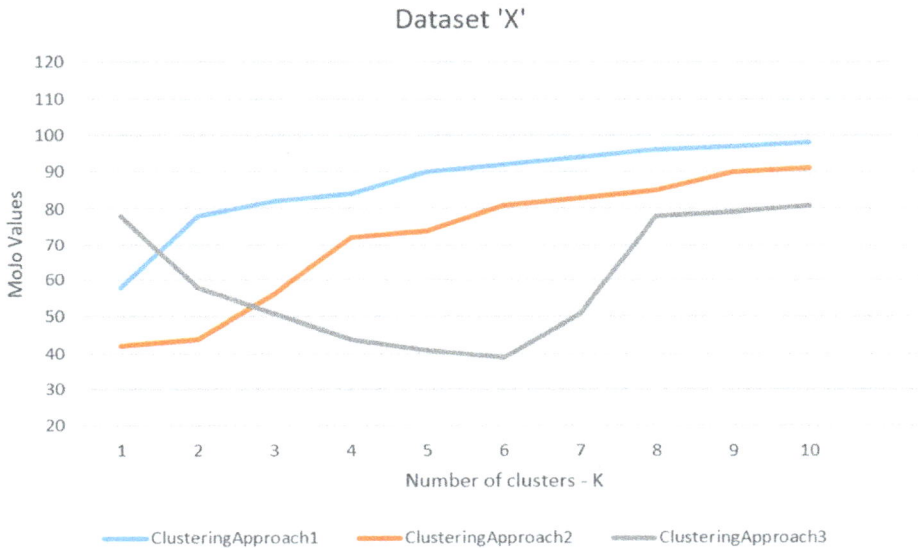

Figure 3.7: Visualization to figure out the right number of clusters across various clustering approach on different datasets.

Single Linkage of HAC works well for cluster hypothesis in traceability. To validate the procedure's established optimal values obtained in the experiment is set as a starting parameter, then run on each dataset's requirements. For every trace query, the resulting optimal number of clusters (which is predefined now) is arranged in the decreasing order of its recall values. Recall values are the number of correct links contained. Then recall values of all the clusters at the same rank averaged over the traces. Results show that when optimal cluster value is set, the traceability links end up being high quality and low-quality groups, high recall and low recall groups. So, this signifies that the IR systems can operate with optimal settings based on clustering, resulting in better performance than the baselines.

Cluster quality is determined by the answers that are available for queries, but in cases where an experiment is not conducted and answers are not available, then automation of this aspect is challenging. Given the cluster, the hypothesis is established, and the quality of clusters is evaluated by dynamic clustering aspects that focus on clustering being based on the query. Since the dynamic clustering was adopted in this study, each cluster will be influenced by the trace query link. Traceability link with maximum tf-idf similarity with the query is considered as the center for the cluster. The entire cluster is then organized based on this center. Clusters are sorted based on the average tf-idf values of the links of clusters to the queries. In place of average, median also would be considered.

Evaluation is done based on recall values; iteratively, maximum, average, and median values of tf-idf as defined above are computed each time the link from the low-quality cluster is dropped. Whichever approach can maintain the high recall values get shortlisted? Between Max, mean, and median, the choice needs to be made based on whichever maintains the high recall rate across several low-quality clusters removed count. Visualization is done as in Figure 3.8.

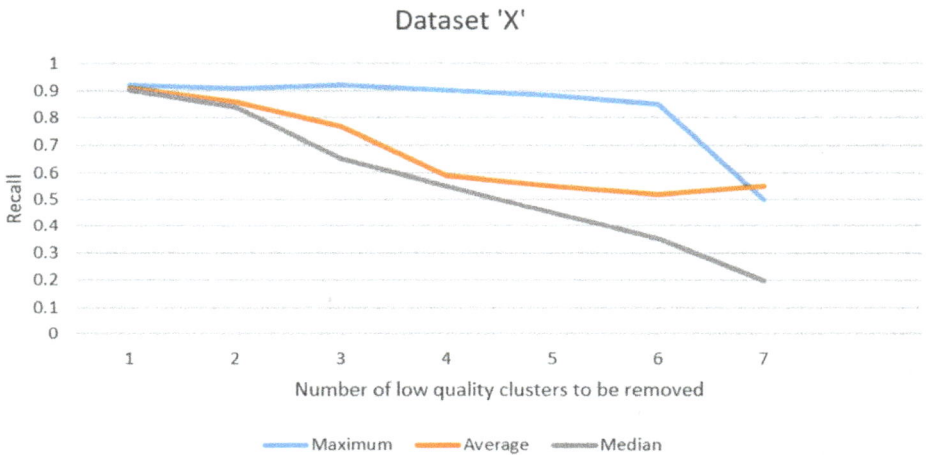

Figure 3.8: Visualization for identifying the number of low-quality clusters to be removed for different cluster configurations. Maximum, average, and median values can be compared.

First test the candidate links generated with the baseline pruning methods. The output is presented based on absolute similarity score, where the resulting link is ordered based on their similarity to query irrespective of the cluster they are present in. Another way is to represent in the context of clusters. Resultant links are shown inside the cluster in relation to the maximum representatives that are located for every cluster. Links also will have ranks based on tf-idf similarity with the query. For the first test where the comparison with baseline is made precision and recall are good metrics, in the second case, MAP is used, which assesses the quality across recall values levels. Higher the MAP values, top links are closer to correct links. While comparing the baseline method and enhanced approach of this study, the threshold is set to cut off the amount of least useful links that are pruning. At the threshold level where the recall and precision value are at optimal, that is taken as a point to cut off the less useful links. This study shows that enhanced methods work better than the baseline method. Mann Whitney test was conducted to test the statistical significance of the difference of the approaches. This test does not make any assumptions about the distribution of data being tested.

The overall cluster hypothesis works well for traceability. A single link with a specified number of eight clusters is a good setting for clustering. Quality can be demonstrated with a maximum similarity of clusters with the trace query. If removed, the three lowest-quality clusters improve the quality of the link's overall performance and indicate the number of false positives associated with them. Selected links to be presented in the absolute similarity format for making it browsable for end users. One of the study's assumptions is the consideration of two clusters as a starting point, which is correct and incorrect links clusters. Test that was done on procedure using the optimal level that was found by previous step iterating over the same procedure is also an area to keep note and validate with. The generalization of these results to industrialized settings is something to explore further. The dataset chosen here is of small size. It may not be scalable for large industrial settings. A variety of domains considered with a focus on requirements to design and requirements to source code trace will mitigate that threat of scalability (Niu et al. 2012).

3.5 Establishing traceability based on issue reports and commits

Links between bugs and commits of version systems are usually missing. These links play a vital role in defect prediction capability and assessing the quality of the software. Methods are proposed to establish a link between bugs and commits based on the text in both. This gets hit when the commit message has minimal information or missing information in many cases. This study focuses on improving the link between a bug and commits message making use of contextual information. Recently introduced ChangeScribe will be used, which produces context-based rich information for the commits utilizing the code summarization techniques (Cort´es-Coy et al. 2014, Dragan et al. 2006). These are derived as features with the bug reports features and input into the classification model that discriminates the relation between both. This approach is validated with the current state of the art method for bug linking which is MLink and approach reviewed here does better.

As part of quality management effort software, developers allow the users and testers to log in to the issues in the tracking system. Later these issues are referred to go back and make the necessary corrections in version control systems. But the link between these issues and their corresponding commits go missing for various reasons. These are important information that can facilitate multiple types of research. This information can be handy to establish the relation between classes and bugs that can reveal the most vulnerable classes and also to study the impact of bugs of various classes across the life span of software development. This also facilitates building an effective bug prediction system. Studies have focused on validating these missing links in the version control system. Validation involves the removal of those links that do not satisfy threshold criteria. These criteria are built based on heuristic methods of exploring various values with training and/or validation dataset. The focus of this approach is to strengthen the existing approach with ChangeScribe helping to build commit messages bases on the contextual information related to code changes done using code summarization techniques. These generated commit messages contain rich information such as structural code changes information, the intent of commits, and the impact of the code changes.

The uniqueness of this study is the use of a classification technique to build a model to discriminate between the bug and commits based on the existence of a link between them or not. Past methods have not used classification algorithm methods. Six major software projects were used for the study. Jira bug tracking system was used as they had more likely linking between bugs and commits compared to other systems. Tenfold cross-validation was run after removing some of the links, and then a model used to predict those links. Precision recall and F-score used for the performance measure of the model. This approach is unique in capturing the contextual information from the commits, such as the intention of commits, code structure, and the impact of changes for establishing traceability.

3.5.1 ChangeScribe

Automatic generation of commit messages based on a comparison of two source codes to perform characterization of commits based on method added, removed, and modified. The output will be an overview of natural language changes explaining what and why of the change. The length of the message can be configured. Two parts of messages are produced based on the changeset type and detailed change information from changesets: general description and detailed one. General one will include details of its initial commit, commit intent, class renaming operations, new module list, property changes details, and so on. Detailed description contains change details organized by packages, removed and new files class goal, and relationship with others, class, method, and statement-level change details. Since the detailed descriptions can be long ones, the length is controlled by learning based on impact analysis. Generally, representative classes are top-listed for descriptions as they have a major influence on the changeset and would be a key reason behind commits. The user can decide the threshold of the impact on the change.

3.5.2 Random forest

Random forest is an ensemble learner with multiple decision trees of weak learners are working together for optimal prediction. Each weak classifier is training on a different sample set of training data. The under-sampling method is utilized in a Random forest to figure out the missing links. Weka toolkit with default settings of random forest is used in the experiment.

3.5.3 Approach

The framework has training and deployment phases. Figure 3.9 depicts Framework discussed in this review. Training has an input of commit to bug link, with the link may be related or not based on if the commit fixes the issue or not. The issue report needs to have information about report date, author name, updated date, priority, summary, and comments list with identification and descriptions of the issue. The training phase delivers the model that will be used in the deployment phase to predict the links of new items.

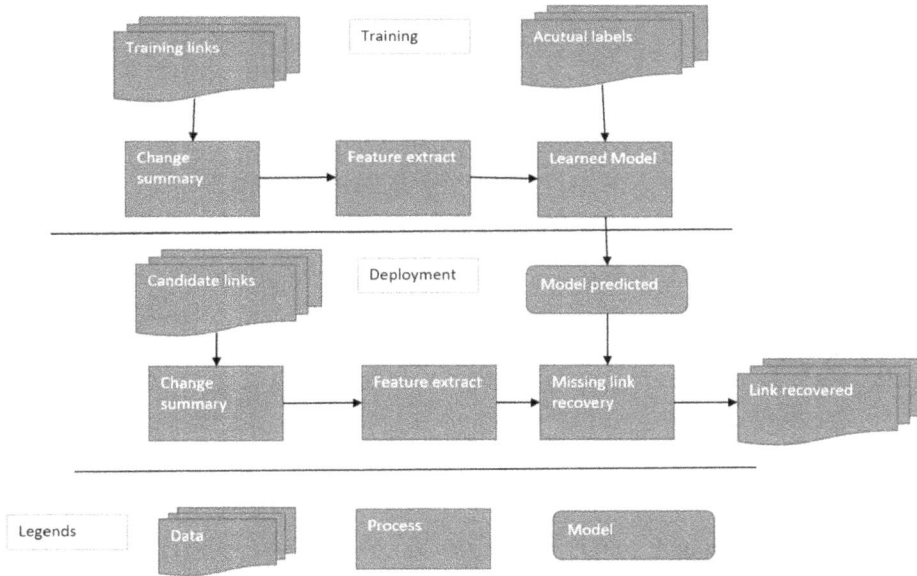

Figure 3.9: Framework discussed in this review.

The training phase involves issues and commits links obtained from the issues tracking system and source code repository. Details for training include textual and metadata descriptions of issues, commit messages of previous and updated ones, the label of true or false link. Change summary, feature deriving, and learning are the phases in model training. For change summarization, ChangeScribe is used on source code change of each commit. The output is then combined with the messages of the developers. Feature retrieval is done on top of this to extract key aspects from the links. The extracted information is used for model building. In the deployment phase, unseen links are taken as input; these are not seen during training. These links are created by linking all the commits with all the bugs. These links are analyzed by the model to validate the correct link. In the deployment phase, as well, the commit messages are extracted by ChangeScribe and combined with developer comments and fed into feature extractor before the input gets fed into the model for link prediction. Finally, the prediction is made on the validity of the link between the issue and commits.

Feature extraction is done on issue commit link and link representatives in two parts, one being metadata features and other text features. For text features, the issues and related commits will certainly have similar text descriptions. Text preprocessing to be done on the text content of issues, developers written messages in commits and ChangeScribe generated commit messages before the feature is extracted. Text normalization helps to increase the possibility of gathering the commonalities across the text. Stemming, stop word removal, and normalization are the preprocessing steps involved. Text normalization includes the removal of special character and punctuations. Later the words are generated from text. In the case of camel casing, each word will be separate, retaining the original form of the word to make sure that it helps in the case of the rare word.

Stop word removal is important as words like "a" and "the" appear very frequently in English but unlike to provide any useful information in the context of IR. A list of stop words can be obtained from the standard sources. Porter stemmer is employed to stem the word into its root form like operator, operating, operated will all be transformed in to its root form "operate."

After text preprocessing, values of features are generated based on cosine similarity. To represent the text content from commit and issue, as document vectors, dimensions based on the frequency of words. Weight is assigned to words based on the term frequency-inverse document frequency (tf-Idf). For a tf-Idf number of times the word appear in the document is one factor across collection of documents called as corpus and individual documents. Based on tf-Idf cosine similarity of the documents is assessed. Tf-Idf provides the vectorized form for the documents. Among various features, one is cosine similarity between the text of commit message from developer, issue, and ChangeScribe commits message. In another feature average cosine similarity can be taken, and similarly, the maximum value can be taken, or the other variants' normalized values would be another feature. These would be key parts of the feature engineering process. Other features would be constructed based on common words shared between texts of commit and issue reports. The ratio of shared words to total unique words would be the feature.

Features are constructed from the metadata as well. Metadata are all the other supporting data beyond text such as report date and last update date. Metadata-related features would look like the relation between issue and commit based on several files modified by commit, as mentioned in the issue. Other features would be the ratio of several changes in the file to the total count of comments and the issue's priority values. The feature would look at the link between issue, commit, and committer, as generally issue may be assigned to the committee; also committer may have a comment in issue as well. The feature is configured based on commit time and issue reported and updated time. Generally, the date of the issue and its corresponding commit would be closer. Boolean features like if the commit date is in between issue reported and updated date would be used. The time difference between commit date, issue date, and issue update date would be featured. Also considered are date of commit and last comment update for the issue, as the committer who fixed the issue will drop a comment in the issue log most likely.

One of the main challenges of model training is the imbalanced dataset, where the actual linked commit and issue are rare compared to false links. This calls for using the under-sampling technique of sampling. This helps to improve prediction as well as training time. In this format, we select several false links for every true link, which are the nearest neighbor of true links based on textual similarity. In under-sampling format, the total number of training links, several nearest neighbors, and commits, and issues related to the link are considered. To arrive at the nearest neighbor for issues and commits, other issue reports and commits are ordered in, decreasing the value of the cosine similarity.

Evaluation metrics used are precision, recall, and F1 score that are classic metrics in case of the classification problem. These are based on the true positive, true negative, false positive, and false negative measure. F1 measure is the harmonic mean of precision and recall. Generally, precision and recall are inversely related, so the F1 score helps strike balance while we evaluate. The cross-validation technique with tenfold is the usual approach taken to make sure that the model is trained to generalize the prediction for the unknown dataset. It was also observed that the issue and commit link from JIRA and other issue tracking systems have an id that was cross-referred between issue and commit. This makes the relation explicit; hence this information was removed from training data to simulate the situation for any general context and make other information in the issue report and commit being base for linking. In under sampling process right number of nearest neighbors needs to be explored and configured.

The study target was to evaluate the effectiveness of the approach to sort out right and wrong links, the validity of the approach compared to best in class approaches, and the influence of several neighbors selected for the experiment. As the number of neighbors is increased for the experiment, it influences the number of negative cases considered. Another study objective is to explore the feature importance of all the features considered in the experiment. Fisher test is the general test for checking feature importance in machine learning. Fischer score of zero indicates no influence of feature on output, whereas greater than zero values indicate a bigger influence. Their importance ranks features based on the fisher test.

The recall is the primary lookout in the experiment, as finding true links is the focus since not many links are produced by developers that can separate false positives. This approach outperforms the best baseline approach by far. The influence of increasing the number of neighbors is positive but at the cost of training time. For the importance of feature based on fisher test, maximum cosine similarity for the issue's text content and commit report is the top one. Among metadata-related features, a Boolean feature of whether commit date is in between issue reported date and issue last updated date appears in the top 10. Combatively text-based features come out as most important compared to metadata-related features.

3.5.4 Threat to study

There would be known experimental errors involved. The solution's generalizability would be a question as the experiment was limited to the bug tracking system of JIRA and SVN as the source control. A limited number of projects were considered, and Java programming language–related project used in the study also would be a limitation.

3.5.5 Other related work

On recovering missing links between issue reports and commit reports, many studies are done (Bissyandé et al. 2013, Nguyen et al. 2012, Wu et al. 2011). Missing links exploration is done based on creating representative candidates by looking at three criteria textual similarity, the time duration between issue reported and commit, and the relation of bug reported and committer. The threshold of several manually created true links is experimented with to arrive at the experiment's optimal threshold.

In another experiment, VSM was explored for missing links for simple IR. This study also agreed to the fact that JIRA has consistency in terms of maintaining the required information about bug tracking. Another study tried a multi-layer approach, first generating the representative links and then filtering out based on criteria such as the similarity between texts based on various features, as we discussed in the approach reviewed here (Nguyen et al. 2012). Also, to commit changes, code changes were included here but ignored information related to committer and comments in the bug report. The approach discussed in this study is unique in terms of Change-Scribe used to commit comment automation, not only for code part that is changed but also for the parts affected by the changes. The classification algorithm leveraged for the link validation is unique in this approach, particularly for the under-sampling technique. The extent of feature engineering done is also extended to account for variable possibilities of the features.

One of the other study target product-specific requirements associated with regulatory code (Cleland et al. 2010). The focus here is to identify the probability of the

terms being related to regulation and in turn, relate the regulation to the requirements. There is an approach to study web mining and associate the requirements document to regulatory requirements. Studies use bug report similarities to assess the
missing links. Another similar study compares bug reports with contextual words related to maintainability, functionality, and so on. (Alipour et al. 2013). Though in the
approach studied here, conceptual information is used, but they come from the code
summarization part.

3.5.6 Classification models application in other areas

Classification methods are used to derive key information from the bugs logged in the
system, like assessing the bugs' severity levels. Priority mapping has also been arrived at
from this information. Severity is assigned based on the user's need, whereas priority is
based on the developer's view. Different families of the feature have been studied in the
area. Work also focuses on classifying an issue in to feature change requests or bug reports (Antoniol et al. 2008). Multi-class categorization of the content into documentation related, improvement suggestion, bug report, refactoring request have been made.
Work has focused on estimating the time needed for bug fixing if it is short or long duration (Zhang et al. 2013). Classification techniques are also extended to identify defect
prone modules using semi-supervised learning and active learning with dimensionality
reduction (Lu et al. 2012, Lu et al. 2014). Cross-project defect data have been used to
predict defect probability in another project (Panichella et al. 2014).
  (T. B. Le et al. 2015)

3.6 Topic modeling—based traceability for software

Traceability in software becomes critical as the processes become more complex,
and artifacts involved keep growing. In this approach topic modeling, machine
learning technique is explored to automate the traceability process. Over the life
cycle of the software development, traceability is built, and for topic modeling,
probabilities of the artifacts are built. The software domain comprises many documents related to requirements, design, and code throughout its lifecycle. Traceability tasks focus on making a connection between these artifacts to facilitate various
engineering tasks. Full automation may be possible in smaller projects. Difficulties
are added on due to the lack of accuracy and false-positive cases of existing links
provided for the artifacts. In this work, the focus is on topic modeling usage and
automated linking information usage. Traceability can be assessed based on the retrospection of the processes and artifacts manually. In the case of automated information, retrieval traceability is done retrospectively. In a proactive traceability approach,

the links between artifacts are established as the process goes on. In this approach, proactive traceability and topic modeling are looked at.

In proactive traceability, the links are created online, and these are refined and improved over time. Developers are benefitted by using these links. This online information also helps assess the users' time-based actions in design scenarios, which facilitates as the link for multiple artifacts that are not of the same kind. IR and machine learning approaches have been employed in this area. LDA algorithm unsupervised learning based has been used for learning semantics from artifacts without historical data. In this approach, LDA would be employed in proactive traceability. Another key differentiator is using architecture as the center of the approach rather than depending on the source code only. Proactive base traceability is closely associated with topic modeling. Though topic modeling has a restriction of application on different artifacts like sound files, since they focus on user actions associated, it plays a linking role. Proactive linking will not cover semantic aspects of artifacts that will be taken care of by topic modeling, as that is its specialty.

3.6.1 Tracing based on retrospection

Machine learning techniques have been associated with monitoring run time and program analysis to create the link between our source code entities automatically. Visualization of traceability is a common approach where clusters of representative traceability results can be projected. Graphics to show term frequency and the tree structure of the requirements are common ones. Topic modeling helps bring out the semantic relation between artifacts and the software ecosystem.

Retrospection traceability is mainly manual and partially automated. The linking is established by referring to the artifacts' IDs across each other, making traceability at the artifact level. Some tools can establish a link based on information from the user. The devices also depend on formal training to enable the capability to express the relation between groups of artifacts. Once this predefined information is available automated tracing is possible in the system. As part of the retrospective method, the users are expected to provide potential issues while checking in the code. Model-based automated linking would use the transforming techniques that have built-in rules. Earlier approaches from the authors of this approach have included capturing users' activities over multiple platforms for using the artifacts, based on recorders. Based on these activities rules are configured. The recorder development and customization of rules include the effort of setting up for the first time. This approach does not account for the semantic nature of the artifacts. If the user is working on multiple tasks of different projects, recorded links may be irrelevant, as the approach looks for sequential accessing of the artifacts. Thus, topic modeling will be the right approach.

It is important to assess the parts that can be automated and those that need human involvement. So, the focus is to reduce the task of document search and

posting of the links. The approach ensures this as the development process progresses. Categorization of the links is done semantically, and visual presentation is
done to carry out link analysis. It also facilitates easy navigation in the editor for
validating the correctness of the links. Performing the traceability should ideally
happen in the background as the users proceed with the work. Ongoing improvising
of the link quality and visual presentation of the links are beneficial for the users. It
is essential to make sure that the traceability process to be scalable, both the link
generation process and presentation of the links. Since the examples available are
too small, scaling will be challenging. One prospective thing is the process being
incremental, will assist scalability. Since the topic modeling based on probabilistic
inferencing is explored, models have been scalable.

The approach of creating the links to each of the documents is infeasible. Trace
must be created for a purpose only if there is a direct need for the link. If the focus
is to create source code lines to concepts, it will make the scenario complex, unman-
ageable, and unusable. Scoping the links to the architecture level is a good scope
for traceability. In this setup, the trace links are associated with the architecture
components, and this goes well with the understanding that the source codes are
inline with the architecture defined. Many system design parameters, such as gov-
ernmental constraints, domain needs, and user specifications, are handled at the architectural level. Since the architecture forms a bridge between requirements where
the problem initiates and code and testing where the solutions hide, the gap between
the space is minimal and restricted to the difference between source code and requirements conceptually. Architecture is an abstraction of the source code which supports comprehension of the overall system and evolution of the software system, and
these are the areas traceability target to improve.

3.6.2 Topic modeling

This is a machine learning approach that derives the semantics of the text corpus
from deriving topics. This approach is discussed since semantics-based traceability
is established by LDA is considered for topic modeling. One of the major expectations of machine learning algorithms is to unearth the hidden patterns in data structures. This has encouraged the progress in the exploration of algorithms that can
facilitate dimensionality reduction. LSI serves the purpose, and, due to its focus on
text corpus, it has its applicability in software engineering. Here, the document is
represented based on the count of the words in it compared to a total number of
vocabulary words. For all the document vectors described in comparison to vocabulary size, the matrix representation of data space is sparse, and LSI plays its role in
reducing such large sparse matrix into lower dimension space. Probabilistic LSI was
an improvisation of LSI version on top of the Bayesian-based model LDA was introduced; LDA has the specialization of overfitting.

LDA represents the topic as a probability distribution of the associated words which are derived from Dirichlet distribution with associated parameter. Also, each of the documents, in turn, is the probability distribution of the set of topics derived from the study. The Dirichlet distribution represents the discrete distribution. The topic variable is compared with each word in each document. Actual word tokens are derived from the topic distribution. Word counts are represented as a sparse matrix as defined earlier. Bayesian calculus is involved in deriving the topics hidden in each word of the documents and word vectors that represent the topic distribution. LDA is unsupervised learning with only inputs of a set of corpora of documents that are represented in the form of a sparse matrix and the desired number of topics to be learned. Improvements in the Bayesian approach has helped to learn the representations in real time with fewer data. Collapsed variational Bayesian inference algorithm is used here; this format of topic modeling makes it scalable to hundreds of documents and operates over distributed computing and efficiency also helps in traceability.

3.6.3 LDA application

Learned LDA model can be run on the words to create a list of words with probabilities in decreasing order for every topic. So overall corpus can be presented with these sets of topics. Also, the document distribution can be validated with topic distribution to identify the topic of the content. Distribution of topic probability for the corpus provides information on the most relevant topic for the corpus. Probability representation of the matter for a corpus can be used to compare the corpus with other ones based on similarity metrics like cosine distance to find out similar corpus. So LDA assists in finding similar documents and visual display documents based on the topics. LDA also applies to identify the topics in source code. Compared to another approach, in this approach reviewed here, prospective traceability is focused on, and topic modeling of text in the artifacts of requirements, design, and other software lifecycle is focused and not on source code.

3.6.4 LDA drawback

One of LDA's drawbacks is the number of topics that must be specified, and, if the topics are too less, then it represents the content with general topics. If the topics are large, then it overlaps the semantic representation. One of the approaches would be to model various topic choices and visualize the optimal count. The hierarchical Dirichlet process can model to learn the number of optimal topics in a nonparametric way. Another drawback is the representation of the top topics based on probabilities and no interpretation of the semantics of the topics involved nor the interpretations for topic labeling.

Traceability will be of significant value with the topic modeling techniques being applied to software development related artifacts. Topic modeling helps proactive traceability by providing semantic details for artifacts. The proactive traceability approach discussed here is based on the architecture. When the developer is working on the architecture aspect, the model assesses the actions about the developer's visit to a page on the wiki, the opening of the requirements specifications, and changes to the issue report. Based on these actions, verified artifacts are automatically linked to the part of the architecture that is worked on. Once these setups are done, the architectural graph visualization will have all the linked information for the users to access. Since the topic modeling also comes up with the topics identified, it helps the user to gather the information necessary from online documentation and other sources. Instead of searching to complete the project repository, this visual representation can help navigate the required information effectively.

With topic modeling, the entire system view is enhanced. Since the collected artifacts are linked with the components of the architecture. When a user looks at visuals with the associated artifacts of the component, the artifacts' topic distribution is also available. So, this creates the reverse linkage of the components to the group of topics based on the semantics. This facilitates describing each of the components as an association of the artifacts through topic distribution. This gives a topic-based bird eye view for the system. This helps gather artifacts and components based on semantic-based topics. As time progresses, the quality of topics improves due to the accumulation of more documents. The contribution of artifacts happens online as well, which makes the artifacts reference base large and diverse. False positives considerably get controlled with this approach. The threshold can be set and beyond which the link can be ignored as noise. Artifacts cluster can be displayed for users to choose from, and also propose other similar artifacts of users interest. This also provides an extension to understand the topic that users are working on and how it has been evolving, which plays a significant role in program management.

Tools associated with the framework are a tool for topic-based artifacts search, proactive capturing tool, and enhanced topic architecture visualization tool. Figure 3.10 depicts Tools involved in the approach discussed here.

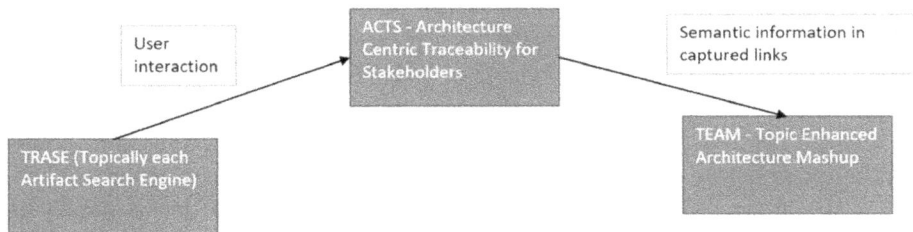

Figure 3.10: Tools involved in the approach discussed here.

3.6.5 Topic modeling capturing tool

This is a topic-based artifacts search engine, which traces through the artifacts in the project repository and learns the topic based on LDA in real time. Key technologies are used here to facilitate keyword-based search. Since the topic model inference algorithm adopted is high speed, this tool learns topic-based models for all the artifacts returned in the search. They came up with high probability words and distribution of these words over topics for artifacts selected. In the visual display, each topic is color coded, and the artifacts will have the color representation to visually highlight the extent of the topic covered by each artifact. So that makes it easy to identify topic distribution in the searched artifacts. The tool displays the artifacts based on topics, and similar artifacts are displayed. So, users can choose the artifacts belonging to a topic or pick the ones with a high proportion of topics. Tool for capturing the actions will record the artifacts visited by users on the displayed tool. The same is presented as links of the trace that will get associated with the functional components of the architecture. LDA operating search engines can work on the web pages of the internet. The linking tool is enhanced based on the topic that makes the traceability process more efficient and accurate.

The linking tool helps to trace the activities of the user and capture links. Activities of the users over the architecture and all other artifacts are tracked. The adapter that is tool-specific for recording user actions is available for MS Word, Excel, PowerPoint, and Firefox browser. Once the recording is done its filers through the rules to remove error links and make association based on semantics. If the user visits any sites from Firefox using the search engine's search option, in this approach, then recorder in the browser captures that URL. At every session, artifacts visited in the session are listed after the rules help refine the search artifacts associated with architecture components. Capturing tools can link to various types of media, including images. This possibility of capturing non-text-based content is advantageous in case of proactive traceability. Figure 3.11 depicts Topical information mapped on the software architecture in TEAM tool.

3.6.6 Visualization tool

The visualization tool aggregates all the links obtained from the linking device and puts it on top of the architecture. Topic modeled information is picked, and visualization is provided for both components and artifacts. The tool gives the ability to visualize the entire system and see which components are related to which topics. To semantically distinguish the components, the topic distribution of the components is studied. If the topic's probability is greater than a threshold decided by the user, that component is associated with that topic. The visualization tool displays all the topics and their probabilities. When the respective topic is clicked, all the components associated are highlighted in color code. This helps to visualize the topical architecture of the whole system. Also, each component

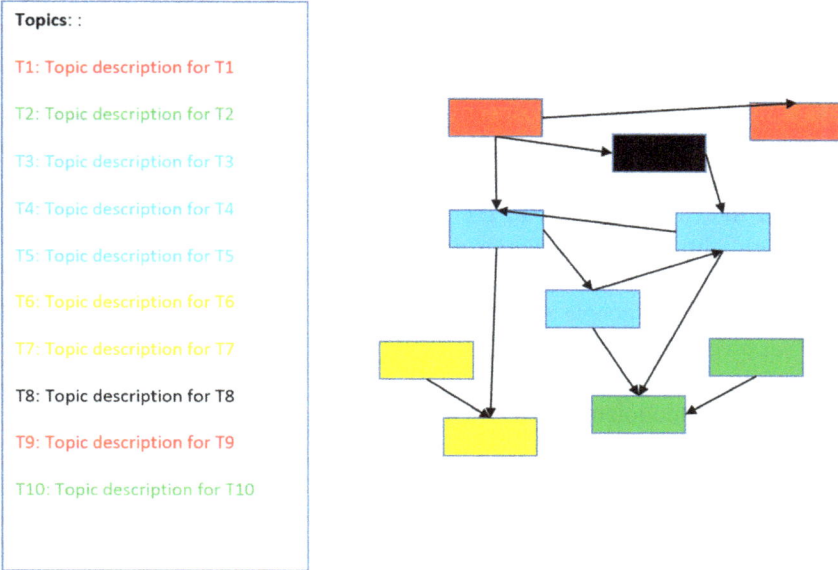

Figure 3.11: Topical information mapped on the software architecture in TEAM tool.

can be zoomed in to figure out the topic distribution related to artifacts. This visual config-uration helps the user to pick the artifact of interest and access those in the native editor. Overall user is helped with the system level and component level artifacts view.

In the study of the tools, the focus was to understand the unfamiliar part of the architecture's base system chosen for the study. To begin, the base tool was explored for various artifacts, websites where the tracking tool captures all navigation for its links. To make artifacts adaptable for topic modeling, they are converted to plain text, and in the case of large artifacts, they are broken down into separate files. Other preprocessing steps for text are automation in the topic modeling model. The search-ing tool was checked by running the search on the studied base, and it was efficient and quick. Once the link was captured by the tool, topic modeling was performed to generate the topic summary for artifacts and components. These were visualized in the visualization tool, then by testing the topic it was linking to associated topics. The visual outcomes studied on base tool was validated with the user group and found useful and relevant. This visualization also facilitated users to analyze if the linked artifacts are right and question the linking if there are deviations. Some of the users' suggestions were to provide topic details in finer details to avoid the need for examin-ing multiple components. Also, the feedback was about making topic detail more rep-resentative to avoid overlap with multiple topics of similar high probability. This approach is unique in the sense of capacity provided to generate trace links across various media. Probabilistic analysis of the semantics of the topic also is a highlight. Visualizing the topics overlapping with the architecture is also unique.

As part of the performance test, the perplexity metric used decreases as the number of artifacts increases. Perplexity is the most used metric for topic modeling. Lower the perplexity better is the model. This indicates that the increase in artifacts base plays a significant role. As prospective traceability captures more documents, helps topic modeling, which are similar with better quality topic modeling helps build the artifacts. Study of time taken to understand a different number of topics shows a linear increase in time with the number of topics to study. The time involved also highlights that the system is quite fast and is scalable. The linking tool is scalable as it worked incrementally. Visualization tools also can be made scalable with the incorporation of google map APIs. A precision-recall study was conducted and found that precision is low as different artifacts are considered in a study made of one set of artifacts. LDA outperforms LSI, as LDA is probabilistic and includes its information, the prior distribution of topics and words, which will regularize the process.

The limitation of the approach is assuming that the architecture exists, as the study is centered around architecture. It is realistic to consider that the system has architecture, irrespective of whether it is documented. In case of architecture not documented or complete, then the source code can be organized into components. In the study, text-based artifacts are included to conduct topic modeling; there is an opportunity to include metadata that are non-text. The study's interesting question, can the proactive link capturing work in association with the retrospect-based work in IR systems be successfully combined. Based on the topic generated, can we develop the features, concerns, and nonfunctional requirements? Also, the question arising is exploring topic modeling on large documents by considering the documents' sections (Asuncion et al. 2010).

3.7 Software traceability by machine learning classification

Tracing the links between all the software development artifacts has been a significant focus of the last decade. It is purposed to establish links between requirements and software classes as an example. This review focuses on establishing a link between software artifacts that exist across in similar forms. This work focuses on creating knowledge that can be used in various experimentation related to the topic. The overall benefit can be expected in areas such as security requirements, which are nonfunctional, regulatory requirements, and architectural requirements.

The focus here is to establish a classification-based algorithm from machine learning. To generate a trace between various documents that frequently occur across software projects. Compared to other earlier methods like LSI, VSM, and LDA where the similarity score was learned based on the trace related query and the target artifacts, which were also probabilistic models. Here in this approach of classification, weights are learned from target documents based on which they can be classified into different classes. In the case of a large corpus of training data being available, this technique can be excelling compared to other trace generation methods.

An earlier version of the work focused on nonfunctional requirements (NFR) (Cleland-Huang et al. 2006, Cleland-Huang et al. 2007). Post that, the solution has been extended to tracing quality issues to architecture and code. Solution also looks at the area of tracing regulatory-related code to requirement. In this review, the approach will be discussed.

One of the key challenges faced on the traceability domain is to trace back between common sources of artifacts with that of the targeted artifacts that are specific to projects. One of the discussed applications is in the quality domain to trace back performance, usability, and security-related goals. The second area is to trace back health-related regulatory requirements to product level requirements for health care products. This work has targeted the need for traceability solutions that do not need much of the users' efforts, possibly zero effort if possible, so that the work of users is not impacted. Also, the focus is on complete automation of the traceability creation process, maintaining the performance and quality. Since the regulatory requirements and NFRs are most common in most of the projects, there are returns on investing the effort needed for this work. In the case of frequently seen requirements, this approach has a considerable upper hand on the previous methods (Wieloch et al. 2013).

3.8 Classification method for nonfunctional requirements

This section explores the classification algorithm used by authors in TraceLab, as discussed in the next section. Requirements sources are highlighted to be getting captured in various sources such as interview notes, memos, and minutes of the meeting. Generally, functional requirements get organized and nonfunctional once they get scattered. That causes concerns about the important needs of stakeholders getting missed. The classifier proposed here targets both structured ad unstructured data. The classifier focuses on the early aspects of the development, mostly related to security, portability, and usability. The classifier's focus is around these early aspects of the system and not on a programmatic level. Focus on early aspects also helps to incorporate these major aspects at the design level rather than digging into code later to refactor. Figure 3.12 depicts classification process for NFR as discussed in this review.

Data gathering, classification, and application are the phases involved. Key indicators are extracted from the raw data, including requirements specifications where

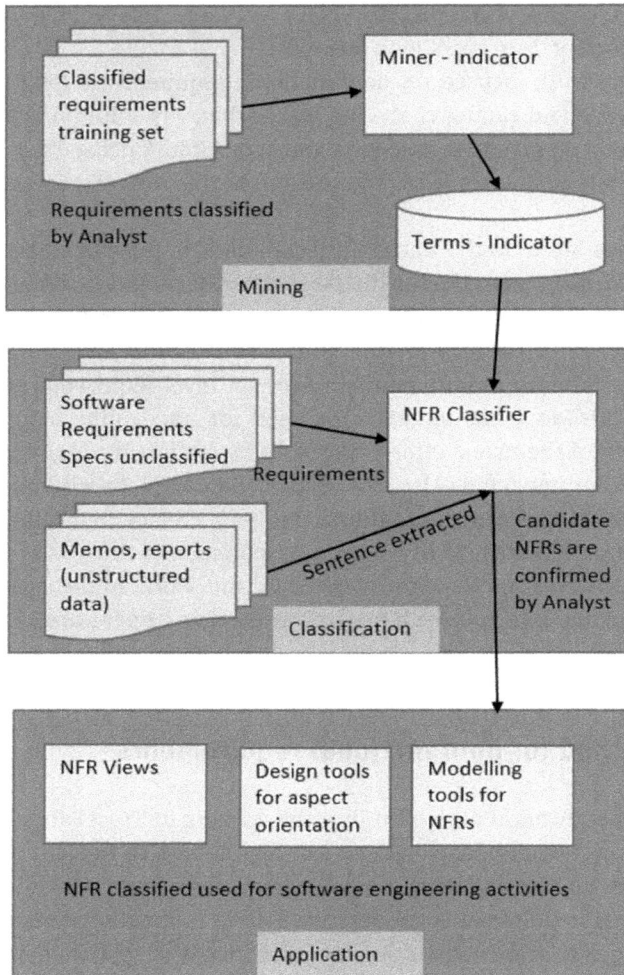

Figure 3.12: Classification process for NFR as discussed in this review.

the NFRs are manually labeled. These terms are used for the classification of NFRs. The classified requirements come in handy for further software engineering activities like architecture design and requirements discussion with stakeholders in the application phase.

3.8.1 A general way of NFRs classification

Elicitation is the method where creative questioning is involved, and checklists may be used. Aspect-oriented programming focuses on determining the granular aspect from design and code and early aspects from requirements. But methods under them are not addressing NFRs. Some of the partially automated processes involve the need to manually labeling requirements to the target classes and then use them for the IR process. This also depends on the good initial reference for the requirements that are providing similar terms across conditions. Another approach which is also semi-automated makes manual identification of keywords, based on which the tool builds the visual presentation about the behavior involved. Using this view, users derive key candidates. This also has a lot of manual effort involved. This also needs the requirements to be grammatically structured. But this approach provides the aspects that are specific to the project and the common ones as well.

Another approach used NLP way to identify the nouns and verbs involved to recognize the viewpoints involved (Sampaio et al. 2005). But the viewpoints need considerable validation from the users for both viewpoints and the target candidates identified. It may also miss out if different nouns and verbs are used for the same scenarios across viewpoints. A validation step attempt was made to develop keywords that assist in classifying NFRs. In anticipation of reducing the training set creation and use. Keywords were extracted from the standard catalogs. These catalogs were the body of knowledge for NFRs. Based on these keywords, NFRs were derived from a set of requirements. Wherever there were multiple keywords associated with the requirements, they were taken as candidates. Multiple categorizations were done to map the requirement to multiple categories if they are applicable. But there were overlapping keywords across and cases of missing keywords as well. The challenge involved gathering the keywords from the standard catalog representing the project under study. So, the proposed classified arrive at weight-based indicators for every NFRs. In that way, the classifier restricts to extract only those for which it is trained. This is not limiting as the overall documentation has hundreds of NFR types documented but in general practice, and only a handful of them come into the picture. This also has a definite advantage on standard keywords derivation methods, as indicators can be generated from already labeled requirements set. Hence, it is easy to customize to the need of the organization.

One of the experimental factors was several top indicators to be chosen; multiple options were experimented with. Threshold values can be managed to remove the system's noise so that precision measure would be improvised. Using a different number of top terms did not make any impact, but using lesser terms will have the potential for overfitting.

Trace classification algorithm works on manually classified documents, like the requirements being classified into security, functional and performance related (Wieloch et al. 2013). Also, reference can be made to the requirements from health regulatory mandating automatic logging off on the software related to health care to protect the patients' information.

As part of preprocessing the data before training, stop words associated with the corpus need to be removed as these do not add any context to the content semantics and stemming of the remaining word to their base form. In the training phase, each category of the training data is provided with identifier factors, and the training assumes that the categorization is done appropriately. For each of the indicator terms or factors, probability-based weights are assessed. Weights indicate the strength of the terms in explaining the types of classes. For example, words like *authentication* and *action* occur regularly in security to take larger weight compared to words like *ensure*. So most occurring words take larger weight compared to least occurring words.

Overall, computation of the weights is based on a type of classification like NFR-related or health care regulator–related. Indicator terms are another factor. The frequency of terms in the documents computed for each document and weights associated with each term to indicate their strength in explaining the class are other factors. The overall probability value for the term over document set is computed based on the frequency of terms across all the document corpus normalized for a specific class under study. Frequency increases if the terms appear regularly in the document set, indicating the term to be potentially one for that set of documents. The second factor of probability calculation is the frequency of occurrence of a term related to requirement compared to the term's frequency across requirements. If the term is more prevalent across requirements, this term gets reduced; this term basically looks for requirement specific terms. The third part of the probability calculation is the term's occurrence in a document for a project in relation to the occurrence in all the projects in training data. This factor provides more focus on terms that are appearing across projects. Overall for every term, the probability is arrived at, and top terms based on probability can be targeted.

In the classification part, using the indicators that have been provided with the probability of occurrence, other artifacts are classified to the respective requirements. The requirements are then bucketed based on the type of artifacts according to the dominant terms representing the class. Conditions that get a term mapped beyond the set threshold will be classed to that term. The rest of those requirements that do not meet the threshold will be classed on functional requirements. The success of classification results will depend on many targets that are getting mapped against specific types. The mapping threshold is set to achieve high recall, which is one of the key objectives of traceability problems. Artifacts classification is done based on probability scores, which arrive based on artifacts belonging to a category. Indicator terms are weighted from training data, which are classified based on categorized artifacts.

3.8.2 Experiment setup

Data for the experiment is taken from two different domains. Experiment is based on, leave one out approach, where the training set is divided into multiple sets, one of the sets is left out from a test and rest of all in the training set is utilized. Several sets were fixed to ten, nine of them used to train the representative terms, and one for testing. In earlier experiments to choose an indicator term, the option was to consider all terms or top n terms or terms beyond threshold values. Operating with threshold works well, so that will be shortlisted in this experiment. Indicator terms shortlisted will be used to classify the documents in other buckets. This entire setup repeated multiple numbers of times until all the artifacts in all buckets are classified.

Approaches taken in earlier experiments by the authors of this experiment considered the top-scoring phenomenon to select type for the requirement. This approach did not do well. The type of requirements here would be security, functional, and others. In another approach that worked well, artifacts were classified into multiple classes if they belong to more than one class (Altman et al. 1994, Antoniol et al. 2002). This was based on the artifacts performing over the threshold for multiple classes. Like, if the requirement maps with security and scalability, then the mapping is done for both, though original labeling as per the original dataset is one of the classes. This approach suits if we see noncentral data around a class and tend to cut across multiple classes.

In this discussion, authors have focused on previous work done, and here the attempt has been made to package this approach for further utilization by the community. End-to-end data gathering, classification training, artifacts classification, leave on out-based validation, and output verification with confusion matrix are packaged.

The experiment is conducted in the TraceLab tool. Starting with classified requirements and get the list of stop words that will be used for refinement. Requirements are refined by preprocessing specifically stop words removal and stemming the words into their base form. The dataset is scattered to multiple buckets randomly with equal size. In each experimental iteration, one of the buckets goes for testing, and the rest used in training. Following classifiers, training terms are created, and top n terms are identified. These identified terms are used for testing. Iteration continues until each of the buckets has played the role of a tester. Outcomes of the iterations are saved so that after the loop completes the results can be represented as a confusion matrix for performance evaluation. The output is extractable to external files. Figure 3.13 depicts Classification based tracing experiment setup in TraceLab.

3.8.3 Outcomes

The confusion matrix results demonstrate that the larger datasets play a major role in the effectiveness of the results in terms of achieving high recall and precision rates. Another key component is how homogenous are the documents in the training set.

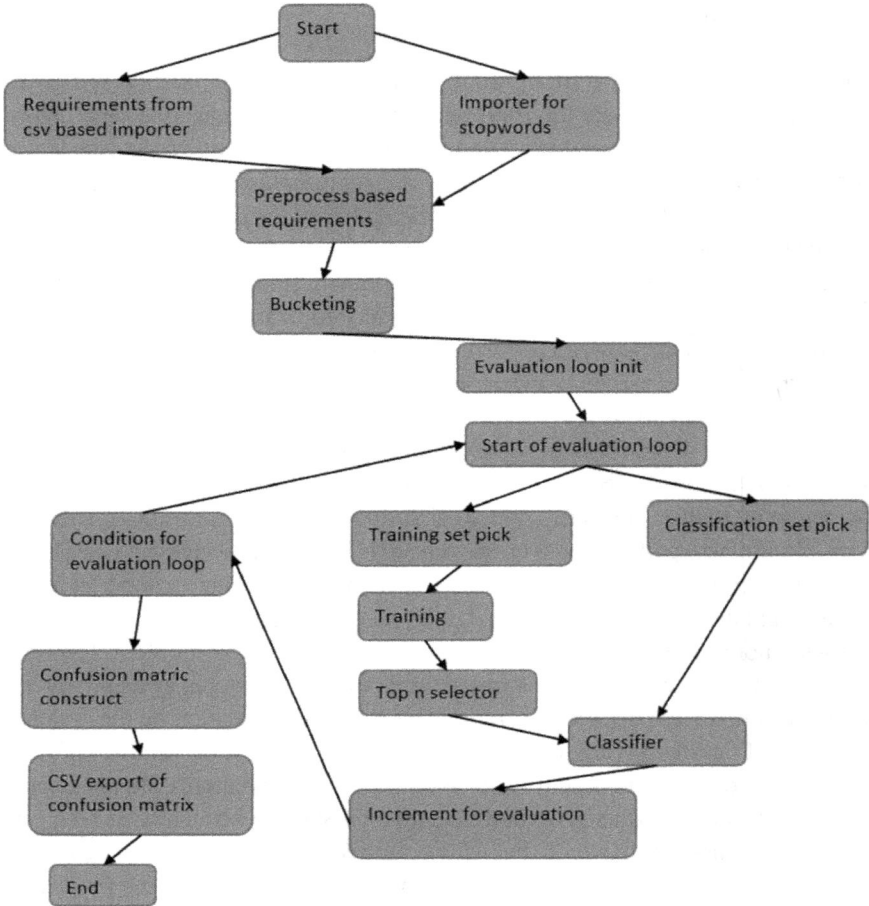

Figure 3.13: Classification based tracing experiment setup in TraceLab.

The more the set is homogenous, it makes way for a classifier to work effectively. In the case of heterogeneous ones, the training content needed is very high.

Earlier work from the author around NFR classification, a dataset was made available to researchers, and experiments were conducted using WEKA to compare the performance of the author's approach against standard classification approaches (Cleland-Huang et al. 2006). Algorithms like Naïve Bayes, decision tree, and feature subset selection were evaluated but were not outperforming the approach from the authors. One another approach from Jalali et al. combining feature subset selection and decision tree got better on the recall and precision compared to other standard machine learning, but recall values were lower than those by the authors here; accuracy was comparable (Cleland-Huang et al. 2006).

3.9 Conclusion and future work

Exploration of the traceability problems was done in detail. It was also noticed that the work done in this area is not extensive, and the need to focus on the actual problem. Prerequirement and postrequirement traceability are explored, and the need to focus on prerequirements traceability was seen. Exploration of the social infrastructure for the sources of information is also key. Future focus is needed for building models that can handle the dynamics of the ever-changing requirements across its life cycle. The research focused on process modeling will help to build a complete understanding of the ecosystem. Works should focus on providing continuous up to date pictures for all stakeholders.

Deep learning methods of building a neural network based on word embeddings and RNN to create trace links have been discussed. Future work focus is to improve the tracing network's precision that was discussed by including more robust training data inclusive of the negative cases. Though the work reviewed focused on natural language text, a further extension to other artifacts such as source code would be useful. Since gathering a large amount of data is the challenge, there is a need to explore the combination of the available datasets and the experience of the experts.

Cluster hypothesis technique for improving IR-based automated tracing was discussed. The optimal settings needed for the experiment were explored. Validation of the experiment against the industrial setup was done to understand the limitation.

The importance of the link between commit and issue report is discussed for maintenance of the software and study of the reliability of the software and predicting prospective defects that may happen. The approach here has considered using context-based information and text classification outputs for creating features to validate the links. Future exploration needs experimentation with various classification algorithms to figure out an effective one. Also needed is the extension of the experiment on a variety of bugs and software projects.

Though the general focus of trace links identification is retrospection based, here we explored the work of prospective exploration. Topic modeling was explored for semantic understanding of artifacts and a further extension to the software's architecture itself. Further exploration can focus on combining retrospective and prospective frameworks. Also relating the topics generated to the concepts of features, issues, and requirements in the software. Exploration of handling long artifacts and to topic model on them is needed.

We explored the approach of combining requirements specifications and text information to classify nonfunctional requirements. Though it is partially automated, it helps a lot for the software fraternity to understand the requirements better. Further work is needed to see if the work done here is flexible enough to accommodate more input data in the future to improve the classification consistency for data across various projects and organizations. Though the work focused on high-level aspects, there is a need for looking at granular aspects like authentication and logging.

Tracing requirements across multiple projects were discussed; the focus is to improve the accuracy of the approaches. Datasets were created for extensive experimentation of the standard classification approaches.

As we focused on establishing approaches for traceability between software artifacts, considering strengthening the requirements management, which is a key element of software development, we now move on to auto-code completion facilitation with the prediction model. This is another key area that will help to improve the productivity of the developers.

3.10 Extension of work to other domains

The ever-changing requirements of the customer are a common problem across various industries. This being the case, exploration done on building machine learning models for this purpose is reusable across various domains. Process model building–based research that is proposed to focus on understanding end-to-end system will also easily help other domains. Overall system architecture is complex and has too many interactions. Another key common concern of various industries is continuously providing up-to-date pictures to all customers and stakeholders. So, all the prospective research areas that are discussed here help to make the system transparent and in an automated way will benefit most of the domains. There is also a proposal to combine the domain expert knowledge with that of the models proposed to build around data. This is also a common theme that will benefit a lot of other domains beyond software engineering.

Cluster hypothesis–based technique for information retrieval–based automated tracing discussion also pointed out exploring the approach with that of various industrial setup. This can also be explored across various domains of industries. The topic modeling–based approach to understanding the artifacts' semantics is also a common theme that can be experimented with that of domains where the unstructured information is captured in the artifacts. The discussion also spoke about making systems flexible enough to calibrate itself as more data is gathered in the future; this is also an element of a common theme to explore in a variety of domains. Approaches discussed here have also focused on providing the datasets for the researches to be conducted. So, the research focusing on all the common themes across the domains can use this dataset because data capturing and processing is a major concern for all industries.

Chapter 4
Auto code completion facilitation by structured prediction-based auto-completion model

4.1 Introduction

Establishing acceptable benchmark evaluation methodology and metrics to provide a quantitative comparison and highlight strengths and weaknesses of the models built for this objective of auto code completion is an area for exploration. Generally, these approaches assess what tokens may apply in the current context, based on rules configured. These rules are configured based on syntactic and semantic aspects of the data under study; even with the work done in this area, managing long-range semantic relationships like suggesting, calling a function many tokens behind is a challenge. One of the research focuses would be to build models that can likely continue a partial token using character language models for auto code completion.

Programming languages have a formal structure, whereas general-purpose language represents a mathematical model of natural language. With this programming language gets created top-down for users by designers. At the same time, the natural language gets evolved bottom-up via social dynamics. Programming language changes in a structured punctuated way, whereas natural language gradually evolves. Formatting can be a key theme in code. The text has a big-time dependency on the environment, whereas the functionality that influences code fades off over time. The execution environment of code also evolves in a fast-paced manner compared to natural language. This formal structure of the code facilitates the code reuse. Code reuse fuels the practice of storing the reusable code in libraries. This approach forms the background for the resemblance of the functions used in the program. The student's project and competition do not have this constraint. In contrast to language, an event can be described in thousands of ways. The code also is explored to have deep patterns compared to text. All these lead the way to code completion as an efficiency provider. Program language finds its transformed format into code, making it essential to be a syntactic and sematic way without any ambiguity. Natural language has the challenge of text ambiguities, whereas code can make the program's underlying structure best. Though small-scale ambiguous code is still real considering the alias used in code. Behavioral flexibility provided by programming language can cause ambiguity in semantics. A structural or syntactic deficiency will cause compilers that do not belong to the standard class.

The recommender system assists code completion tasks. Many of the techniques focus on data extraction and machine learning on the software life cycle documents. Probabilistic modeling of source code is based on recommender systems. This capability assists developers in maintaining the code. In the case of the integrated

https://doi.org/10.1515/9783110703399-004

development environment (IDE), code completion is the most used feature. All the primary IDE has this feature. Code suggestions usually are written alphabetically rather than the relevance for the context. Statistical based suggestion aims to learn the importance of the suggestions based on probability and to provide the rank-based list. Initial statistical code completion suggested constructors that were later improved by a graphical model built on Bayesian theory. Source code–based models used for code completion. Techniques like utilizing the tokens of the source code to figure out the next context. Work on providing comment completion based on code information, topic model graph to provide contextual information and recommendations of methods, and class names based on code tokens.

Machine learning for code synthesis and code completion are the most explored areas. Sequential n-gram-based model and structure-based prediction models target the same area, but there have not been good comparative frameworks to compare both and bring out good and bad in them. This calls for setting appropriate benchmarks, computer vision, and natural language processing (NLP) advancements to focus on this area. Challenges from the lack of information or partial programs are contributors to lack of experience by program editors. Possibly this problem would be resolved by semantic assessment but understanding the intent of the coder becomes key. Advancements in information retrieval mechanisms, by synthesizing the code from part of codes extracted from big code will help code completion, without even exploring probabilities.

This chapter is organized as follows; it starts with a discussion on the code suggestion tool for language modeling, then explores the neural language model for code suggestion exploring context setting, neural language model understanding, attention mechanism, and pointer network. Next, we look at code summarization with a convolutional network, exploring convolution-based attention model, features of attention, and model copying based on convolution attention. Neural Turing Machine (NTM) is explored in the next part, discussing neuroscience and psychology background for the topic. The discussion continues to NLP used for program comments prediction. In the software naturalness, part n-gram model is explored to capture regularities in software (Miltiadis et al. 2018).

4.2 Code suggestion tool for language modeling

Usually, the suggestions are based on the information available at compiling time-based on the variables that are in scope and other factors. Code pattern recognition based on large code corpora's statistical modeling has been effective in providing a context-based suggestion.

Code suggestions drastically improve the productively of the programmers. The basic logic is to arrive at the tokens that fit in current syntax based on the information learned from semantic and syntactic structures and knowledge configured in the rules. So, this brings the challenge of valid semantic and syntactic based suggestions but not so guaranteed in real-world situations. Models based on n-grams have been successful in providing improved suggestions based upon the most globally applied context. This contradicts the software context of the more local-focused view. Since the code is modularized, the format gets local to the module and repetitive as well. N-gram model specializes in capturing the patterns associated globally and fails to capture local patterns. An improvement over this model was to bring in a cache possibility in addition to the global n-gram context, where cache provides the local focus of pattern. N-grams cover the mean probability of corpus linguistic, but the local probabilities vary across the different code locations. Cache component model these local variations and provide mean probability for the local context.

The discussion here is around the cache-based plugin of eclipse IDE. This plugin is specialized in code suggestions based on the cache language–based model. General code suggestions are composed of built-in eclipse code suggestions based on local context looking at semantic and syntactic structure. An alternate approach is an accounting global n-gram model where suggestions are based on the global context and local context derived from the local cache model. An alternate approach with both approaches combined have proved to be performing better than inbuilt general code suggestion.

This plugin operates within the eclipse editor and can be switched on as needed. Cache model utilized helps the editor with features as follows. Suggestions are provided even without type information available making it feasible for dynamic and static languages. Probabilities associated with this plugin suggestions will help refine the built-in suggestions and rank them in order. With the concept of a recently used token having a chance of being used again, this plugin focuses on short term probabilities of terms. The built-in tool does not get the format of changing the order of suggestions based on the project context, even if the situation is the same. But this plugin facilitates that situation by sorting the suggestion based on the statistics derived from the corpus's statistics.

Tokens used by coders are more likely to be used again in a shorter duration. An example would be a method identifier having the possibility of being reused in the code in a short duration of time. Plugin specializes in studying the probability of use of such terms in local context based on the frequency of use. Cache model helps to double the probability of suggestion compared to the n-gram model alone. Plugin banks on the recent frequency of call for a method to suggest the correct one. The possibility of revalidating the ranks based on use is a key feature of the plugin. General code suggestion does not get granular but just depends on the high-level aspect for ranking based on types. Eclipse built-in feature banks of the kind of information, which makes the situation difficult in the case of a dynamic language, thereby the type information is not available. In those cases, suggestions based on the semantic and syntactic structure will be difficult. Figure 4.1 depicts CACHECA mix model architecture.

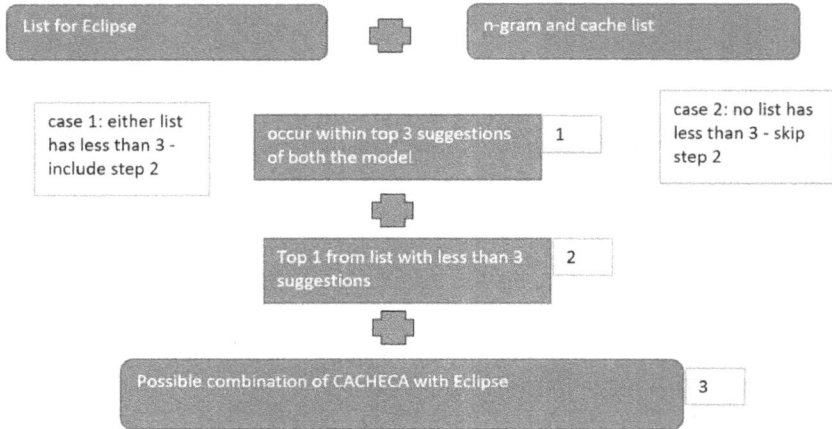

Figure 4.1: CACHECA mix model architecture.

The plugin uses information from eclipse provided relevance score and the cache model-based probabilities ranking. The plugin monitors the opening of the file or the files brought to the forefront. At that time, n-grams are referred from the file that is created based on word count. The empty cache file is also generated. The plugin starts with the eclipse's suggestion, then combined with the cache model proposal into a common list. The selection of the top words is made based on different scenarios. If one of the lists has less than three suggestions, the top one is taken from the list that has less than three. If both lists do not have less than three, then consider all the proposals. The plugin also has a sorter customized to maintain the order of suggestions while combining the lists from both proposals. The final list is displayed to the user via eclipse. Cache model generates the lexical-based code tokens it does not provide parametric information, so combining eclipse provided a list that does not have great ordering. This list is used as a reference to check the list generated by the plugin that is associated with it the ranking reference. The plugin provides a robust code completion system. The plugin is written in Java. To make the plugin compatible with eclipse, the plugin needs to gather the initial list provided by an eclipse that is not ordered and sort it in line with how eclipse does internally based on relevance value. Later this is used in the mixed model to arrive at the final list.

Testing was done on the corpus of Java projects. A tenfold cross-validation was done for each project with 90% of data used for training and 10% for testing. The training process included counting of the tokens, and these tokens were tested on the testing set. The testing objective was with the knowledge of historical token identifying the next token in the test. Symbols parenthesis was removed from training as these would easily be predicted and is an unusual overhead if used in training. For every test scenario, the next token prediction based on a set of previous tokens is expected. This list of suggestions for the next token is ranked order, which would be cut off at ten suggestions. The rank of correct suggestions from input to compute the performance of the model. One of the measures is the percentage of context in which the correct suggestion is ranked for top n index or less. Various values for n top index can be tried out. Another measure is averaging of reciprocal of the rank of correct suggestions over total suggestions made by the tool. If there are no correct suggestions, then the score is zero. The relevance score of eclipse computation is based on multiple weights configured, based on all possible suggestion properties. The model used for mixing of eclipse suggestions and cache model suggestions focusses on improving the accuracy.

Some of the approaches have a similar base but topped up by inputs from a group of programmers involved in crowdsourcing (Mooty et al. 2010). There are methods to use the user's action monitoring and feedback for performance improvement (Bruch et al. 2010, Franks et al. 2015).

4.3 Neural language model for code suggestions

Recent works have propagated that the model based on language can get better by referring to software repositories. In this review, we look at neural language models that focus on capturing long-range dependencies. A large repository of Python code is gathered from the GitHub as a reference for the study. In this study, the neural language–based model provides good performance from local aspects suggestions. But it does not account for tokens that have existed in long-range past. So the neural language model is assisted with a pointer-based network that is good at taking reference to predefined categories of identifiers, which makes the performance better. This combo works better than LSTM baselines as well.

The probability of a sequence of tokens being observed in the Python program for code suggestion using the language model can be represented as follows.

$$P_\theta(S) = P_\theta(a_1) \, .^N \Pi_{t=2} P_\theta(a_t | a_{t-1}, \dots \dots a_1) \tag{4.1}$$

Representation is the sequence of tokens, parameter "θ" is an estimate from the training corpus. In the forgiven sequence of tokens, we need to predict the next required number of tokens to maximize the above equation.

4.3.1 Context setting

One of the key features for the coder who is new to the code base is the autosuggestion, where for a given context of code part, next sequence is suggested, which would be API calls, functional calls, etc. Static language has a good amount of support compared to dynamic language such as that of Python. The recent past focus has been using NLP to study the code repositories and train the data for the code suggestion system. N-gram models find their place in terms of ranking possible suggestions. Neural language models' recent focus has been addressing long-range dependencies. These models have been constrained by the hidden state block requirement to have the same vector representation of fixed size for all context-based information. This makes the model get stuck in local phenomenon rather than call a function that was defined long-range dependencies ago.

In this study, these issues are targeted, and high-quality Python code is gathered from GitHub. Exploration of the attention mechanism for the code suggestion context is done. Though this makes a significant improvement in accuracy, but there are challenges associated with errors in the output. So, the format for long-range dependency learning by selectively focusing on identifiers introduce based on abstract syntax tree (AST). This is a pointer network model, which is in a dynamic mode and selects syntax activated points for taking in to account long-range dependencies and also account for the local phenomenon.

Overall this study provides code suggestion corpus from GitHub for Python, with sparse attention mechanism that is based on tracking long-range dependencies for dynamic language. These language are efficient and helps the analysis for demonstrating the capabilities of these model to study long-range dependencies.

4.3.2 Mechanism

The neural language model needs to be understood before exploring the possibility of extending the model to the attention mechanism. Sparse attention network configuration for pointer network benefits from AST of Python. This will be done in the context of code suggestions.

4.3.3 Neural language model

Code suggestion in the context of the language model is about the probability of finding the next set of tokens in the context of past tokens. Here neural language model is built based on recurrent neural network (RNN) and long short-term memory (LSTM). The token's probability is measures based on the output of the LSTM network at a time step. Every token will have a parametric vector associated with it. Theoretically, the neural language model is supposed to capture long-range dependencies based on the token sequence associated with internal memory. Since the internal memory is constrained by fixed size, and there is a mechanism of this memory getting updated at every time step, this learning gets localized. Contrastingly the lookout of the study is to recognize a function introduced long back in the time. If a function defined right at the top of the file, it may come in to use at the end of the file.

4.3.4 Attention mechanism

The best approach is to use attention mechanisms on the last few sets of tokens that are in the form of vectors from the language model. Earlier implementation of attention mechanisms has been in areas of syntactic parsing, question-answering systems, and machine translation tasks. Taking care of the bottleneck of fixed output vectors and creating the possibility of looking back at earlier output vectors are a key. From the language modeling context, memory is fixed to past "n" output vectors created by the LSTM network. To forecast the distribution of the next token context vector provided by the attention network combined with vector generated by LSTM on the projected matrix that can be trained. This final vector takes in the next word distribution and encodes it, and generates a certain size vocabulary. Softmax is applied to generate the probability of the next token. The issue with the attention mechanism is it becomes a complex computational scenario as the window keeps moving to the next set of words. In the beginning, as the memory is passed in a lot of noise is seen when the network is just attempting to optimize. As in the initial phase, LSTM vectors are basically random; this calls in a pointer network that picks up the memory based on the efficient picking up of the identifiers from historical code.

Attention mechanism with memory at the fixed point represented by $M_t \in R^{kXK}$ for K vectors; $mi \in R^k$ for $i \in [1, K]$, which provides attention of $\alpha_t \in R^K$ for each time step of t. Also, $W^{m}, W^h \in R^{Kxk}$ and $w \in R^k$ can all be trained, 1_K represents k dimensional vector of ones, and c_t is the context vector for attention mechanism.

$$M_t = [m_1 \ldots \ldots m_k] \tag{4.2}$$

$$Gt = \tanh(W^m M_t + 1_K^T (W^h h_t)) \tag{4.3}$$

$$\alpha_t = \text{softmax}(w^T G_t) \tag{4.4}$$

$$c_t = Mt^T \alpha_t \tag{4.5}$$

4.3.5 Pointer network

Attention mechanisms that can filter out the specific view from the history of tokens can be done. At any point, representation includes that of previous "n" number of identifiers from history as a memory. So that creates an opportunity to learn long-range dependencies based on the identifiers. This covers the possibility of capturing the identifiers like a class that is declared hundreds of lines ago. Next, word distribution is generated from a weighted sparse pointer network based on the identifiers as a reference from historical Python code, also using a neural language model. A controller decides weights of sparse pointer network and neural language model. A sparse pointer network operates on a fixed memory from the identifier's representations of fixed numbers at any point in time. Now context vector is generated using attention mechanism for memory restricted to an identifier that is recognized from historical data. Based on this, sparse distribution is obtained on global vocabulary. Neural language model assists in identifying the next word distribution. Final distribution covering language model and pointer network is done by the controller to arrive at weight-based next word distribution. A controller can decide either to refer to the identifier or from global representation based on the information, the controller holds. The information controller is about the next word distribution encoded by the neural language model and an identifier weighted based on attention.

Earlier studies have focused on static language and small corpus. Hence this study picked a large corpus of Python code, which is a dynamic language. Python has been one of the topmost popular languages (Pierre et al. 2016). Python also makes it to the top of the list regarding its availability as an open-source code repository. The expectation was to have the code of significantly high quality as the model intends to learn how coders write the code. But it is not practical to get access to high-quality code. Hence the assumption that popular projects will be of high quality is made. One way to gauge this or a kind of metric for this on GitHub is to look at stars that are more like bookmarked parts. We can also look at forks, a mechanism for users to make a copy of the code repo, and use it without disturbing the original code. So, stars and forks were referred to as a base to pick top-quality projects with top quality code. The code that was compatible with Python3 was shortlisted, and the entire dataset was divided in to train, test and validation.

Long corpus of identifiers also had rarely occurring ones. So, to achieve the generalization of models, we must refine and normalize the identifiers. Normalization includes classifying identifiers into groups. Grouping is based on if they are functions, class, and so on. An identifier that is also grouped will have numbers associated with making it unique. Replacement of identifier is only restricted to ones local with the file, where are those referring to external libraries are left unaltered. As per best practice of creating code suggestion repository, numerical constants are replaced with standard identifiers; comments are removed, also replace tokens appearing less than a certain number of times as standard out of vocabulary terms.

Experimentation was done to ensure that the neural language model operates better than the n-gram models for code suggestions. The pointer network of sparse representation is trained with Stochastic gradient descent (SGD) with a specific batch number. Backward propagation is truncated overtime to restrict to a specific number of identifiers only. Learning rate with the decay of rates used as a strategy after every epoch. Also, the LSTM network is tried with and without attention. Attention language models were tried with the different combinations of memory size and batch size. TensorFlow is used for training with cross-entropy as the monitoring metrics. The last state of RNN gets fed to the beginning state of the same file's next sequence. Models use the hidden and input size parameter, bias state set at forgetting gate, norm clipping used for the gradient. The dropout rate is also fixed for input representation. Sampled softmax is used with the sampling distribution of log uniform.

Models are evaluated using perplexity and accuracy. Code suggestion neural network works superior to n-gram models; performance goes further high as the attention layers are introduced. The attention layer's performance is mainly due to a good prediction of the identifiers. Improvements are further bettered by using a sparse pointer network that looks at smaller memory representation for a smaller set of top identifiers.

LSTM network certainty about the next token is not strong, but the attention network makes a meaningful prediction with the sparse network. It provides not only the correct suggestions but also good top alternatives.

Base work done in this area recommends that true program can be represented in much smaller space than the flexibility provided by the programming language (Abram et al. 2012). Language models have been used to demonstrate the programs' repetitiveness and represent the programs using language models statistically. Cache mechanism helped to retain the local aspects breaking down the modules. Properties of source code utilization are the target of using the cache, which is in line with the attention mechanism discussed here. Language model with lag attention format where fixed-length memory tape was used instead of the memory cell. There is an approach where memory blocks of various languages are used, and it performed better than the neural language model and n-grams (Tran et al. 2016). This makes sure to represent all possible vocabulary rather than the sparse view created in the approach discussed here. The approach discussed here was lexical, whereas an approach is based on probability free of context and grammar-based (Miltiadis et al. 2014). This approach bank on grammar specification and well-defined parsers of source code. This would not handle context-based scenarios like a declaration of variables before using them. This made them include context-based variables. Work has focused on deriving code from natural language (Wang et al. 2016). In the work discussed here, inspiration comes from this work, particularly when the decision is made to choose from the language model or choose from the sparse network's identifiers. But in that work focus was on shorter description whereas the work targeted here had to focus on longer-range dependencies based on memory being chosen in filtered fashion based on earlier identifiers representation.

This approach focused on utilizing the code suggestion model only for the Python code repository, further exploration of the various project would be done. The approach can be made scalable to other projects and collections. This possibility can be looked at to integrate this approach into the existing IDE. Also, further exploration of completing code with partial code using the character language model can be looked at (Avishkar et al. 2016).

4.4 Code summarization with convolution attention network

The neural networks' attention mechanism works well if there is no constraint for fixed maintaining of the dimensions. These models can bank on the features which are not changing locally and can be focused attention on, but the work in this area has not made use of these features. In this approach, the convolution network sits on top of the attention network that works on tokens taken as input to identify the features that do not change with time and those localized. These features also should have long-range topic cased attention based on the context. This setup summarizes source code parts into shorter and more details covering output is done. These features feed into the model in a sequenced way producing a summary based on a couple of attention mechanisms. The first part finds out the next token in a summarized way, where weights are applied on input token based on attention. The second part takes in the tokenized code into a summary without modification. Experiment here is applied on multiple Java projects; from that perspective; this approach leads all other attention-based model approaches.

A scenario where sequence based on input is to be used for prediction or more complex inputs are to be predicted, it presents difficulties using deep learning. One fact of difficulty is the high dimensional complexity of the input and output data space. Also, to add to the difficulty, the dimension range is not fixed. These specific problems are handled in the works that have looked at image captioning (Xu et al. 2015) and machine translation (Cho et al. 2014) problem. The specialization of the attention model is in handling the focus areas as separate items. One to identify the inputs that are most relevant to outputs based on their position. Secondly, find out the actual output position based on all the positions from the input that are identified to be relevant.

The focus of this work highlights the fact of the existence of the features in multiple domains, where these features do not change as the translation happens. These features can be the ones that can be focused on by the attention mechanism. To provide an instance, in literature if there is a phrase, "in the work, we suggest," it is a clear indication that the next part of the phrase is vital. As another instance, if the neural network is working on identifying method names in a programming language, based on the body of language. In case the method starts with a word and contained another word, then it is logical that the part that exists in between these words play a significant role in identifying the name of methods. Earlier networks have not focused on learning these features which are characterized by being non-variable during translation.

The structure of the approach holds a convolution network within the attention network which is also convolution-based. Convolutional networks have the model to go in case of using fewer features and for the domain where features that are not varying with translation are involved. So, they see their success in the non-attention model such as text classifier and image processor. In earlier work, these have not been used in models specialized with attention (Krizhevsky et al. 2012, Lecun et al. 1998). Convolutional layers used in this approach discussed here uses layers without pooling. With the focus of identifying patterns that indicate the existence of the location which is worthy of attention. This setup is used for summarization problem. One of the expectations is to arrive at the short and meaningful name for part of code based on the tokens provided. Source code focusses on being a messenger to CPU to indicate the computation needs. It also is a key communicator to a developer who needs to understand the ecosystem for code maintenance. This makes it important to summarize the code well for the efficient use of developers.

Summarization also plays a key role in software engineering to help analyze the code and understand it well. Since generally code is a well-structured and convolutional way of modeling fits well. This work here is encouraged from the work that focuses on generating names for functions based on features that are predefined (Allamanis et al. 2015). These predefined features are based on classes and methods exiting. But the features may not be available for the case of code snippets and for the languages that are dynamic. Work reviewed here is a more generic one where hardcoded feature are not used but a part of code is used to arrive at the summary which represents method name.

This is a case of summarization task where the method name also represents the summarization. The context of natural language and code summarization is different in the sense that natural language is more complex whereas the code has a structured format. A summary can be said to be an effective one if the summarization is capable to explain how the code provides instructions to get composed into higher-level meaningful components. This summarization cannot just be dumb focusing on just explaining what the code does. To this extent of work, it is needed to understand deep patterns hidden in source code which focusses on the structure of the code and depends on identifiers that are key recognizers for the structure of the code. Code summarization is like the translation problem. The major difference is the size of a sequence of input and output involved in code summarization. In this experiment, the input sequence is large enough to generate a very short summary. While setting up the input sequence it needs to learn features that do not vary with time and learn topic-based sentence-level features. This work also shows that neural translation models do not provide great results.

Another challenge with source code is the words that are not part of the regular vocabulary. Every time new software comes in, a new file of source code comes in, which brings with it the new vocabulary related to software domain and various data structures. Now, these new aspects of the vocabulary do not appear in training data. To specifically address this issue attention mechanism which is convolution-based recognizes the most important token even if they are not part of the vocabulary. This mechanism is a copy machine. As part of decoding, there may be a decision to take over the input token directly into the output token.

The primary focus of this work is a convolution-based attention network that does code summarization, an approach that considers working on all drawbacks of machine learning approaches and from software community issues in the area.

4.4.1 Attention model based on convolution

Input for the models comes as a sequence of code tokens and output the summarized methods name. Neural network models each of the input token sequentially and models the output probabilities. The neural network also takes into account the subtokens that were created earlier. Based on these inputs network create a state vector.

Convolution based attention network now uses the state vector created and uses the group of convolutions generated from the set of tokens which are embeddings. This helps generate the attention feature in the form of a matrix. So, representation is one vector for each position of the sentence, this vector is the attention vector. The outcome vector is normalized, which represents the distribution of the location of input tokens. Each of the tokens in this representation will have weights. Now based on the weight context is figured out that is used to find out the distribution of the target expected. This distribution is probability-based. The output provided by this model is a summary of code parts and is generative with bimodal characteristics.

4.4.2 Attention feature

The first part is to learn attention from input tokens and hidden states from the previous part. The general part of the network is convolution which targets to get the position of the features based on the context. Now, these attention-based features take in code tokens sequentially and are of a certain length. Each location is a matrix representation, where the matrix is formed based on attention features. The size of this matrix attention feature is dependent on the length of tokens used as input, a constant and several features used to represent the location. Constant used here is the fixed amount of padding needed for sequence. Attention is based on an input token; convolution computes several features for every location. The previous hidden state is used as a gating mechanism to multiply, facilitating retention of only current relevant features. The final layer does normalization of these features. This helps to take into account the input sequence of any length. Names and codes are converted as a specific dimensional embedding. Convolution kernels are two in number that has window size parameters for convolution. ReLu is a rectified linear unit used for transformation. Pooling is not involved in the convolution, so the output sequence is of the same length as that of the input sequence. Now, attention weights must be derived for the normalized representation of the embedding and a kernel of convolution. Figure 4.2 depicts Convolutional attention network architecture.

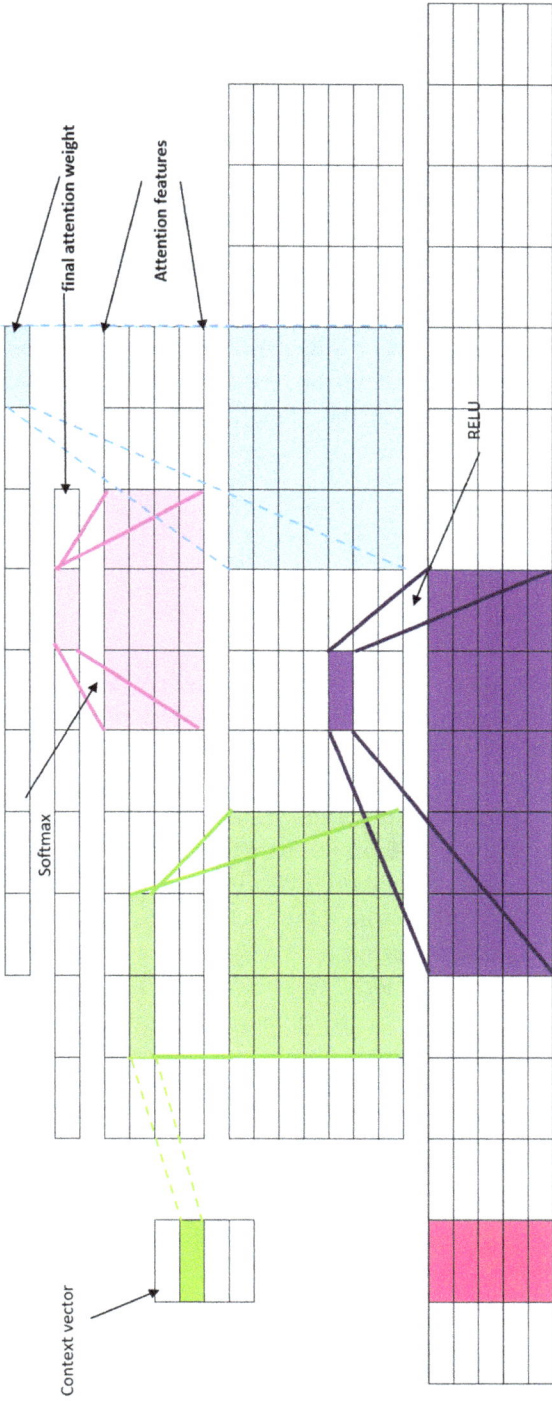

Figure 4.2: Convolutional attention network architecture. Input sequence and context vector are used as a reference for attention features to learn. Based on these attention features attention weights are derived with the utilization of softmax and convolutional layer.

Since the prediction is to be a summary, it is supposed to be a sequence prediction problem. Here the predicted sub-tokens are based on sequential order input considered as tokens from earlier states. Gated recurrent unit (GRU) is used at this stage. Testing is conducted by passing in current token embedding, previous state vectors inside a GRU.

Components briefed above forms blocks of the network. The convolutional attention model builds convolutional attention. This utilizes the attention vector coming from weights generated for attention. This is done to provide weights to the input tokens. Ultimately for prediction of target embeddings. All the tokens in the vocabulary will have distribution. The model is trained based on a maximum likelihood objective. Model generation is done by starting with a special token and a previously hidden state-input at every time step to generate the next token. This generation is based on the probability distribution coming from the attention of convolution. Now with the new token, the next state is generated using GRU, and the process goes on until the last special token is reached.

4.4.3 Convolution attention model copying

This convolution attention is further processed with a vector to identify out of vocabulary terms. These are copy mechanisms. The large part that is one-third of output is in line with input. This encourages the network to copy these from input representation to output representation directly. In the process of predicting a new token with a probability "n," it will copy from input representation to input with "$1-n$" probability based on the convolution attention vector. For copying from input to output probability of copying is based on the weights of attention. Both parts of attention are combined with overall potential vocabulary. Copy signal is initiated with vector passed to copy attention, where the vector elements are one, in case, if the input code token is like that of output code token.

For predicting method names, start with special token sequence and generate the highest probability tokens, group. Each time the highest probability token is generated based on which next token is generated and the previous one is pushed into the group. Once the end token is reached then move on to the suggestions list. Since overall interest is to pick top n number of suggestions at every point. This suggestion list must be pruned based on ones with lesser probability than the one that is the best at that point in time.

4.4.4 Data

With the objective of generating function names from code set, such kind of dataset does not exist for a code snippet. So next best is to consider functions as the code parts and the method names used by coders as the target summaries. To get good dataset quality GitHub Java projects were explored. The top popular project was chosen based on the count of viewers and forks on the project, where the Z score of both was summed up. The focus was to get diversity in data and to have large and maturity of the dataset. Files are cleaned to remove overridden methods, class constructors and abstract methods. Static analysis helps to find overridden methods based on inherited relations. The reason for dropping overridden methods are for their repetitive nature and ease of prediction. Cases where the method name and tokens are the same, in that case, a special identifier is created. The method name and code-based tokens are split and lowercased into sub-tokens.

4.4.5 Experiment

The experiment is measured with two metrics for gauging the predictions' performance, Exact match and F1 score are used. The exact match is the metric on method names correct prediction percentage. F1 score is recorded for every sub-token. As part of the prediction, a list with ranking is produced as suggestions. Each project is used separately for training and testing purposes. This strategy is due to the project belonging to a different domain and the learning cannot be transferred from one to another. This is mainly due to code reusability. An attempt was also done to use all project data for the model, resulting in poor results. Each project data was split into train, validation, and test set. Bayesian optimization is used for parameter tuning. The performance was compared with a model based on tf-idf (term frequency-inverse document frequency). Tf-idf focuses on the vectorization of the code parts and suggestions are based on the nearest neighbor which works with a coined similarity-based distance. The experiment was also done with the attention model with two RNNs having all connected components, which work perfectly for machine translation. SGD with parameter RMSProp was the basis for optimizing optimization to training copy attention network and convolution attention. Dropout is used on all parameters. Gradient clipping also employed to control the gradient passed back during the learning of SGD. Mostly parameters are initiated with a random noise around zero. For copy and convolution, attention parameters involved are the kernel size of two kernels, attention window sizes, dropout rates, and vocabulary dimension.

4.4.6 Evaluation

F1 score is the competition metrics across all methods referred to with the current experimented method. Performance against all the projects is averaged. Comparison is done for tf-idf based model, standard attention model, and convolution attention and copies attention networks. Tf-idf better performance indicates that the bag of words is a good representation between code parts and its names. The standard attention model would not lead the tf-idf model; overall copy attention and convolution attention perform best. Convolution network has similar performance to tf-idf, but precision-wise convolutional network excels. Though code snippets are like natural language parts; the structure within code makes it unique. Also, code is varying across longer lengths.

Measurement also was extended to figure outperformance of copy attention on out of vocabulary parts. Its measure of out of vocabulary words percentage. Standard attention model and tf-idf models are not configured to handle out of vocabulary words. This makes the copy mechanism useful for projects in this domain, and for small projects, with higher out of vocabulary word. Accuracy of the out of vocabulary words also differs due to the style of code being different.

The difference in standard attention and copy attention model points at what aspects of learning of copy attention gets its place over standard attention. One differentiator that can create this possibility would be copied attention's possibility of capturing long-term dependencies for long inputs. To validate these thought sub-tokens are mixed up to remove out the tokens that follow sequence-based information. With the absence of a local feature, the performance of the models to be like tf-idf. With this change in setup, copy attention degrades understandably due to the input sequence now resembling that of the natural language. Standard attention also worsens and hints at the inabilities to handle long-range tokens.

To start with, the attention is on the complete method to pick the first token. If the attention mechanism is confident about copy attention based on the probability generated, then sequentially part is picked from code to name. It demonstrated that the copy mechanism has focused vector compared to the attention vector; the reason for this is the difference in the signals gathered from training by respective mechanisms. Copy attention learns features that do not vary across time and topic-based features.

Though convolutional network-based work has been in image processing, text classification has also been an application area. Similar work on document representation with convolution utilizes activation of the network in place of attention. Works also have used only attention encodings without convolution for attention to arrive at summarization (Rush et al. 2015). Pointer networks use a RNN for attention (Vinyals et al. 2015). Work on a syntax-based convolution neural network that can learn the representations without depending on the actual names.

Understanding the source code and representing them with modeling approaches may assist various areas of software engineering. This area of exploration of code summarization is a good starting point toward the area of source code being represented in machine learning language. This representation will create a probabilistic representation for code that can create software engineering tools that will facilitate maintenance and building of software. The entire setup of software code and its associated artifacts provide a different genre for the machine learning compared to its conventional data structures of image and natural language. This makes the exploration of software engineering more interested to explore different possibilities in line with those of image and natural language (Miltiadis et al. 2016).

4.5 Neural Turing Machine

Work extends to external memory from a neural network where interaction will happen with attention. This system will have an analogy with Turing Machine but being end to end will be its uniqueness. This end-to-end nature will help to learn the gradient descent in an inefficient way. Initial understanding of the projects are that, the NTM will assist in basic models which are applicable for sorting and copying, between input, and its respective output.

Computer program capability is in three areas: the logic of flow of control, basic operations like that of arithmetic ones, and memory externally in which information can be written to and access from, in the context of computation. Though machine learning has been extensively explored, less focus is seen on external memory and logic control flow being used as a base. RNN stands out for their ability to track over time the pattern and use that acts as a base for complex computations. So RNN is quite capable of executing the procedures in the simulation if properly structured. Theoretical possibilities will have their challenges in the practical realm. It makes sense to focus on refining and better the RNNs for making algorithms operations simpler. This improvisation will be using large memory targeting the finite state of the machines.

NTM will be a facilitator of training based on gradient descent which makes learning a program practical. Working memory shares a larger similarity to algorithms operations, based on human perception. Though working memory is tough to comprehend, basically it is the capability of storing short-term information and the ability to operate on that information. Rules needed here are arguments stored in programs or maybe a simple program. This system focuses on solving tasks based on approximated rules, which is like a memory system under work, to create variables quickly. These variables that are created rapidly are the temporary variables that are created to get an output in the context of a conventional computer. For getting a product of 50, you put in 10 and 5 variables to be used for multiplication operation into a memory.

NTM's closeness to operating memory is its ability to pick the content to write and read to memory based on attention. NTM can make use of this memory information to apply the operations and use this information for learning purposes. This is unique compared to models built on working memory. Context setting is done with exploration of working memory and its application in the neural network, artificial intelligence, neuroscience, and other areas. On top of this, the current research discussed in this summary will focus on the attention-based controller and memory structure that will help execute and induct the programs. Problem areas are picked to demonstrate the effectiveness of these architectures.

4.5.1 Neuroscience and psychology

Psychology makes use of this operating memory in the context of tasks that are short term based on the purpose of information handling and processing. The general thought is the core of the memory being attending to and acting upon the memory's information. The capacity of the working memory is also defined in psychology as the parts of information that can be called back, this is a limited capacity. This is a clear indication of the limitation in human memory, but the work here will focus on extending that limitation. Neuroscience attributes operating memory to the prefrontal cortex. In terms of experimentation, it is the signal from a group of neurons for tasks performed, where there are observations on the cue, with a lag period, further there is a response that is happening based on information from that cue. Working memory usage modeling focuses on how persistent neuron firings are learned by biophysical circuits and use it to solve tasks.

4.5.2 RNN

RNN is a representation of a multistate machine that is dynamic and depends on the input and current state of the network. Markov models are similar but RNN has considerably larger memory due to its distributed nature and makes that a computationally good. The importance of the network's dynamic state points back to the ability of a signal to make its influence at different points of time from the point of it entering the memory. This nature helps to perform context-dependent computation. LSTM was a breakthrough in RNN. This came into existence to tackle RNN problem of vanishing and exploding gradient descent also can be considered as vanishing and exploding sensitivity. The whole problem tackling capability is due to the perfect integrator that works within memory storage. General integration is a signal of the current state created by adding an input signal to the current state signal. But the input signal will be associated with a matrix that will control the signal from exploding and vanishing off. If the integrator is associated with the program, it takes care of understanding the input and decides when to listen to it based on context. This will be a computational gate attached. This provides the capability to store the information selectively based on context for the required length of time.

$$x(t+1) = x(t) + i(t) \tag{4.6}$$

$$x(t+1) = x(t) + g(context)i(t) \tag{4.7}$$

Example of LSTM is $x(t+1) = x(t) + i(t)$, where $i(t)$ is the input to the system. To avoid dynamic vanishing and exploding of the gradient, identity matrix $IX(t)$ is utilized. With this integrator network can choose to decide on action to be taken on the input. So transformed equation would be $x(t+1) = x(t) + g(context)i(t)$. So, the information can be stored for any period.

This provides RNN capability to process the signals of variable length without making any changes to signals. In the problem where inputs are coming in sequentially, they are also coming at different points of time, this creates the situation of variable length or complex structures being available to be processed on several steps. This ability to handle variable length of inputs makes this apt for text generation, speech recognition, handwriting generation, and machine translation. This nature makes it less priority to build parse trees to represent the complex structures.

4.5.3 Neural Turing Machine

NTM constitutes the controller for neural network and memory bank. Figure 4.3 depicts Architecture for NTM. The controller contacts the external entities with a vector way for input and output; this is a normal neural network. This controller is also featured with a memory matrix that can make read and write a selective

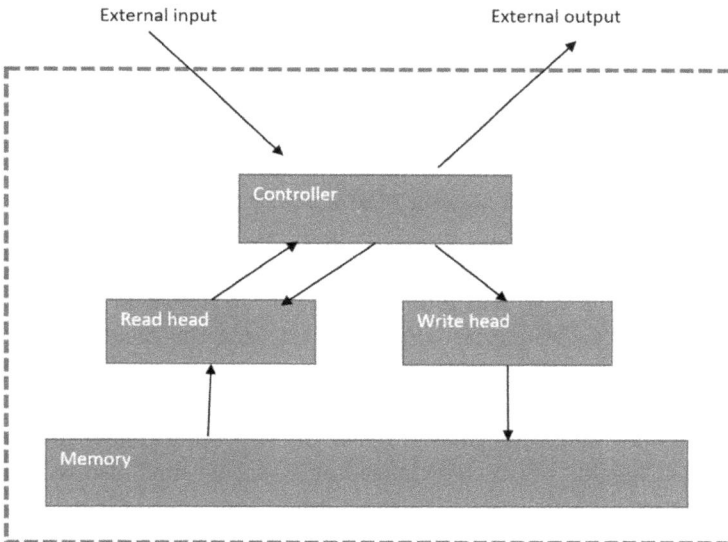

Figure 4.3: Architecture for NTM. The dashed box is the differentiator between the network and the external world. The controller receives input from outside during every update and provides the output response. The controller also writes to memory and reads from it. Read and write heads are utilized for this action.

process. In comparison to Turing Machine, this part of external interaction can be considered as "Head."

All part of the network is trainable with gradient descent. This ability is due to the network's ability to read and write with all components of memory but to a different extent. In contrast, in Turing Machine it is restricted to one element. The degree of consideration is influenced by the attention mechanism where the focus is restricted to the required portion of memory, ignoring other parts. Since this creates sparse data storage that pushes network to save data without any disturbance. Heads provide the specific outputs based on which attention picks the location of memory. This output is specialized in generating weights to the memory content for locations of the memory. Every read and write heads will have the weight of its own. This weight is the information to decide for the head to pick up the required amount of information from the memory location. This provides head to decide if the memory at a location must be focused specifically or general focus is needed across different memory locations.

Content will be stored in the matrix of memory at a time. The size of this memory is based on several memory locations and the vector's size at every location. The weight matrix represents weight distribution across memory location at a point in time. All these weights are normalized, so the overall weights across locations will sum up to one. Read vector provided by head will have a specific length of a vector at every memory location; this will be a combination in convex fashion for all row vector memory content. This will be unique compared to memory and weights.

Reading of the part of the NTM can be represented as:

$$\Sigma i \, wt(i) = 1, \ 0 \le wt(i) \le 1, \ \forall I \tag{4.8}$$

$$rt \ \leftarrow \ \Sigma i \, wt(i)Mt(i) \tag{4.9}$$

where, contents are represented with M_t with NXM memory at time of t. M is vector size for any location and N memory location count. For read head and time point t, for N locations, w_t will be vector of weightings. Due to normalization of the weightings all N elements $w_t(i)$ will follow constraint as shown in equation $\Sigma_i w_t(i) = 1, \ 0 \le w_t(i) \le 1, \ \forall i$. Read vector rt with length M from the head will the combination of convex nature Mt(i) which is row-vector in memory represented by $rt \leftarrow \Sigma_i \, w_t(i)M_t(i)$

In the spirit of forgetting and input gates of LSTM, the network will have to add and erase components. Weights provided by ahead at a specific time also with a vector for erasing, where all the elements of this vector will have entities in a range of zero to one. The memory state of the previous is then decided based on these components. If the weight and erase are one at the location, then memory at the location gets set as zero. Alternately if the weights or erase is zero, then memory remains as is. There is no specific order for erasing in the cases of multiple heads that are writing, the reason for this is multiplication being cumulative. Write head

also adds on a vector of specific length to memory after the erase function has operated. Again, the add function sequence is irrelevant. All the heads that are writing
will have combined erase and write instances that in combination creates the memory
at any point in time. Writing operation that creates a composite outcome can also be
differentiated since they are built from erase and add which are also differentiable.
Elements that are modified by erase and add are fully under the control of these respective functions, as they operate with their own set of independent elements.

To understand the creation of weights, one part is similarity-based attention between values at current state and values coming out of controllers. This facilitates easier
retrieval. The controller can pull out a rough match with stored information and then
use it to compare with memory to create a replica of stored value. All scenario does not
facilitate this content-based addressing. Some of the tasks involve random variable, but
it needs an identifier. It is like arithmetically two variables a and b can take any values
but the function of a and b generates a product of a and b can be defined. These variable
values will be stored in a specific location by the controller as needed for the operation
they will be gathered back. The reference here will be the position of the variable and
not an actual constituent of the variable. Location-based storage will be specific as location information is involved and content-based one being general. It was seen in the
experiments that the location-based information was key, so both methods are used.
Figure 4.4 depicts Addressing mechanism flow.

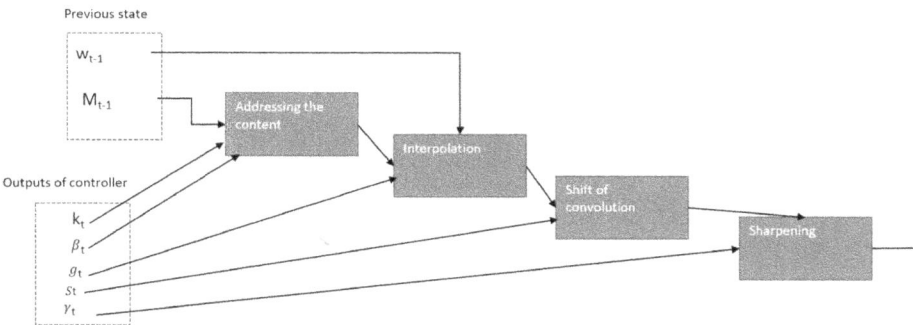

Figure 4.4: Addressing mechanism flow. Memory matrix M_t exerted to content based addressing
with key vector k_t and key strength β_t. Interpolation gate g_t helps interpolation of the content-
based weights with previous step weight. Extent of weighting rotation is decided by the shift
weight factor s_t. Weight sharpening is done by γ_t.

4.5.4 Content focus

All the heads of write and read generate a key vector of a specific length. This is
compared with all vector inputs and they will have a similarity measurement for

this. Based on the similarity and key of strength weights are created, which decides on increasing or reducing focus. Cosine similarity is used for measurement.

Cosine similarity representation is as shown below:

$$K[a, b] = (a.b))/(||a||.||b||) \tag{4.10}$$

4.5.5 Location focus

This mechanism has a combination of randomly accessing the memory and iterating across the memory. This random jump access is based on the shift of weights, positive and negative concerning the current location. These weights will decide the direction of the move to pick the next location. At every jump, the interpolation gate combines the contents. Blending gates combine weights from the previous timestamp with that of the current time step. These weights are generated by heads. In the case of the gate being zero current weights are not considered, only previous weights are accounted for. In the case of the gate at the value of one, the previous run is ignored, and content-based weights are generated. At the point of interpolation, there is a weightage for a shift that comes from a normal distribution of specific integer. The degree of shift will be over −1, 0 and 1 if the allowed shift is −1 and 1. The softmax layer at the controller with a rough size can be a good definition of the shift's weightings. Weights from interpolation, location-based addressing, and content-based addressing are complementary to each other. Content system–based weighting for a shift can be chosen without touching the location-based system. Other options can also be explored to make weights iterate through addresses in sequence maintaining equal distance every time.

$$w_t^g \leftarrow gtw_t^c + (1 - gt)w_{t-1} \tag{4.11}$$

Weighting w_{t-1} produced in previous time step by the head and weighting w_t^c at the current time step are blended by value g producing weighted gate w_t^g.

4.5.6 Controller

Parameter of the NTM is the memory size, the range for shifts, read and write head-counts. The controller's neural network type has a significant bearing on the architecture. The feedforward network and RNN are the choices. LSTM helps with the internal memory that can complement with a matrix of large memory. Memory matrix is equivalent to RAM, the controller is a central processing unit (CPU) equivalent, and hidden activation of the controller resembles the processor's register. Hidden activation of controller help combines information across multiple time step. Feedforward-based controllers are like RNN, but they read the memory's content and write it in the same location. This nature adds to network transparency as reading and writing to the same memory location is straightforward than RNN's internal state. But feedforward network is constrained with several parallel reads and write that can be done, this gives lower hand on the computational capability that is needed in the case of NTM. Based on several reading heads, either one or two feedforwards can only perform unary or binary operations. Whereas RNN controllers can store the memory from time steps.

4.5.7 Experiment

On top of solving the problem, NTM can also learn the program in the objective to validate. This will enable generalized learning for algorithm rather than getting stuck with training data pattern. A simple check is to validate if the network trained to copy 10 sequences can work for 50. Three formats experimented with was LSTM controller–based NTM, feedforward controller–based NTM, and normal LSTM. The dynamic state of the network was reset at the beginning of the network as tasks were episode based. The previous hidden state was learned bias for the case of LSTM. Similarly, the controller of NTM took previous read vectors and memory as bias values. Supervised learning problem with bi-target, had an output layer of logistic,with sigmoid and cross-entropy optimization as the problem setup for the network.

The copy experiment was to explore if the network can save and get back the long sequence of random information. Input comes in the form of a binary vector randomly sequenced with a flag for the delimiter. All networks historically have had the challenge of handling information for a long period. Here the focus is to see NTM's capability of doing better than LSTM on long-period information. Sequence length was random between 1 and 20, and the sequence of the copy was an eight-bit vector; this was the training format of the networks. The expectation was a simple copy of this input sequence. When the target sequence was generated by network no input was provided in anticipation of getting a complete sequence without any intermediate support. NTM with feedforward or LSTM performed faster learning compared to normal LSTM. There is a significant difference between the learning rates of NTM compared to LSTM, which demonstrates the difference between both the model operation.

The focus of the study was to see if the network can generalize beyond what they have learned during training. The experiment clearly demonstrates the wide difference in the performance of NTM over LSTM. NTM works to cope effectively even with the increase of sequence length whereas LSTM loses track beyond sequence length of 20. The analysis shows that the NTM has been able to learn from beyond that LSTM demonstrated. To get a feel of this aspect experiment was done for controller and memory interaction.

To summarize the algorithm, initialize and move the head at the starting point. Receive input vector until delimiter for input is seen, post seen input is written to head location. Keep count of head location by the counter. Head return to starting location. Read the vector of output from the location of the head and generate output, keep head location count. This is a vague representation of low-level programming language to perform this task that the network does. In terms of data structure, the array is learned and created by the network. The format of content-based and location-based is combined with the network. Since the head jumps to the beginning of content it is content-based and since the head moves along the sequence its location based. Ability to generalize long sequence learning is influenced by a shift that happens between the previous read and write weights relative to each other and the weights sharpening their focus helps to maintain the precision over time.

Repeating copy is a task that keeps generating the sequence until a certain number is reached, or the end of the sequence is reached. Nested function reading possibility of NTM was the motivation of exploration. To be specific to see if NTM can execute for loop based on learned subroutine. Scalar value for several copies and binary vector provided in a random fashion which are in random length sequence are the inputs. These appear on a separate channel for input. The network must keep count of several sequences copied and figure out any extra input so that the end marker can be produced at right time. No input was provided to the network except for a number of times to repeat and initial sequence like the task of copying. Sequence length and a number of repeats will be a random choice of one to ten; eight random vectors of binary will be expected reproduction from network based on its training. Repeat numbers were generated in the normalized form of zero mean and one variance. Repeat copy tasks also were solved fast by NTM compared to LSTM, but both completed the tasks correctly. The key difference in the network structure comes into play when looking at the extent to which the generalization can be done beyond the training. In this case, generalizations were needed for both perspectives, which is for the length of sequence and number of times to repeat. Figure 4.5 depicts The learning curve for copy mechanism. Figure 4.6 depicts visualization of the associative recall's generalization performance in case of longer item sequences.

An experiment was done for the variable of sequence length and number of repetitions on both the network variants. LSTM was not able to succeed in both, NTM was successful in handling multiple repetitions as well as various lengths of sequence. NTM was only struggling to recognize the end of the sequence. It ended up putting an end marker after every repetition. This may be due to NTM trying to represent the count of repetitions numerical way but has not been able to generalize on that element.

As the earlier experiment demonstrated, in the case of a simple experiment, now time for a complex scenario such as one data point to others. A list of items is created, and the expectation is pointing at one item needs another item to be returned. The binary vector sequence with the delimiter symbol at both ends will be the representation of the item. After such sequence has been exposed to the network then a random item is presented to the network to guess the next item. Every item is represented by a 6-bit vector of three, which is 18-bit representation. The training was done for two to six range of items for every scenario. Within 30,000 episodes

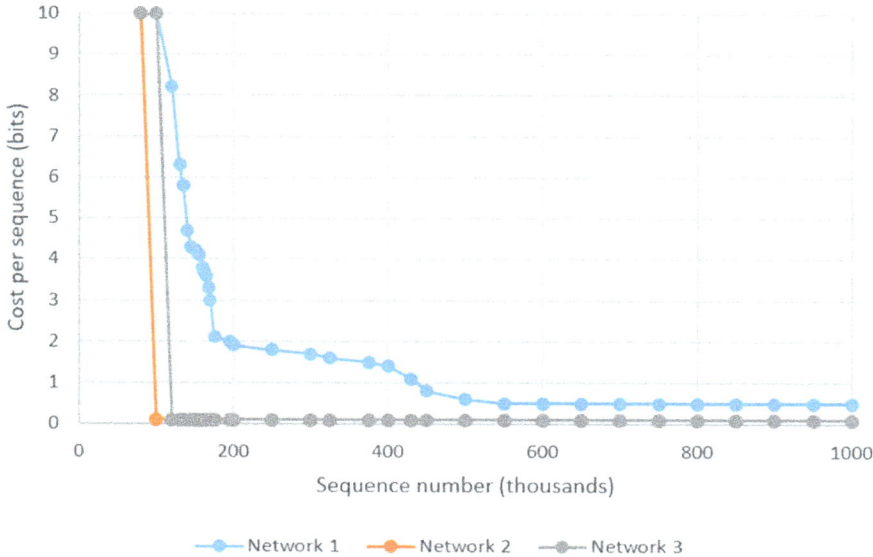

Figure 4.5: The learning curve for copy mechanism. Visual representation to review the rate of learning to converge to a lower cost. This visualization helps to compare the performance of various network architecture.

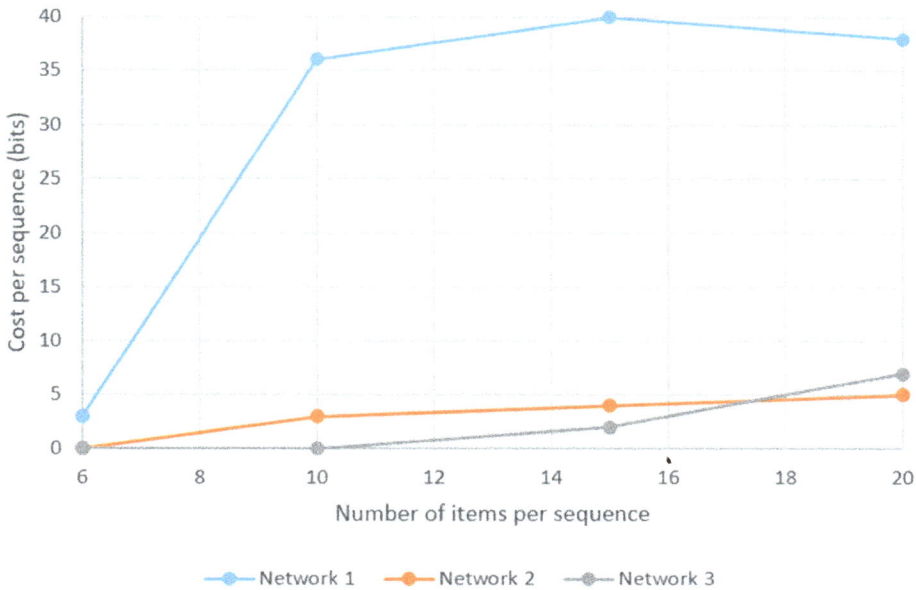

Figure 4.6: In the case of longer item sequences, associative recall's generalization performance can be visualized for the various network as in the figure. The depiction is to get an idea of at what length of the sequence does the network operator and its associated cost per sequence.

NTM learns the tasks faster than LSTM and gets to near-zero cost. LSTM does not reach zero cost events after millions of episodes are tried. For this scenario, NTM with a feedforward controller does well compared to one with LSTM. LSTM's internal memory would not do well compared to NTM's external memory. LSTM fails to generalize for the longer sequence where NTM does good. For sequence up to 12, NTM with feedforward does well. This is twice compared to the training sequence. The average cost also is within one bit for a sequence for a case of fifteen bits.

For single head for the test, one head LSTM in NTM, inputs are single bits item delimiter. After passing multiple sequences of items delimiter gets ready for query item. Query expects the next item in the sequence. Output seen in the experiment shows that the next item is produced, for weights on the reading controller picks up the target item from the stored location for the last three time steps. This is interesting as the controller has specifically jumped to the correct location to identify the target item. Write weighting explains this as the memory was written to the location when input provided the delimiter in mid of items. Adds confirm that the data was written to memory when delimiter came up. Also, its seen that vector addition of different nature every time the delimiter comes up. After query, network reads location based on content; this comes out on the analysis of memory. Content-based lookup is done based on weighting that shifts the item by one. There is a key assisting in the content lookup relative to the vector that was added. So, after the memory is accessed previous three-item slices compressed for are written by the controller when item delimiter is shown. Query items compressed form is computed by the controller. This is used to do a content-based lookup which helps to find first representation that was written earlier. Post this pick a subsequent item by shifting one place in lookup. This is a combination of location-based and content-based lookup.

The next scenario is dynamic n-grams which helps to validate, if NTM can recognize the new predictive distribution. Usage of memory as a reusable table keeps stats of transitions to play like normal n-gram models. Six-gram distribution on the sequence of binaries is considered. Based on five binary histories of various lengths and providing the probability of next bit being one with a table of 25 that is 32 numbers, this is a representation of six-gram. All 32 probabilities are randomly and independently drawn from beta distribution to generate six-gram probabilities for training purposes. The current lookup table was used to draw two hundred back to back bits for training purposes. The next bit is predicted based on one by one lookup of sequence from the current lookup table. Bayesian estimators help find out an optimum estimator for the problem. The Bayesian analysis will consider five-bit previous context, the value of next bit, number of zeros and ones seen up to that point after five-bit context. This provides a chance to compare the NTM with optimal predictor similarly with LSTM as well. Two hundred sequences with a thousand length are used as validation data and it comes from the distribution like training. In this scenario, NTM has a slight edge on performance but will not converge to optimal cost. NTM memory is used to count ones and zeros encountered in different scenarios so that it can estimate in an optimal way.

In the next scenario, the attempt is to sort the data. Scalar priority rating of each vector and random vector of binaries in the sequence is the input to the network. Target output is the binary vectors arranged based on the order prescribed by priority scalars. Sixteen highest-priority vectors were produced for twenty binary vectors and their priorities that were given as input. Priorities are used as a base to find a location to write in relation to input; this comes out as the objective when memory is analyzed. To test this objective a priority-based linear function was fitted to the write locations that are observed. Observed write locations are figured out by the linear function fit on priorities on returned locations. The sorted sequence also is traced by the network by reading memory location in increasing order. Observations show that NTM with feedforward and LSTM controllers both outperform the LSTM. The feedforward controller would excel in this task with eight parallel writes and read head. That highlights that operations of unary vector alone will have a tough time sorting the vectors.

RMSProp algorithmis used, with all LSTMs had three stacked layers. Quadratic increase in LSTM parameters as the hidden units increase. This is caused due to hidden layers of recurrent connections. In the case of NTM, this does not happen. Element wise clipping of the gradient seen during the backward pass of the training.

This discussion was about NTM, neural network architecture inspired by digital computers and biological memory. Gradient descent–based training and differential nature are like a conventional neural network. The demonstration of the network's capability to extend its learning from the simple dataset and generalize the learning was seen (Alex et al. 2014).

4.6 Program comments prediction with NLP

Natural language documents are analyzed and described well by the statistical language model. Works have focused on predicting code using language model focus has not been on coders' comments prediction. Java source files are targeted in the study from open-source projects. Topic models and n-gram models will be explored. The performance of the model is validated with varying amounts of data being fed from the project. Comment completion capability is checked with context like code completion tools that operate with code editors. This demonstrated large part, close to half of the code commenting time be optimized.

4.6.1 Introduction

Software source code represented from language-based statistical modeling has been a recent research trend. Research work shows that the code being the output of humans will have repeatable and predictable nature. Documentation plays a key role in software maintenance in the form of developers' online documentation, code tutorials, and online comments of developers. Code segment use cases, methods, classes and variables kind of identifiers are the examples of a high-level description of the code which is covered in the documentation. This kind of good documentation helps a great deal in the long run to maintain the code but needs extensive task of maintaining the comments which generally gets ignored. Code comments help, as covering a significant amount of information of relevant code and concepts of code comes out as code comments. Code categorization, locating code related to a certain topic, and querying over code base are linguistic-based tasks facilitated by good code comments prediction. Natural language captioning and image and video commenting are other areas where NLP tasks are well used.

Class comments have been targeted in this study using a statistical language model. N-gram models specialists in this scenario will be explored. Language and speech recognition have major success with n-gram models. Document expansion tasks ignore syntactic dependencies at the locality and extract import aspects of code only. From this context, link latent discriminant analysis and latent discriminant analysis (LDA) will be explored as a term extraction scenario related. Separating code and text in code will help get greater performance on topic modeling.

$$p(\theta, z, w | \alpha, \beta) = p(\theta | \alpha) \, \Pi w \, p(w | z, \beta) \qquad (4.12)$$

for set of N topics z, for topic mixture θ, for N observed word tokens of source code document, with observed word tokens given as d= $\{\omega_i\}_{i=1}^{N}$, with Dirichlet parameter α, β, joint distribution can be defined as $p(\theta, z, w | \alpha, \beta) = p(\theta | \alpha) \, \Pi_w p(w | z, \beta)$

Combined code and tokens of text sequences, from training data, set, combinations of source code documents are used to train n-gram models. N-gram of 1, 2, and 3 are tried. Absolute discounting methods with back off strategy to lower n-gram order with Berkley language model package are used. The same data is used to train the LDA-based topic models. Various combination of topics is tried. Several topics and topic mixture parameters form joint distribution from the source of code that has word tokens that are observed. This uses the Dirichlet parameter to distinguish between code and text which makes it unique compared to other work. Code and text tokens are two different entities representing documents. Text words cover string literals, comments, whereas programming syntax and identifier represent code part. Link LDA models are tried with the various topic combination. Topic mixture, topics, and words form mixed membership for joint distribution. Joint topic distribution, words observed in the document, the topic associated with code and, those associated with the word, are used for link LDA training. For the purpose of topic inferencing, link LDA and LDA use Gibbs sampling and entities per document either being single or multiple.

Tokens of Java class comments are to be predicted; these precede class definitions. The next comment token gets its probability from all the models. The previous set of tokens forms the basis to arrive at the probability of the current word, in n-gram modeling. The comment that is to be predicted and class definitions are the separated tokens of documents for the purpose of topic models. Class comments are taken as text tokens; class definition with both code and text token is the rest of the tokens. The text would be comments and string literals from source files. Rest of the token provides a posterior probability for document topics. The probability of comment token can be derived from the document distribution estimate.

4.6.2 Data and training

Source code of nine open-source Java project data is considered. Each project is divided into train and test sets. Three training scenarios are considered. Documented code data of the same projects can be used to predict comments of the mid-development phase of the project. For prediction at the beginning of the life cycle, a similar natured project to be referred. Out of the project's training data is configured for this purpose where eight other project data from the source for training. Compared to text, code forms a major part of the data in source code. The third part of training data is predominantly natural language text combined with code parts, as the focus is to predict the comments. Stack overflow online portal content is considered for this, where people have technical conversations in. Complete dataset of the portal was made use of. Java-related questions are only used as a reference from stack overflow. Testing set of projects are used to test the models; average performance across all projects are considered. Tokenization of code files is done to separate identifiers and code-based tokens. Tokenization has considered reducing camel casing format to normal format for identifiers, such as methods, variables, and classes. Single characters and non-alphanumeric content are removed from code. Mallet statistical natural language process package (Mc-Callum, 2002, McCallum et al. 2002) is used for the tokenization of text from comments and strings. Stack overflow tokens are parsed and tokenized with the Mallet package. In stack overflow, all the content marked with "<code>" was considered.

4.6.3 Evaluation

Vocabulary used in each of the multiple data sources is diverse, making the model comments likelihood diverse, as the model is trained on these various data sources. The reason for this is the out of vocabulary tokens being of diverse nature in models. Character saving metrics are used for evaluation of the model. In a word completion kind of setting, several characters that can be saved by using the model in terms of percentage are used. This is like a code completion tool built-in code editor. Two most likely words are predicted for a comment word from every model filtered by zero to a total number of characters in the word. From the top two words predicted, the top few token characters are selected, the rest of the characters that are not selected are saved characters. Measurement is the average percentage of characters saved using each of the models. Overall results are taken across all the projects considered in the dataset.

4.6.4 Performance

The model that is trained on in project data for predicting in midterm development performs way better than those trained on other data sources for predicting at the beginning of the development. About almost half of the characters are saved using the trigram model. This can be expected as the project's data will have a similar identifier and comments appearing across various classes. Within the project, data should be given first preference while training. In the case of other projects or out of project data referred cases, a model that used large text from stack overflow performed well compared to the model that used more of code data from other projects. This must be traded off on time taken to run as the large vocabulary of the stack overflow has to be processed. Projects from outside the associated programs which influence on the target prediction performance, were not considered in scope for the study. For all data sources, the trigram model performed the best. In LDA model where all tokens are considered, the text was low in performance compared to link LDA that considered text and code tokens separately of tested topic models. There was no correlation found between several latent topics learned by topic models on its performance. Data sources gave optimal character saving for, different number of topic models used for modeling.

Topic models used in work are unigram, so the results are comparable with the unigram model and will not have benefit from the back off strategy used by trigram and bigram. Link LDA model performs way better for comment prediction than compared to models that do not separate code and text. N-grams model used without back off strategy do worse than any other model. Instances show different model types as useful in predicting different comments and hints at using combined models. More words are completed better by n-grams, the topic model also completes a considerable portion of words better that hints at the use of the hybrid solution.

A mixture of code and text tokens contained in the source code file is used to predict the class's comments using language models. The language model saves up to half of the characters as per the experiment. In the project, data has a larger advantage of being used, compared to out of project data. If more text compared to code are available, an event without project data is better. Multiple models to be combined as different models have shown good performance for different comments (Dana et al. 2013).

4.7 Software naturalness

English is rich and natural, delightful use of the language can be an inspiration. Due to the way of daily life the language gets simplified and way easier to predict and repetitive. This gives room for statistical modeling of the language. A simple code completion model for Java is worked upon in this work that can improve eclipse's code completion capability which is built in.

The naturalness of the code is in terms of it being a natural outcome of human effort. Naturalness is picked from NLP where the intent is to process the texts into natural language like English and use it for other key purposes like machine translation that is translating from one natural language to another. NLP had its slow early start with dictionary, logic and semantics-based processing improved the field. In the 1980s corpus-based focus kicked in which involved statistical analytics. Multiple language-based translations, online corpora of text, and computational resources availability have contributed to greater progress in the field. Google translate is one of the classic demonstrations of advancement. Though the complexity of natural language is high, people's extent of use is regular and predictable. A similar analogy for the software as well. Open-source code available with the notion of the repetitiveness of the use of programming structures will make statistical approach utilization a possibility. Applications can be widely around error checking, program summary, code search, mining of software.

This work provides a simple statistical language model that can be widely used, which is based on estimation techniques with corpora of software as the base. Statistical regularity is demonstrated by models with software, which is evident with cross-entropy and perplexity metrics, based on n-gram levels. Language model utilization is demonstrated with a simple tool that can significantly improve the inbuilt eclipse editor's code completion capability. Various software engineering tasks can be facilitated with the vision of establishing usage of the software in its natural form, for modeling purposes.

4.7.1 Motivation

Natural program statistics can be made use of in many ways. In the case of speech recognition, if there is a noise that leads to incomplete information, a good system should be able to get the context and fill up the noise. Similar context of the IDE where the developer when it comes up with part of code the system completes it meaningfully based on context.

The reason why these contexts seem easy is due to the predictable nature of the contexts. It is possible to order with rank on the next part, with a good amount of prior knowledge with a good statistical model, which is in line with this style. It is notable that its repetitiveness characterizes natural language. These have been revealed in a number of studies adding on to the fact that the large part of code parts repeats themselves. Developers can be assisted with the completion of code if we can figure our most possible next part of the code. It is all about identifying the probability of code sequence distribution by making use of the needed amount of code corpus. It is easy to guess the probable part of the next part of the code if we can figure out distribution with the lowest entropy and confidence of the guess also will be high. Language models are the distributions in NLP that cover the form of the distribution that we are talking about here and all associated parameters of the distribution.

The utterance is attached with a probability by language models. In the case of software, utterances will be programmed, it is about program sequences and its allowable program tokens. Probability distribution provided by language models is over the systems, where systems are assumed, possible implementations. In the case of a chosen parametric distribution from a corpus of program, for this distribution lookout is to find out maximum likelihood estimate. The language model comes out of this process. Easiness of estimation and use decides the choice of the language model. All these contexts provide a nice setting for n-gram models.

$$\text{Language model representation: } \forall s \in S[0 < p(s) < 1] \land \Sigma_{s \in S}\, p(s) = 1 \qquad (4.13)$$

Utterance of the language is assigned with probability by language model.

N-gram statistically predicts a token's possibility based on the sequence of its past tokens, coming from a document. Group of conditional probabilities and their product provides the probability estimate for a document. Markov property is base for n-gram model, where its last five tokens influence the token if it is a case of five-gram model. Token sequences are counted for frequency based on maximum likelihood while we estimate on the corpus. N-gram model computation are complex due to the richness of the model versus sparsity of data available. In the case of 102 token vocabulary 106 coefficients are to be estimated for a trigram model. Situations of trigram not appearing in the training corpus may lead to a probability of zero trigram. This trigram may still exist somewhere else and be relevant. Entropy values in this case will be infinite, smoothing technique comes into rescue making the results statistically usable. Techniques also take care of handling the coefficients in extremes where they are smaller than expected or larger than expected, those are smoothed. Options of choosing bigram model in place of trigram can be explored in these situations of technical issues. New modeling and estimation technique that is apt for software will have to be explored, as this is a new area.

4.7.2 Good model

The extent of regularities in the corpus that the model can capture indicates the goodness of the model. A new document will be predicted with high confidence, to have come from the same population by the model derived from the representative corpus, which will be a good model, in that case, the new document will not surprise the model. Cross entropy is a measure of this, no surprise for the corpus. Perplexity is another metric that is a log-transformed version of cross-entropy. Entropy computation takes into account the sequence of tokens, several tokens in the sequence, the language model is chosen, and the probability of the token, given the previous set of tokens. This probability is close to 1, if the token is most frequent and will be close to zero, if the probability is low. IDE should be made intelligent with the highly efficient hypothetical model. This should facilitate programmers to complete the code parts, with very high probability for most of the program. This creates a situation where the program can be codded in with just hit of tabs. But a lot less can still take us too far.

$$\text{Cross entropy measure: } HM(s) = -1/n \log pM(a1 \ldots \ldots an) \qquad (4.14)$$

For a document s = a1 . . . . an, with length n in a language model represented by M, the probability of document as estimated in the model is assumed to be represented by cross-entropy measure equation.

4.7.3 Method

To assess the question of how natural the software is, the experiment focused on natural language and code corpora. As a comparison of naturalness, cross-entropy is a measure to be used. Various combinations of code corpora comparison with English text and code corpora comparison with other code corpora will provide further insights on similarities and differences. Gutenberg and Brown corpus, which are famous, are used for natural language study. Ubuntu application collection and Java project collections are code corpora chosen. These are broken into the application domain. Lexical analysis was done on projects to produce token sequence as per language syntax; these are used to find n-gram models. Extension to another language would be easier as the corpus selected for this work are C and Java. Java focused language model with code suggestion as a plugin for eclipse; and cross-entropy studies were the target for experiments. Tokens are extracted from each of the files. A unique number of tokens provide a glimpse of the amount of supremeness in the corpus.

The distribution model is representing the corpus, and when a test document is presented, how surprising is the document to distribution is the cross-entropy. For testing purpose corpus must be divided away from the train part. Average cross-entropy is measured, train and test are divided as per standard norm, and a tenfold cross-validation is used for validation. In the case of unseen tokens in the corpora, it is attached with small probability to an unknown token for smoothing, which is an open vocabulary model. Cross-entropy is measured on the parameters of the distribution model over training corpus. Figure 4.7 depicts Visualization for comparing cross entropy of various languages for different order of n-grams.

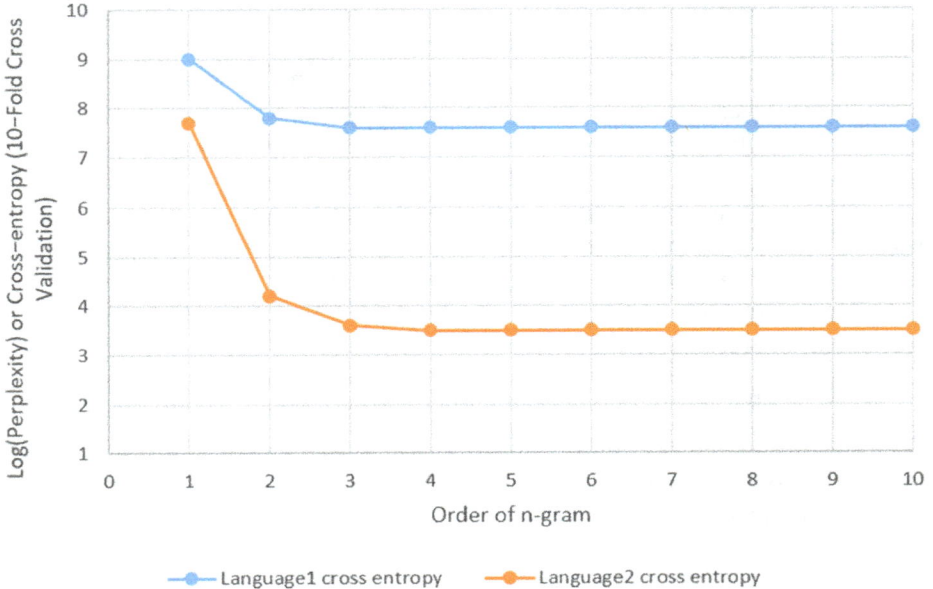

Figure 4.7: Visualization for comparing cross entropy of various languages for different order of n-grams.

4.7.4 N-gram language model to capture regularity in software

N-gram models are tried for various *n* values over Java language projects and English corpus.

Cross entropy is computed for the same project test data on which it is trained, so it is a self-cross entropy. Each project is combined as a single document for the sake of performance measure.

Uniform distribution of unique tokens has a higher entropy compared to unigram entropy because unique token frequencies are quite skewed. At three- or four-gram cross-entropy saturates rapidly. This decline of cross-entropy is surprisingly common across English and Java language. This shows language models' capability to capture the repetitiveness in Java language on similar lines as that of English. This should be promising as local context regularity modeling for natural language has been of great significance for statistical modeling of natural language. Now on the same lines in parallel, if it can be demonstrated to software that will be the crux of the study.

Many instances demonstrated that software is very regular in comparison to English with cross-entropy going down at a much higher pace. One concern to account for is doing the difference between Java and English gives the situation of high similarities that are noticed in software. Structured syntax of Java makes it much simpler compared to English. There should be consistency of results across projects if the statistical similarity was about the simplicity of Java. The language model's local context capturing statistical similarity is what is looked at here. The idea is that if the model is trained with Java language, on its testing on another project, the output should be reproducible.

So, we need to figure out if the statistical language model's local regularities demonstrated are language-specific or project-influenced. The experiment would be to train on the individual project first and then test nine other projects. Cross-entropy loss needs to be used for measurement in the trigram model scenario. A tenfold cross-entropy of self's average needs to be used as a base for comparison. Trigram choice is influenced by its simplicity of memory usage and can provide less entropy.

On visualization, the x-axis shows all the Java projects. Each project box plot shows cross-entropy performance across nine other projects. Figure 4.8 depicts Visualization for cross-entropy versus self-entropy. The average performance of self-entropies of each of the projects is used as a base reference. The project is represented as a document. Cross entropy of projects against itself is low with tenfold cross-validation technique applied; there is no possibility of overfitting. Apart from Java-specific nature, every project has its uniqueness, which comes

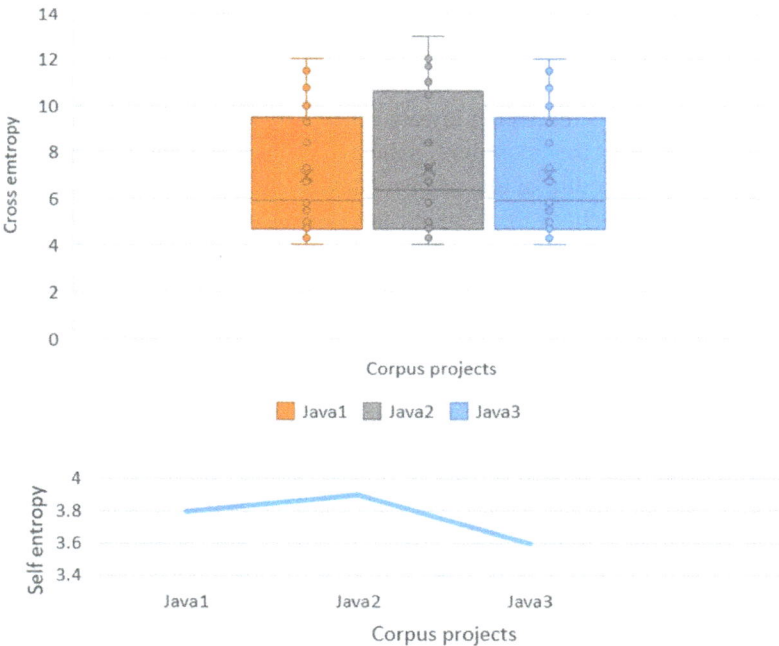

Figure 4.8: Visualization for cross-entropy versus self-entropy. The x-axis can list all the Java-based projects, Box plot shows the range of entropies for each of the projects against the rest of the projects under comparison. Trendline graph at the bottom provides self-entropy of the projects with itself.

out in mode. The experiment demonstrates its ability to figure out tokens that are not specific to Java, about half of the time. This indicates that the simplicity of Java language alone does not influence low entropy. The concept of product line comes from the fact that the projects are not completely independent but contain similarities within the domain. This demands the question of n-gram possibility of making out similarities within the domain of project and differences across project domains. Different categories of a project within a project was picked to assess this. As shown earlier, self-entropy was computed for each of the categories and cross-entropy against other categories. It shows the regularities within domains and less of regularities across. Some domains demonstrated high regularities with low entropy that needs study. Projects with large data show that new projects' ability to make use of corpora from common domains.

Language models figure out application domain-specific and project-specific local regularities. Beyond Java language simplicity, projects and domains have a large play in these similarities. Project-specific regularities can be used to enhance IDE-specific code suggestion engines. Code summarization and code searching also can be enabled with this local regularities' nature of the code. Semantic properties also generate out of these local similarities.

4.7.5 Next token suggestion

Smoothed n-gram model that created the lowest entropy for the token sequence locally shows the regularities at the token-level sequence. The next token in sequence can be guessed with just a handful of tokens in sequence.

Eclipse plugin is built to check this idea. Based on context-based type information IDEs suggest the next token with their auto-suggestion feature. Natural token sequence-based prediction of the token can be done by improvising the eclipse's existing code suggestion feature with an n-gram model suggestion engine based on the corpus. This model uses the last two tokens to predict the next one with the trigram model built from the project corpus, the static corpus of code source is used. Likely next tokens are proposed based on rank order formed of probability estimation. Both eclipse-built feature and the add on a feature from the approach discussed here have proposed a list of options for next token, based on several permissible options to be proposed, best from both the list are presented. Plugin from the approach reviewed here was good at the short token sequence and eclipse built-in one for a long token sequence. Merge algorithm is used to handle the output of both. Specific length is seen as cutoff after which eclipse outperforms the n-gram model discussed here. Minimum length of the sequence is set to six; if eclipse provides a suggestion of long token within the top selected number then that is taken; else half of the choice from eclipse and n-gram is done.

To identify the key differences in the performance of the eclipse inbuilt feature and n-gram model, a good amount of study is needed. Suggestion engines can provide a wide variety of output, which may be few or more, all longer suggestions, and so on. Focus here is not on the best user presentation with the merged model, but to explore the language model capability that works on the corpus. Suggestions less than three characters will not be useful and will be ignored. When combining lists at least one from each will be picked. A trigram is a good option compared to four or five grams as it provides the right amount of context and perfect cut off beyond which the entropy reduction declines.

4.7.6 Advantage of language model

For every top number of suggestion different visualization is done. The graph is plotted for suggestion length in the x-axis and, percentage gain over eclipse in the y-axis. One trend for percentage gain and the other one for raw gain count. Figure 4.9 depicts Visualization of suggestion gains after combining suggestions of eclipse and n-gram. Count of correct suggestions reduces for eclipse as the length of suggestion increases. Merge algorithm has substantial performance over eclipse, though with an increase in suggestion length this reduces. Performance up to six-character range is quite substantial. Tokens that are generated by n-grams when assessed, it provides a sense of why n-gram models do well with shorter tokens. Most frequent n-grams

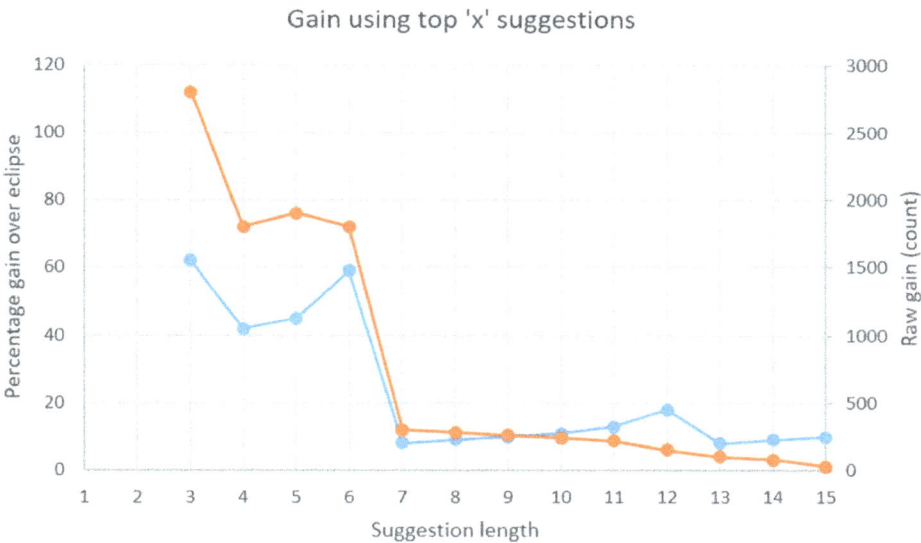

Figure 4.9: Visualization of suggestion gains after combining suggestions of eclipse and n-gram. This visualization can be done for different top suggestions. Percentage gain and raw gain count are visualized together.

are from across all the files and the language model built here is based on all these files. N-grams pick up the short words with a stronger signal, and the most frequently used words are shorter in the corpus. More than half of the language model's suggestion is based on project tokens and not based on Java keywords. So, the statistical model captures local regularities in projects that are not specific to language. Between the merged model and eclipse, several keystrokes saved is another measure of performance.

The effort here was to optimize the software tool of code suggestion using one specific language model. With a combination of scope, type, and syntax, related data language model can be more sophisticated, and much lower entropy can be recorded.

4.7.7 Other work

Code suggestion is suggesting applicable token of code and, code completion is in the case of partially typed in code being completed. IDEs provide a unified way of code completion and suggestion capabilities. The current syntactic context is explored by eclipse to arrive at code completion based on inputs sent with keyboard short cuts. Based on Java specification multiple semantic and syntactic rules are configured in eclipse. The basic lookout is to assess the surrounding code to get the context, based on which list of expected tokens will be derived. Rules-based on "type system" kicks in if the list has reference type; this generated a list of type names that are applicable as a list of completion. Likewise, for variables as well, finally lists are combined based on defined rules. An approach that is reviewed here complements this format. N-gram model captures the most often occurring ones rather than just focusing on immediately available context and semantics. The language-independent approach that was discussed in this review is more flexible. This approach's precision is to do with the practice used space being much smaller compared to possible space applicable for language. Further refinement of language model output can be done using statistics of the corpus. The extent of improvement shown by eclipse code suggestion after using the n-gram model demonstrated the code's naturalness.

The advanced approach of Bruch et al. (2009) looks at predicting method calls based on earlier usage in the calls' context and frequency. Review of work in this discussion provides a broader view for software tools utilizing code corpora by language model, it is not restricted to method calls. Recorded human action and code knowledge are used up collectively to come up with advanced IDEs by Bruch et al. (2020). Work in this discussion had emphasized the naturalness of the code, which fuels this big vision of next-generation IDE. Hidden Markov models are explored for generating token from short tokens by Han et al. (2009), where backtracking and suggestion are used which are done by dynamic programming trellis. Matching code clones based on the partial program task was the n-gram model proposed by

Jacob and Tairas (2010). Based on the partially completed code tasks, the candidate code clone was presented, based on clone groups that were generated from the language model. In this case entire corpus was not used instead code clone was used.

Using names to improvise the meaning of artifacts is the line of exploration done by others. Apart from finding the influence of name on the meaning of artifacts, the possibility of improving the same needs to be seen. Method naming is targeted at entropy-based measure and static analysis on the semantic property distribution of the corpus methods to determine methods' ability to discriminate. This provides information about names that are inconsistent with the corpus.

The natural language description of the code is another area of exploration. Static analysis is used to explore and asses the semantic property of the code. Different approaches compared to identify the regularity of natural code based on a statistical model. A statistical model of software corpus can be enabled from the static analysis if it is done efficiently and to a good scale.

Software mining is another key area of research to extract information from software repositories. Error pattern recognition, application programming interface (API) usage, change handling and topic extraction to name few. Approaches used may vary but the naturalness of code that is provided as based in this review will provide some good implementation approaches. For representative software corpus where known information is identified, useful information can be figured out. This comes out as the implementation approaches are reviewed. The availability of useful and simplified information from software can be used to establish facts and required information that has a statistical relation between the information available and the information that is looked for. This relation helps to arrive at information from new programs that are in line with the corpus.

The corpus-based statistical model helps many other engineering tasks. The extension can be done on the language-based model discussed in this review. A large body of code can be subjected to analysis. It can build on the semantic, type information, scope, and syntactic information coming out of this extensive study. Data sparsity is another problem that comes from large data used for making the model richer. In that case exploration of IDEs to handle the sparse data, techniques like smoothing can be looked at.

Programmers with visual impairment having difficulty in using a keyboard would be enabled with speech recognition. This scenario suffers from high error rates, and there are no attempts to use statistical language models trained on corpora of code, in this area. Language models can significantly reduce the error in this scenario. Confidence about this comes from the fact that the code corpora largely represent the software's reengineering and maintenance area.

Summarizing the code parts or code changes in natural language as a task. Alternatively, extracting code part like method call from English description. This is related to statistical natural language translation (SNLT). Input for SNLT is the set of sentences that are provided in more than one language. In the case of translation between two languages, the joint corpus of two languages is used as a base reference. The conditional distribution of one language output sentence concerning other language inputs based on the Bayesian process is used as a reference.

Statistical estimates from several corpus can derive to handle summarization task. English and code-based corpora can be generated. English log message in the case of commit messages, program version history, and code changes are the variety of sources. Code and nearby inline comments form a good pair. English in the design document, code, bug reports, and discussions associated with the project can give good English corpora. These sources provide code and English content to arrive at a model that can facilitate translation.

The naturalness of the deep properties of software is what provides the naturalness of software. These are computed by powerful software tools. Semantic properties of the code manifest with the repetitive usage of the code syntax by coders. These semantic similarities bring out the easily identifiable properties that are costly for static analysis to do, or sometimes not even possible (Abram et al. 2012).

4.7.8 Conclusion

A linguist reveals theoretical complexities of the natural language. But statistical modeling of the regular and predictive natural utterances is a possibility. Even a simple statistical analysis can bring out the software's regularities, which is perceived to be complex. Such a simpler statistical model can influence current state of art IDE. The naturalness base provides the hope of language translation approaches being parallelly explored for code search and code summarization. Software analysis tools traditionally have been computing deeper properties of the software that comes from the code's naturalness.

4.8 Conclusion and future work

Future focus would be to create a plugin for eclipse which can combine the default code suggestions with that of the suggestions from the tool. Further exploration is recommended for dynamic programming language. We also reviewed the neural language model application to dynamic language like Python, with attention mechanism to see that the accuracy of identifier prediction was top-notch. Sparse pointer network was reviewed that would easily capture long-range dependencies, by taking filtered memory of earlier identifiers that were represented. Further point of exploration is extending the study beyond a common file, like used in this discussion. Further point of exploration is extending the study beyond a common file, like the one used in this discussion. Further to this scale the approach for whole code projects has to be devised. Further experiments would be needed for integrating the code suggestion tool in the IDE. Work can also be extended toward the completion of partial code tokens with the character language model as a base.

In NTM discussion, we looked at an interesting neural network architecture that was combining inspiration from biological neural systems and digital computer design. This architecture was differentiable like traditional neural network and can be trained using gradient descent. The language model was explored on a source file that had code and text tokens, for predicting class comments. It was also noticed that the training data related to the same project will have a significant influence on better performance compared to data from outside the project. But interestingly for data from other projects with more text than code, the performance was also high.

The naturalness of the natural language and the possibilities of exploring statistical language modeling were discussed. In the context of software language, significantly higher patterns exist and the same can be modeled statistically, with ease. This opens the possibility of building a powerful suggestion engine. Further exploration can focus on the natural language translation approach being utilized for code search and summarization.

Chapter 5
Transfer learning and one-shot learning to address software deployment issues

5.1 Introduction

Deployment is another prominent area of software development that needs attention. One-shot learning and transfer learning can be subjected to advanced research to handle critical challenges. This module addresses the problem involved in the concept drift resulting from code ecosystem's drastic changes. It results in the need for machine learning models to evolve with this change, which costs time and money. As frequent training of the models for this concept, drift becomes a need and adds complexity.

5.2 History of software deployment

History of the influence of various dimensions on software deployment and anticipation of the prospects are explored. Deployment is part of postproduction activities for the software. Multiple aspects of the software will need services to and from various other features. Adding to the complexity is the distributed system on which these parts are deployed and their heterogeneous nature. The discussion here concentrates on standardizing terms used for deployment and other supporting activities. Deployment problems and approaches are discussed by looking at the case studies of the deployments done with the latest technologies. Also, we deep dive into the deployment issues and the direction needed to act on these.

5.2.1 Deployment process

Deployment process marks as the bridge between execution and takeover of the software. Personnel in charge of this exercise take over the software, prepare for the execution, and complete the execution (Object Management Group 2003). This time of deployment is when the customer needs are taken care of at the granular level. The entire release process is the combination of various subtasks at the end of the software development life cycle: managing the configuration of the software, installation into the targeted environment, and activating the same.

There is a need for deriving common terminologies that hold good for most of the software deployments. The component is the common terms as per UML2 specifications (Object Management Group 2005), representing the system's module. Its existence gets replaced within the environment as per (Szyperski 2003); a component is

https://doi.org/10.1515/9783110703399-005

a unit of integration that explicitly defines prescribed interfaces and dependencies of the context. The performance of the component is influenced by interfaces that are needed and those that are available. Assembly is the gatherings of the components integrated. Interface requirements are provided for components associated with the assembly or by the system's environment. The resource is another key term related to the supporting artifacts: software, hardware, or other artifacts. The collection of these components is the application, which is intended to perform some functions. Versioning of the component brings in time dimension on the revisions on the components related to the platform's functionality associated aspects.

Metadata and assemblies also form the part in some cases of the systems. Different implementation done in the package helps to take care of the need for software and hardware. Thus, choosing the most suitable component will be necessary when the deployment of the package happens. Metadata would be the description associated with the package for the components involved. There are models where the components are to be operating in specified environments like containers. Containers play a crucial role in protecting the environment's components, policy application on the components, managing the life cycle, and provisions to identify other and components to bind with them as needed. In the case of binding components, figure out their counterpart required for the component's efficient operations. Based on the time of deployment, packaging, assembly, the building of the components' components and bindings are influenced. The naming of the components also plays a vital role in the binding process. The component's get their generic naming convention in web services where the Uniform Resource Identifier (URI) is under use.

5.2.2 The life cycle of the deployment

The release is a point of software development lifecycle where the person in charge of the release will be the center of system for all the stakeholders involved. Installing process is where the handover to the customer happens and activation is done with required configurations. The process of activation of software involves starting the software or configuring the trigger that will start the application at the right time. These processes are handled by the interfaces designed graphically or with scripts in many cases. Deactivation is the reverse process that is needed before the changes are incorporated into the system. In case of new versions or changes to the part of the software are involved then the update process kicks in. Like installation, the update is a special case where the software must be deactivated before the process kicks in. Adaptation is another key element where the installed software changes to make it respond to the environment in which the software operates. Also, another aspect is to remove the software from the deployed machine.

Specification for software process requires composing software and metadata needed to customers from the producers, setting up the software into the deployment environment before the deployment is proceeded, understanding the needs of the execution environment, planning the infrastructure and procedure to deploy the software, configuring various parts of the software into deployment space, and launching and monitoring the software.

5.2.3 Virtualization case study

To understand the issues associated with software development, we will explore the case study of the deployment. Using one technology as a reference to understand the deployment landscape helps as overall there is a wide variety of operating systems, hardware, and technologies for implementation.

Most of the deployment issues are the outcome of the issues associated with the products' interactions that go for installation, execution-related issues, and deployment environment. One way to handle these dependencies is to handle interdependencies with abstraction. Also, these issues can be handled by creating a perfect environment for the installation of the components. Though this was a tough job, now virtualization makes it possible.

The virtualization trend in software deployment will be of major importance. The conventionally single operating system will handle single hardware and provide a consistent environment for all programs. In virtualization, a single piece of hardware that runs a layer for virtualization represents multiple hardware forms on which heterogeneous natured multiple operating systems can run. That is how multiple versions of the windows can run on common hardware. This is how the partitioning technique works. VMWare (2006) virtualization is set up to combine the operating system with its applications in a common virtual machine. The single image file will encapsulate the CPU, file system, a disk known as a server earlier. These images are put into a virtual environment and virtualization software executes this.

The operating system and application working hand in hand have a greater influence on deployment. The deployment process with virtualization is said to have five steps to create, generalize, test, distribute, and update as per Schumate (2004). In the first step image is created with data, operating system, and other components. Physical hardware will have image baselined; also it will have a software image, which is a file generated from the respective tool. Together these copies form a master image. Then users' needs are looked at and respectively the image is customized. Testing also needs to be done for the distribution. VM images are executed by the host, after being shared by the distribution step. Network boot technologies are used for this purpose. For updating the software new images must be created from the master that we created. This helps to test the entire distribution without depending on the hardware or software, as the image represents the whole of the environment.

5.2.4 Deployment issues

Binding of the components may happen at different points of time, like compiling time, package assembly creation time, before run configuration time and run time. Any of these binding times used will have its impact. In the case of the compile-time binding, it is only about binding between the code pieces and there are no components involved. Code fragments like class are what are denoted by name in this case. Classes will be closely coupled during this instance. In this context code-created data objects and code itself will occur together in the environment and are containerized.

In assembly time, while the reference between the components is resolved, abstracting the names with some language mechanics and concrete instance part of the assembly can be done. The factory pattern (Gamma et al. 1994) uses this to support the exact need. This also just involves code fragment bindings and not on components.

At the pre-run time to resolve references, a wide variety of the bindings may be created, after the instantiation of the components. Reference will cover other components in the container, pure code, local components, and extant code references. During run time a similar context of bindings is possible.

Run time environment decides on bindings in the resolution context. Based on the type of environment whether local components or other arbitrary components can be addressed accordingly.

Now talking about issues related to run time and container, ability to abstract over the local environment that covers the operating system and hardware has been demonstrated by .Net and J2EE. In these cases run time environments that are machine-independent are employed and use of containers are facilitated. Technology buys in trade-off are done by both technologies to ensure the components are isolated from their complexities. This also brings in the constraint of developing the application components in the language specified and maintaining conventions around semantic and syntactic writing the applications as needed for the container. Hardware and operating system dependencies mitigation are supported minimally by the containers and environment introduction. This setup's key influence is to reduce the complexities involved in the interaction of hardware and operating system.

This helps the application run on the environment, which facilitates their need, which is the advantage of virtualization. This will cover all the required entities like components, the application needed, operating system, and hardware.

We looked at the container technology and general technology conflicts involved. The lightweight container concept helps to take care of this conflict. Techniques such as inversion of control (IOC) or dependency injection are used in this context. This technique is like "don't call me I will call you back" pattern of design. At the base, it is about removing the dependencies on other containers and components. General context components identify other dependable components, which is a service locating pattern. Infrastructure calls the needed components instead of components looking for ones, this is how IOC facilitates. IOC has its major contribution in deployment compared to support on component identification. In some of the cases, the required components are supplied when the component gets created, which is constructor injection. In other case setting methods are provided to get components, that are capable of setter injection. Some cases' interface decides the interaction between various components or with the container itself.

There is a commonality between the deployment needs and the way IOC operates, which encourages IOC to be a good facilitator to take care of issues. Some of the deployment models and IOC format both have the format of infrastructure resolving some of the references that are not resolved earlier for the components. Source code is free of maintaining the dependencies between components in both cases. In some cases of lightweight frameworks a variety of XML configuration files come in place of the code being freed from handling the dependencies.

IOC seems a good fit across the life of software; components coupling together can be alternated with this, during programming. Assemblies created with bounding components can use this technique. In some of the frameworks, controls are injected by node manager along with required components port. This will also include other components and IOC will be handy in case of those deployment situations, IOC will be handy. During the run time components can be bound and unbound using the IOC concept. Fractal (VMWare 2006) component model houses this concept.

Given reducing the cost of design, deployment, and maintenance, the fractal model plays a role as programming language–neutral. Recursive data modeling ability is the key contribution of the model. The model also helps to retrospect components and hence it is reflective. Aspect-oriented programming (AOP) principle of separation of concern is the base for the fractal model.

Throughout the lifecycle of the software data metadata must be maintained; this is the next area to look at a deployment issue. Maintaining the trace of design decisions throughout the life cycle is assumed to be a key need. Workflows, version history, languages used, source code, data on interfaces, and component description may all be metadata. The entire gamut of software management across acquisition through evolution and retiring, including deployment, is governed by dedicated application lifecycle management (ALM), a new field. So, deployment becomes the scope of this. ALM intends to have an architecture for software-related documentation that covers nonfunctional and functional requirements. UML designs independent of the platform are modeled as software architecture, which facilitates a refined framework of architecture. Domain-specific modeling framework, rather than a complex all domain-based framework, is the intent of model-driven architecture (MDA). As a fractal approach to modeling, domain-specific language can be used to abstract at a different level, to cover different aspects. Meta-object facility (MOF) provides integration across different languages to help interchange of artifacts across.

Metadata repositories are needed to hold the metadata with tools to extract those metadata and schema for the same, all these to keep by the deployed system. Frameworks assist in developing models and code and are kept in line with each other. There is the hope of this extending from component focus toward complete life cycle focus across development covering deployment.

5.3 Issues and challenges in deployment

We have explored the issues in deployment and the techniques involved for deployment. We will explore the deployment issues that will impact the development processes.

5.3.1 The granularity of components and environment

It is not consistent in terms of the granularity of the components and environment. Granularity varies with a large scale. Some of the systems have the finer granularity of objects of programming language. Some other systems will have larger file system objects that are binary executables. The difference also extends to typed objects dependent on programming language and untyped binary object-based language-independent ones. Granularity plays a role in terms of the dependencies between the components. The large granularity of components is seen in service-oriented architecture; otherwise, it may have to be broken into a distributed object model. Assemblies of components that are packaged also do not have a standard granularity agreed upon. Users will not get a chance to decide on the package components whether to repair, uninstall, or install, in the cases of installers for Windows.

The granularity of the environment is another key aspect and its variable as well. Implementation can be specific to a server instance or would be machine-wide. The virtual machine (VM) image creation process will be inline with assembly creation. VM image will be unique in its self-containment and independent from the machine for its image execution.

Replication and concurrent execution in different sites will be possible once the image is created as it acts as an assembly. Assembly can be black box based tested or white box based tested before it is delivered, that the advantage this approach brings. This also provides a higher degree of confidence to the customers, for its accurate. That points to understanding how the association of external world assemblies work.

The deployment cycle gets simplified after deployment with the application of the VM approach. This approach has its own cost and complexities. With granularity colocation also was a key factor; VM helps break the barrier of having applications in a common location. Every application can run its hardware on their need base. Traditionally multiple processes were run on the hardware compared to that, if multiple VMs are not running it will result in its performance implications.

The VM provides the self-contained environment for applications. Configurations are needed for such application even with this setup; these configurations focus on details of web services needed, local registry details, database servers, and so on. Also, this facilitates integration between externally running applications and ones on the VMs. Controls need to be injected for this purpose. The configuration management system then would need metadata for configuration purposes. Such scenarios will be to identify if the VM invokes the information need or web services. This calls for addressing modeling and meta-modeling relate issues.

5.3.2 Deployment based on the distribution

A deployment that is distributed brings in a lot of issues that must be addressed. Security and trust model's inconsistency contributes in case of distributed deployments related issues. Deploying and executing harmful, malicious agents to be prevented by strong security mechanisms. It should also help avoid accidental interference between the components due to malicious activities. Since it is a trade-off between the interaction between the separation and sharing of the resources' components, it plays between the behavior of resources and the infrastructure.

There is an approach where code can be easily and safely deployed on any outer machine (Dearle et al. 2004). Code and data duo called bundles were included in the assemblies that are self-contained and the same will be deployed. Small kernel helps to establish the authenticity before deployment this of happens on each of the nodes and that ways bundles are signed off.

Inter components bindings' maintenance and monitoring is the ability that is needed for the case of distributed third-party-server-based deployment of applications and its components. Lightweight frameworks can be used for these purposes. Distributed IOC environment also will be like IOC. With components being the same inter-component bindings are managed by methods. Permission is needed to allow such bindings between local components, between machines and local spaces (containers), which is a need for fractal bindings. This kind of setup will be an extension of the fractal model where first-class entities of the model would be the nodes and address space of the containers. Intra- and inter-container and internode update-level updates would be needed to expose the bindings and interfaces. To achieve this interface of binding manipulation of the components and applications are to be exposed Brebnar et al. (2005) criticized this inability of applying the application and components in some of the frameworks. Multiple levels of support are needed to expose binding manipulation interfaces, including middleware support.

5.3.3 Specifications for architecture

Dynamic changes are not well explained and do not manage up well in many of the architectural specifications, architecture description language (ADL). The future intended state of the system is explained here in these specifications. To make this a generally applicable one, architects must use ADL to define software systems. Problems that are related to the deployments come from the ability of ADL to be modified for its software specifications. Some of the scenarios included are as follows. First of all, can the components be created and destroyed, second, can topology of the architecture be changed followed by the unnesting and nesting possibilities of the components. It is suggested by Medvidovic (1996) that there should be a possibility for including new components, change in mappings of components to machines, and reconfiguration of the architecture. He proposes such capable language as architectural construction notation (ACN).

Though the problems related to autonomous management are claimed to have been solved there are co-ordination-related issues. Updating will generally be planned during a time when demand is low, which includes temporal complexities to the update task. Failures noticed in the deployments are the same kind of failures that are prevalent in the management systems that govern the deployment. Failure resistant collection of managers, are good fault tolerance method to be built into the system. These introduce issues such as reliable event distribution between application to control system. This also helps to have information on time when the instructions provided are executed. Management infrastructure-based fault tolerance engineering and other issues are also addressed.

As part of the discussion on software deployment history, we explored the deployment and its associated aspects. Internet domain has been the key focus for academia and the industry. Mobile and sensor-based network domain is significant for both fields. Deployment issues get complex in the mobile domain due to heterogeneity and number of devices involced. In case of IP-based protocols, the network also connects as an add on. The mobile device is with individuals and lack of central control also adds on to the situation. So, this area provides greater scope for further research.

As part of the discussion on the history of software deployment, we explored deployments of the key fields. Deployments in case of sensor-net are simple and the field is also widely grown. Mostly these huge programs are directly connected desktop programs. As extensive use of the same has been seen and there is a need to update the node. Wireless network–based components' deployment will then be needed. In line with the deployment issues in the mobile domain, senor-net brings in the challenge of conservation of energy and communication bandwidth being low (Dearle 2007).

5.4 Artificial Intelligence deployment as reference

After understanding the background of deployment, we will know to explore how deployment is handled in the world of Artificial Intelligence to explore the possible best practices that can be taken into software development processes.

Mobile applications have found extensive use of deep learning techniques. There is a need for centralized guidelines on the best practices of developing solutions related to deep learning solution deployments, particularly for real-time applications and smartphones. Open source tools are widely used for the purpose though. Deep learning inference networks for smartphones based on the best applicable tools of deep learning will be explored. For Android and iOS smartphones consistent flow of implementation is looked at. Improving the real-time throughput using the multi-threading approach is also looked at. For validation, throughput happening real time, consumption of central processing unit (CPU)/graphical processing unit(GPU), and accuracy parameter–based benchmarking frameworks are looked at. Deep learning models are being used as real-time smartphone apps, due to the use of well developed deployment approaches. These can be done with ease for the publicly available software tools of deep learning and smartphones. Related convolutional neural network (CNN) is explored for the approach mentioned above and the outcomes are validated against metrics that are mentioned above for benchmarking.

Machine learning has had its major progress with deep learning approaches (Lecun et al. 2015). Speech recognition (Deng et al. 2013), object detection (Krizhevsky et al. 2012), and localization of object (Object Management Group 2005) areas have gone beyond the traditional approach due to the progress in the deep learning area. Progress also has its contribution from the exponentially growing computation capabilities and the availability of the large data.

Deep Neural Network (DNN) which are part of deep learning has layers of nonlinear processing units that facilitates the abstraction of the data at their deeper layers. This will contribute significantly to feature engineering, as it avoids the conventional need for building hand made features and provides the feature extraction capability. Also, the need for data cleaning is minimized as the raw data can be used effectively for feature extraction. Natural language processing (NLP), speech recognition, and computer vision areas have seen a great accuracy of operations by the DNN approach, due to their ability to learn the features well.

Smartphones have made their mark for implementation-related platforms in mobile computing space outreaching 2.5 billion owners worldwide (VMWare 2006). Sensors are a key part of smartphones on top of CPUs and GPUs capabilities; these sensors do not need hardware to interface with like in the case of Raspberry Pi (Gamma et al. 1994) and Arduino (Schumate 2004). Application programming interface (API) for smartphones adds on top of other capabilities.

Voice assistants, augmented reality, automated text prediction, and many other solutions based on deep learning have enriched and made smartphones the top users. They are also becoming the platform of choice for the research in areas such as hearing studies looking into voice activity (Sehgal et al. 2018), jaundice diagnosis (Mariakakis et al. 2017), concussion detection (Mariakakis et al. 2017), and schizophrenia (Chu et al. 2017). In these cases, offline data simulation or server-side processing of the data collected from smartphones is employed. Literature does not provide much of the information for deep learning solutions to be deployed on a smartphone. If DNNs are to be manually coded for real-time operations on a smartphone that will be a complex scenario. The discussion here will further look into centrally providing all the resources needed to enable deployment of the deep learning model on smartphones to address the gap between deployment and development, with software tools that can be used publicly.

Mobile implementation for the deep learning solutions approaches has been worked out in literature (Han et al. 2015, Howard et al. 2017, Lane et al. 2016, 2017). Smartphones are enabled with on-the-edge learning for deep learning capabilities. Kirin 970 neural processing unit is from Huawei and A11 bionic chip has a neural engine from Apple. To address the challenges of cloud computing and the need for internet, processors are being dedicated by the smartphone industry to enable deep learning on the device. This also comes up with the advantage of secure data that was a challenge for server data. Multi-thread operations become a reality with multicore CPU-based smartphones, making the operations efficient, and enhancing throughputs during real-time operations.

Exploration of deep learning tools and their progress has reached a maturity point considering the libraries that have come into the area. Since there are no guidelines for usage of deployment tools for the smartphone learning curve is steep.Thus, the Deep learning model operations which are deployed as apps on Android and iOS are looked at with a unified approach. For this all required knowledge is in a commonplace. This makes smartphones moving platforms a good area for the research.

With these backgrounds we explore further best available libraries of deep learning for deployment, operating system–based deployment mechanics, software tools needed, and the device involved itself. Benchmarking criteria and various DNN models are looked at. Multi-threading operations for throughput improvement in real time with multicore CPUs on smartphones will be explored, wrapping up with the validation results for the same.

5.5 DNN model deployment on smartphones

Publicly available software to deploy DNN models is explored here. Approach for tuning the deployment methods for Android and iOS based apps, using DNN are explored.

5.5.1 Software tools for deep learning

Public library's availability has been revolutionary in deep learning making its mark and at an accelerated pace. CNTK (Microsoft Cognitive Toolkit) from Microsoft, Google's Tensorflow (Abadi et al. 2016), Berkeley AI research's Caffe (Ren et al. 2017), and Facebook's PyTorch(PyTorch) have been the major libraries. Training and prototyping of the models are done by these libraries in Python. It has been a confusing state of where to start from for researchers even with a wide variety of options to train the deep learning models. From deep learning libraries three key ones are looked for the task under consideration. Active support from developers and portability to mobile apps are the factors considered for the selection of libraries.

TensorFlow being a programming library is focused on the dataflow. Users are enabled with a graph to operate with on data streams of input as the TensorFlow provides stateful dataflow graphs. TensorFlow gets its name for multidimensional arrays of data or tensors' way of data representation. Graph-based neural network representation helps manage multi-threaded operations possible on GPUs and CPUs, covering various hardware platforms and facilitating efficient memory use and computation. TensorFlow and Android being Google products help ease integration providing module TensorFlow for mobile as a dependency. TensorFlow Lite is in experimentation for the purpose, but supports lesser operations compared to TensorFlow mobile. In both cases deployment flow is the same.

Python has a higher-level library called Keras (Keras) that can work with CNTK or TensorFlow as its backend and is widely used. Providing building blocks for the DNN layer, simplified coding syntax, and data preprocessing tools are Keras' facilities to make model building faster and efficient. Keras models can work on Android as TensorFlow is being used as a backend for Keras.

To run machine learning models on iOS devices Apple has developed CoreML (Core ML) framework. CoreMLTools(Coremltools), which are Python-based tools, are available that get CoreML-supported models from existing machine learning models. CoreML model can be derived from Keras model, due to this capability and implement this as an app on iOS devices or iPhones. CoreML model can be extracted from the TensorFlow model by using a converter such as tf-coreML (tfcoreML).

5.5.2 Process of deployment

Figure 5.2 demonstrates the activities involved in deep learning models deployment on a smartphone.

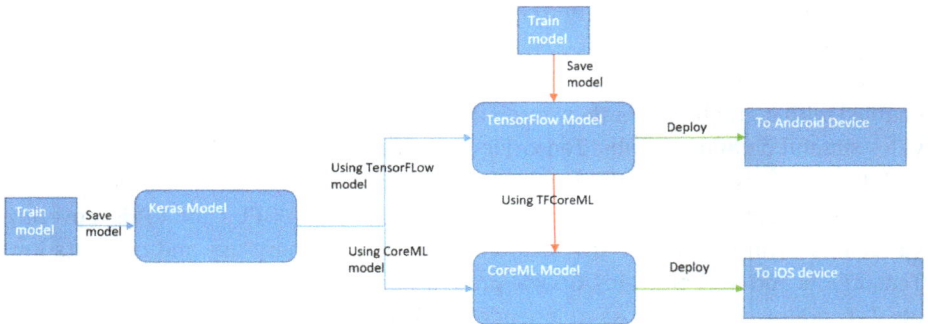

Figure 5.1: Figure depicts deep learning solutions are deployed on smartphones using the deployment tools available publicly.

For iOS conversion to CoreML models and TensorFlow backend to Android conversion can be possible with Keras being the prototyping library primarily. Tf-coreML converter helps covert to the CoreML model for iOS implementation, in the case of the Tensor-Flow model. Publicly available converter tools (deep-learning-model-convertor) are used if any other libraries are to be converted into TensorFlow or Keras base. For now, Keras-trained deep learning libraries like CoreML also provide required conversion tools.

The network's feedforward path is only executed, removing all the training-based layers and retaining only inference-based ones. Optimized running on the platform is possible due to the computational graph-based representation of the models. Figures 5.2 and 5.3 provide the flow chart for deploying Keras of TensorFlow model on Android or iOS smartphones respectively. Figure 5.4 shows iOS or Android app creation steps from the models that are converted.

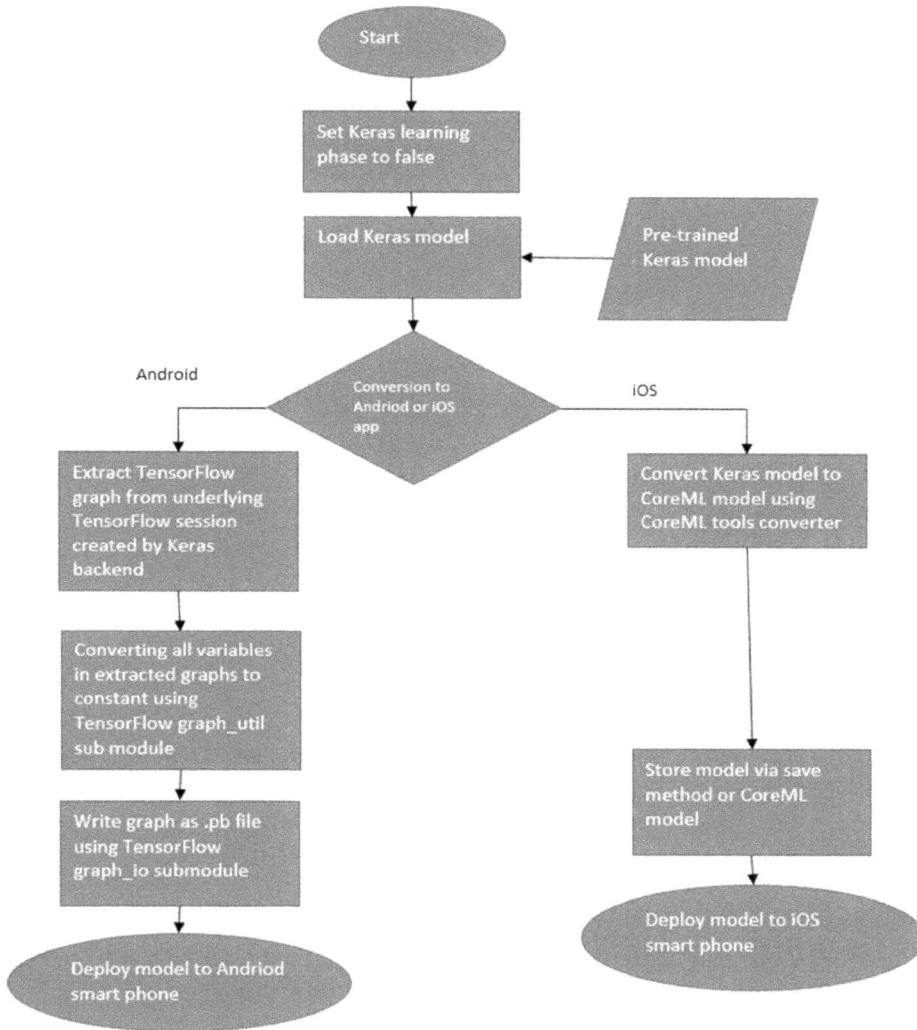

Figure 5.2: Flow chart for the smartphone deployable version of the Keras model in Android and iOS.

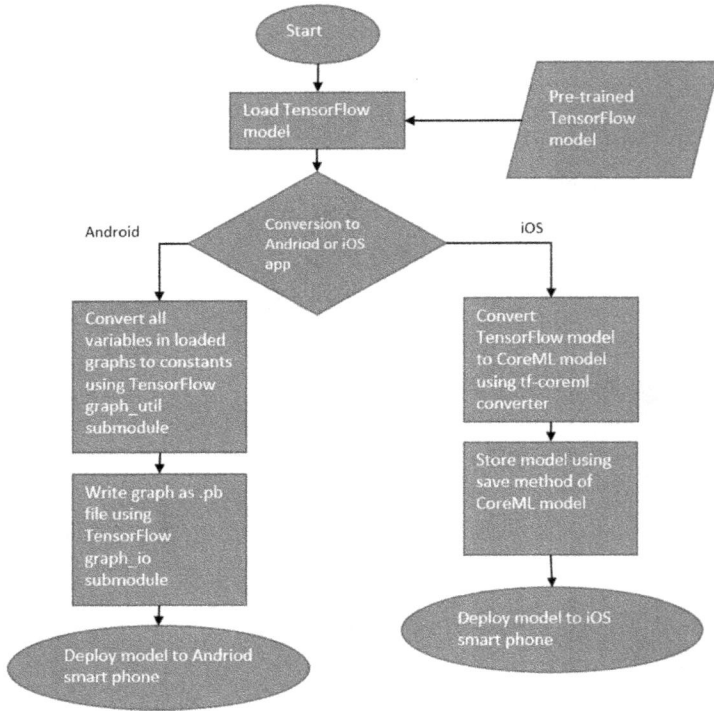

Figure 5.3: Depiction of the Android and iOS deployable model for smartphone converting from TensorFlow model.

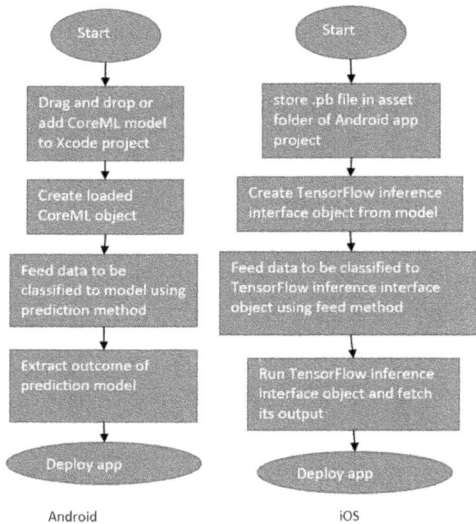

Figure 5.4: Android and iOS app creation from the converted model.

5.5.3 Model generation for iOS

Pretrained model or Keras model needs to be trained first for the CoreML model. Hierarchical data format (HDF) also called .h5 is the format in which Keras is stored. The different backend can be trained for the same model using this format. The numerical array is the storage format for this file format for graph architecture and the graph's tensors' weights. The CoreML model can be retrieved from the Keras model using the Python library's CoreML tools, once it is loaded into Python. Multidimensional array or image is the format provided for input using the converter. For the model preprocessing parameters are defined by the converter using image inputs. The model's ability to be used on raw images and ease of switching without depending on preprocessing is the benefit, especially with preprocessing being of different variations across models.

.mlmodel file format is used for CoreML model conversion. This makes MLModel available as a class for direct use and makes it a plug-and-play format. Neural network building from scratch is not needed as CoreML API can compute for DNN. Instead of implementing the neural net users will focus on deploy and test aspects.

5.5.4 Android model generation

Gradle-based build dependency is provided by TensorFlow mobile (Gradle Build Dependencies for Android Studio) and the trained model can undergo inferencing with just a predefined function. Protocol buffer file format, or.pb, is the format in which the trained model of TensorFlow gets stored. The architecture of the model and trained weights are also stored in a format like HDF. Execution on the Android phone with dependency is accounted for by the predefined functions, with a model created using .pb file from TensorFlow for the interface of mobile inference.

Variables of the model need to be changed to constant as the precursor for Keras model to generate the TensorFlow model. The backpropagation input update needs variable tensors. These tensors are supposed to be constant in the case of inference-based models. TensorFlow has graph_util, which is a utility for the graph; this can be used for the purpose. The TensorFlow session created by Keras needs to be accessed for this graph creation. Keras has a backend module for this purpose. Converter function graph_util constant function can get the session graphs generating graphs with weights of constant. .pb version of this constant graph using a submodule of input/output of the graphs; the Android app imports it with the inference interface of TensorFlow. All required DNN computations are handled by sessions like the TensorFlow session of Python. The session's output model can be extracted with input data or images being fed in the session. Exclusive preprocessing of the images is needed when compared to CoreML. In CoreML only the model's output can be extracted whereas any intermediate layers can also be extracted from TensorFlow for mobile. But for TensorFlow for mobile, it needs manual setup of the model by feeding input, output node names, size of the input and .pb file. But that is not the case with CoreML.

.pbmodel can be converted into .mlmodel with tf-coreML converter with DNN having TensorFlow as the primary framework. The reduced set of computation is generated by this converter tool; in the case of changing the model for unsupported computation, the converter GitHub page has a reference setup.

5.5.5 Software tools for smartphone

The uniform approach is created using iOS and Android jointly to show deep learning inference on the device; 96% of smartphone shares are with Android and iOS (Mobile Operating System Market Share 2018). Organizations of Android and iOS have maintained the developer's tools well and are freely available online.

Android Studio IDE (Android Studio) available for all operating systems that can be used for Android app development. DNN models are run on Android smartphones with Java being the language of choice for Android development. For deployment purposes, Android Application Package (AAP) files can be used with a package of Android apps.

macOS machine with Xcode IDE on iPhone or iOS device is the possible deployment for iOS applications (Xcode). Objective-C or Swift is a programming language used for apps from iOS. Apple developer registration is needed to develop iOS apps on the iPhone.

5.5.6 Processors for smartphone

Sample Android and iOS phones Pixel 2 and iPhone 8 respectively are used for experimentation purposes in this work that is reviewed here.

Google-built Pixel 2 has Qualcomm Snapdragon 835 64-bit octa-core system on a chip running on the Android operating system. 1.9–2.35 GHz of CPU clock speed for core is used. 4GB LPDDR4x RAM is involved as the internal memory of the smartphone. Adreno 540 GPU also is involved. This GPU does not get utilized in the TensorFlow mobile.

In case of iOS iPhone 8 it had hexa-core SoC (system on chip) having a 2.39 GHz CPU clock rate and A11 bionic 64-bit Hexa-core for Apple. 2GB LPDDR4x RAM for iPhone 8 internal memory. The dedicated neural engine provided in the A11 chip helps with efficient modeling for machine learning models just not depending on GPUs. Six thousand billion operations per second is the capability of the neural engine. The latest iOS version 11.4 runs on the iPhone used in this experiment reviewed here.

5.6 Benchmarking metrics for DNN smartphones

5.6.1 Models of DNN

Six popular CNN are applied for smartphone apps of DNN models and the benchmarking framework of the same is discussed. Convolution-based computing is done primarily using the convolution class of DNN. Image processing type feature extraction kernel is created for providing a higher level of abstraction by convolution layers (CONV) in the case of CNN's. The image of a matrix input can be tracked by the convolution layers designed on CNN. The audio processing system, speech recognition, computer vision, and image processing have state-of-the-art models based on CNN. MNIST(MNIST) dataset, an entry point dataset for deep learning with data from ImageNet; large-scale competition for visual recognition (Jia et al. 2014) is used for top CNN representative models. Let us review these networks.

For MNIST dataset LeNet (Le Cun et al. 1989) was used for digit recognition purposes. The network architecture is designed to classify 28×28 grayscale image into two CONV layers of single digits output, followed by fully connected (FC) layers of three numbers. A 2×2 average pooling intermediately helps reduce the dimensions. The sigmoid activation function is used in LeNet.

To bypass weight layers with the use of identity mappings, CNN based Resnet (Residual Network) (He et al. 2015) will have "shortcut" or "skip" connections. Identity mappings and output of the weight layers are then summed up. Identifying mappings helps prevent the shrinking of the backpropagated errors with ResNet's skip connections, which is also nothing but avoiding the gradient from vanishing deep in the network layers. Two 3×3 convolutional layers are part of ResNet weight layers, and 1×1 filters based bottleneck layer is used by ResNet to optimize several parameters involved 1×1, 3×3, and 1×1 layers replace the two layers; the bottleneck layer does this. A 1×1 filter manages a decrease and increase of weights. ResNet-50 model used here has sixteen bottleneck layers after the CONV layer and an FC layer.

Google Net that is Inception model based, has CONV layers with specifications of 1×1 CONV, 3×3 CONV, 1×1 CONV, followed by 5×5, 1×1 CONV, and a 1×1 CONV, wrapped up by 3×3 max pooling. Various combinations of inception modules are devised in InceptionV3 to reduce the number of parameters, reduce the number of weights and in turn computation cost, and promote sparse representation of high dimension.

In the case of limited memory systems SqueezeNet (Szegedy et al. 2015) CNN model is utilized, this has accuracy like AlexNet (Krizhevsky et al. 2012). Squeeze layer and expand layer are stacked in SqueezeNet which is also called as "fire" modules. The 1×1 filter configured in SqeezeNet, a squeeze layer, helps optimize several input channels to the module. Concatenation of the post-activation layer with a mix of 1×1 and 3×3 layers is fed into the module. Baseline accuracy is maintained while reducing the parameters involved due to the sparing utilization of the 1×1 layer.

Maintaining the accuracy size of SqeezeNet compressed with deep compression (Han et al. 2015) results in 510 times reduction compared to AlexNet. Existing deep learning software tools cannot manage with the compressed model. CONV layer is followed by fire modules of eight numbers and then CONV layer at the end. CONV layers in comparison to the FC layer have a higher number of parameters, so FC layers are not used.

MobileNet (Howard et al. 2017) CNN model is used for embedded vision and mobile applications. Depth-wise convolution and pointwise convolution are the two divisions of the CONV layer in the MobileNet module, resulting in memory and computation reduction. Depth-wise separable convolution is this representation. The channel width of one is maintained for the filter in case of depth-wise convolution. Output channels of the depth-wise convolution are expanded by pointwise convolution. MobileNets modify model shrinking hyperparameters, which are width and resolution multipliers. At each layer to reduce the network's channels uniformly width multiplier is used, which helps reduce overall parameters of the layer. Computational cost of the model can be reduced, with a change of input resolution using resolution multiplier.

The residual learning framework of the ResNets is extended by DenseNet (Gradle Build Dependencies for Android Studio) that are densely connected CNN networks. Every layer relates to the feature map dimension that is similar in the case of the DenseNet block. Input is concatenated as compared to ResNet, where it is added. Bottleneck architecture of ResNet is like CONV architecture of DenseNet, to reduce the number of channels of input with a 1 × 1 filter before the 3 × 3 CONV layer gets its feed. The transition or compression layer between the DenseNet block helps reduce the number of feature maps flowing into the next DenseNet block. To maintain the same accuracy fewer parameters are enough for DenseNet in comparison to ResNet. Compared to ResNet50 DenseNet-121 provides 1% less accuracy but operates with three times lesser parameters. In this case, the overall architecture has a layer of CONV, DenseNet blocks which are four in number, three transition layers come in between and following up with a layer of FC.

For Inception and SqueezeNet total number of CONV and depth are dissimilar as parallel CONV layers are inbuilt in their layers. The computational expense of the CNN model is represented by several floating-point operations (FLOPs). The built-in profiler of the TensorFlow helps computing FLOPs of the model. Keras' pretrained models are used on the ILSVRC-2012 validation set to compute the model's accuracy; 50,000 images are part of the dataset here Multiple crops of the images are used in literature usually whereas here single crop image is used to compare the accuracy. Single crop accuracy is relevant for real-time operations. InceptionV3 in comparison to ResNet has a smaller number of parameters but FLOPs are higher due to the CONV layers. Transfer learning can be used to apply the models available here with Keras that are pretrained, and an extension of the model is possible (Yosinski et al. 2014). Figure 5.5 depicts visualization for models' performance.

CNN model parameters

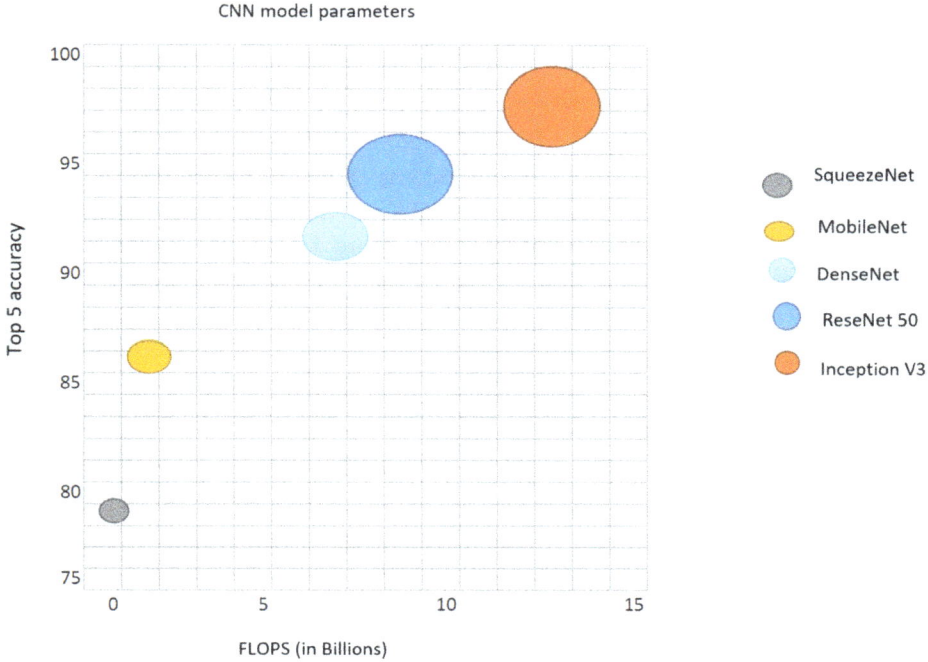

Figure 5.5: Models' performancecan be evaluated with visualization as shown in the figure. The top five accuracies and model sizes with FLOPs (Floating Point Operations) are evaluated in the visualization. Circle size represents the size of the models.

5.6.2 Multi-thread format

Computational expense is seen in the case of DNN. Delay is caused due to the execution of the model on the main thread in the case of real-time smartphone operations. This delay causes a drop in throughput and is due to the delay in the capture of frames. Frames per second (FPS) also get reduced. Multi-threading works when a slower rate of operations compared to the frame rate of the app is acceptable in the case of DNN models applications. To create computational bandwidth, for the main thread for desired FPS, the DNN model can be run on a secondary thread. This is in the context that the latest smartphones operate with multicore. DNN model running on a parallel thread will prevent skipping of the audio frames from the main audio thread resulting in removal of computation expense.

5.6.3 Metrics benchmarking

CNN model validation is done with the ILSVRC-2012 dataset. Thousand objects categories from 50,000 images form the set for validation. Single crop image with top-five accuracy metrics on smartphone and PC platforms is the setup for DNN model validation. Literature experiments use multi-class classifiers with multicrop images showing high accuracies. But it does not work for real-time apps on the smartphone. But it does not work for real-time apps on the smartphone, but works for single real-time crop and is more relevant. MNIST test set–based validation is used in the case of LeNet. MNIST has 10,000 handwritten digits images.

With respect to about CPU/GPU consumption, the operations of CPU/GPU majorly influence battery utilization; this becomes critical for smartphones. Only CPU consumption is measured as smartphones have CPU only. For parallel computations of the CNN model, using GPU, the neural engine of the iPhone 8 is being used. The majority of the computation is handled with the iOS app, so its GPU consumption is measured. CPU/GPU consumptions are not measured for LeNet as it is treated like a non-real-time app.

Looking at real-time throughput, number of consecutive screens displayed works with the measurement of screen time per second. Twenty-four FPS or more are needed for video-based apps where video data with the smooth perception of visuals are needed. Measurement also includes several frames processed per second. There is a direct relation between FLOPs and the number of frames processed; an increase in FLOPs results in a reduction of several frames processed. The model's efficiency is deciphered with this measurement. If the throughput is an important aspect, application with higher number frames processed per second models do well. Accuracy goes up with an increase in FLOPs. So, for better accuracy frames processed per second can be reduced. Frames processed per second and the FPS will remain the same if the model is running at frame speed. Several frames are processed per second differ compared to FPR in the case of multi-threading.

For throughput metrics, time taken per image that is scoped for classification is possible as LeNet is not the real-time one.

5.7 Discussion on results

The developed approach was used for Android and iOS applications with LeNet. The validation set for a smartphone on the iPhone showed 99.43% accuracy, which is like a PC's platform. For Pixel2 and iPhone 8, 5.66 ms (Mili second) and 5.20 ms of processing time were noticed. Figure 5.6 depicts ILSVRC challenges data validation comparison with PC, Android, and smartphone.

To check the ImageNet dataset's accuracy for a challenge with 50,000 images Android and iOS apps were implemented from ILSVRC models. All the apps' performance came out almost with the same accuracy, with about 0.5% difference only.

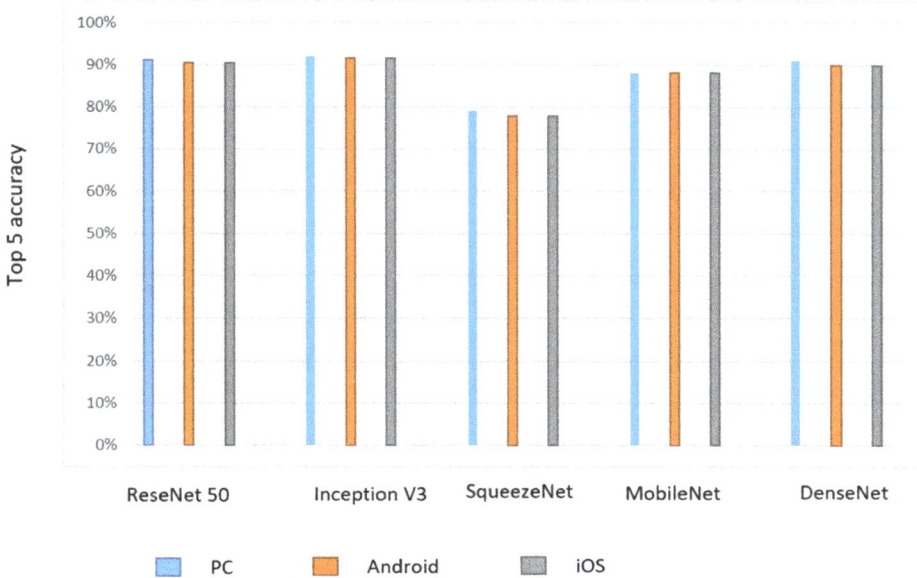

Figure 5.6: ILSVRC challenges data validation comparison with PC, Android, and smartphone. Single crop and top 5 accuracies looked at. Since the floating-point numbers are handled differently in ARM and Intel-based processors, there are small variations and inaccuracies.

Smartphones ran the models in real time to monitor the throughput and consumption of CPU/GPU. Android profiles (Android Profiler for Android Studio.) of Android studio IDE help monitor CPU consumption for Android. XCode IDE's performance analysis tool (Instruments for Xcode IDE) is the instrument used for GPU consumption monitoring for iOS. Instrument performance analysis also helped to monitor FPS for iOS and Android OpenCV camera API (Android – OpenCV library) assisted for Android. For multi-threading operations, frames processed per second were measured for the app periodically.

SqueezeNet that ran on 11 FPS provides the best processing rate for Android. The number of FLOPS used by the app was in proportion to Android apps' CPU consumption. SqueezeNet alone achieved 24 FPS and more in the case of iOS. iOS app also would enjoy the benefit of GPU compared to Android, resulting in improved throughput. The number of FLOPs used by the app was in proportion to GPU consumption. Due to higher processing consumption, a faster battery drain was seen for the DNN app due to the higher number of FLOPs.

FPS was constant at 30 FPS for multi-threading deployment in the case of Android apps but for these apps the number of frames processed was different. While running model at a lower rate and in parallel, for visual perception of nature, 30 FPS were possible due to multi-thread running the GUI. Since GPU was used, iOS version apps operated at higher throughputs for frames processed per second and FPS. In the case of

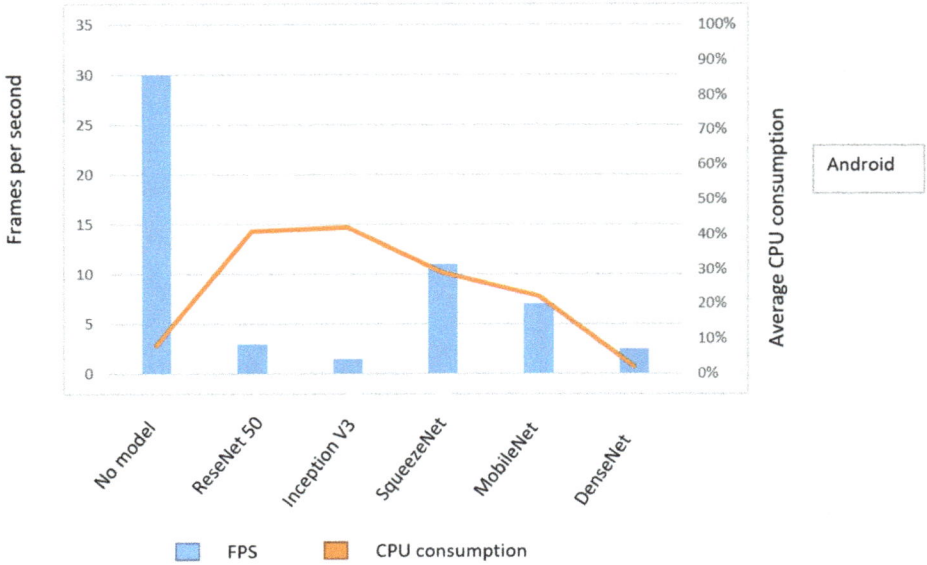

Figure 5.7: The figure depicts the CPU/GPU consumption. The primary y-axis shows for DNN Android average frames per second, in secondary Y-axis average CPU consumption is provided.

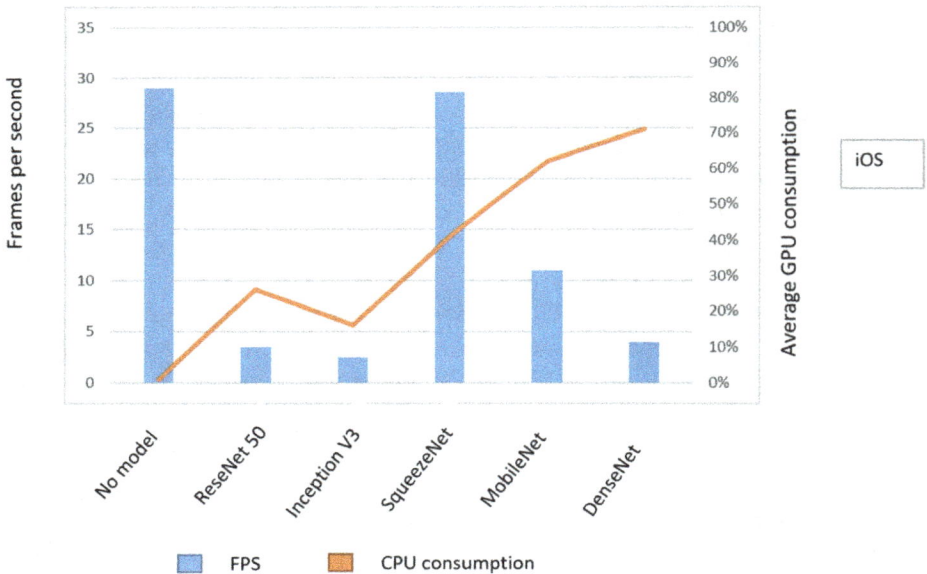

Figure 5.8: The figure depicts the CPU/GPU consumption. The primary y-axis shows for DNN iOS average frames per second; in secondary Y-axis average GPU consumption is provided.

multi-threading, due to concurrency when using multi-threading, there was an increase in CPU and GPU usage. Figure 5.7 depicts the CPU/GPU consumption for DNN Android and Figure 5.8 depicts CPU/GPU consumption for DNN iOS. They also provided the benefit of the video stream.

DNN models implementing apps in comparison with SqueezeNet with iOS smartphones showed high throughput and high energy efficiency provided at thecost of the accuracy when compared to other models. Applications that need higher throughput will be the model to go for. With lower energy efficiency and throughput, for multi-threaded settings, high efficiency was provided by MobileNets. For accuracy being the focus, InceptionV3 with a multi-threaded setting would be the approach of choice. For comparatively higher energy efficiency and throughput ResNet50 and DenseNet 121 are good choices, but accuracy loss of 1–2% has to be sustained.

The study provided a consolidated view of deep learning inference network deployment for Android and iOS smartphones. Exploration also included deep learning software tools that are publicly available to make sure smartphones run deep learning models as apps. Also, exploration involved the app's enablement of real-time operations. Metrics like CPU/GPU consumption, accuracy, and throughput are devised to benchmark the framework. Popular CNN models were used extensively in the application of deep learning where the validation done by these was the benchmark used to validate the process discussed. Benchmarking showed that without accuracy being compromised, models of deep learning were implemented. For achieving real-time throughputs multi-threading leads have been useful. The overall discussion focused on providing smartphones' steps to have a deep learning inference model being deployed as the apps.

Neural network compression methods also can be an extension of the work. Methods used here helps matrix decomposition, which converts weights of the model to integers with lower bits from numbers of floating points, which is quantization; this helps in reducing computation time by reducing the size of the model. Smartphone hardware supporting these methods after their full development and these being available in public provide an opportunity to use the deployment approach discussed here seamlessly.

5.8 Automated software deployment survey

After looking at deployments in the world of artificial intelligence to derive inspiration for software development deployment, especially for extensively used mobile applications, now we will explore automation possibilities for software deployment, which opens the gateway to using machine learning.

Cloud technology is making its mark and automated deployment becomes important for cloud service providers. Cloud services are interesting area of exploration as they provide the option of utilizing resources based on need and release the infrastructure when not in use. It makes it more interested areas. This review aims to build the controller that facilitates effective multi-cloud deployment of applications, which will be an additional facility on top of cloud services provided. Only that these controllers will have to operate within the options of cloud services provider APIs. Exploration will focus on the process involved in the usage of the controllers for the purpose specified above.

Cloud services bank can improve uptime and reduce cost based on load providing scalable server. Up to 59% of Fortune 500 companies experience a weekly downtime of 1.6 h minimum (Vision Solutions 2014). So, for a company with 10,000 employees with $30 per hour salary (Vision Solutions 2014) a downtime loss of $480,000 weekly would be expected not considering the loss of sales and negative impacts on customers. It becomes of utmost importance to have services being installed and run for companies, faster, and with consistency, reducing the cost. This creates the need for deployment and configuration that are automated. The pace of adoption of cloud services is not the same in academia as it is in industry. The reason for this slow pace of academia is due to the lack of enough information and the nonavailability of the resources and time, whereas industries have been investing significantly (Cosmo et al. 2015).

Amazon, one of the cloud service providers, hosts 50,000-server capacity data centers for a zone and has more than a center per zone of availability. Faster and easier deployment becomes an important need in the case of these big centers in service providers or big companies. Specialized deployment methods become important going beyond simple bash scripting which may not feasible.

This review's focus is to look at approaches that help one-click implementation of software and deploying infrastructure into complex environments of the cloud. The overall target is to look at a common framework to reach and configure a variety of environments of the cloud.

The host system facilitates scripting and deployment of N-nodes from this framework to different frameworks of various cloud offerings provided by major cloud providers, this also applies for in-house private cloud provisions. The host will gather the choice of application to be hosted first, followed by the lookout of the need for single or multiple servers. Common operating system options and the quantity of nodes selection will be the next option from the framework. Infrastructure and cloud building and software deployment will happen after the initial selection of basic items that are also key for SSH (Secure Shell) user credentials. From the list of available software, required preselected software are used by the specific operating system that is on clusters of servers, created by users. Figure 5.9 depicts application deployment controller for a case of multi-cloud.

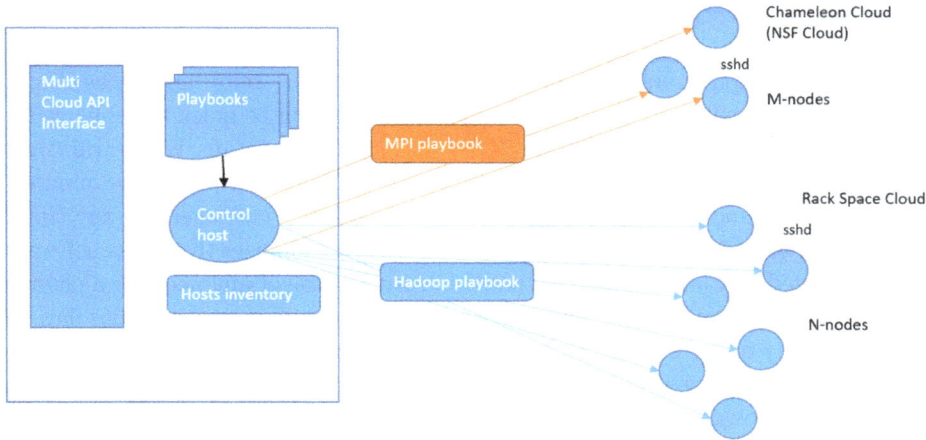

Figure 5.9: Application deployment controller for a case of multi-cloud.

Works earlier have focused on simplifying cloud services for researchers and educational purposes. Blender (Cosmo et al. 2015) is one such program that simplifies the process of cloud infrastructure building and program installation on that. These approaches are interesting vigorous testing and adoption are needed like other commercial zone products for infrastructure deployment. Some of the products include OpenStack heat, or Juju, and for software deployment Chef, Ansible, and SaltStack are available.

"One size fits all" does not work in educational research such as scientific and engineering computation, as the need of researchers varies quite widely and depends on the domain of exploration. So, aspects like computation power, program requirements, storage needs, and system configuration needed by the researchers must be studied by academic researchers. Running predefined scripts, setting up the required infrastructure, necessary software automatic installation, and configuration will become possible after identifying the needs said above. So, to get the custom cloud implemented researchers will find it optimal and effort optimized. Since work discussed here uses commercial products for automated software packages deployment they stand different compared to Blender.

Different domain-based use cases are looked at in this review. Areas like chemistry and data analytics are looked at. Research domains chosen will assist in assessing various deployment models to figure out the speed of deployment, ease of use and effectiveness. Code required to execute, time to execute the clean install, and time to execute a single versus double node with existing installation are the metrics that provide evaluation state.

5.8.1 Images and scripting

Imaging and scripting are the standup server infrastructure setup approaches. Both have their benefits and disadvantages. Imaging a server will be quick and reliable for deploying a preconfigured system once the image has been set up, which will cover potential tough infrastructure of service and authentication. In just a few minutes, rightscale.com kind of companies can bring up images for customers on any cloud provider services. Limitations of these setup are outdated images that make software parts outdated and missing security patches. Also, in case of any need for a change from the customer, it calls for the image replacement. Imaging sees its major flaw as the inflexibility of the images. New images need to be saved for any change needed. Large space is taken by the set of images that are only slightly different from each other. Further slowness will be seen for patches, customizations, and software updates.

In the case of complex tasks, though scripting is not instantaneous like imaging it resolved a lot of issues. Installation of all needed software, updates, and architecture-related activities are all made flexible due to this approach. To make required files work well, which is not easy when using imaging server setup, distributing custom files can be made dynamic with scripting on to each of the N-nodes. Distribution of customized hostnames, software license files, and IP addresses are some examples.

Lesser space needed will make scripting an advantageous task. In the case of scripting, only one version of the operating system and software is enough as the scripts are the only kilobyte sized, so no duplicates are involved. Figure 5.10 depicts Space requirements of images and scripting. Figure 5.11 depicts Agentless node management, which is standalone management.

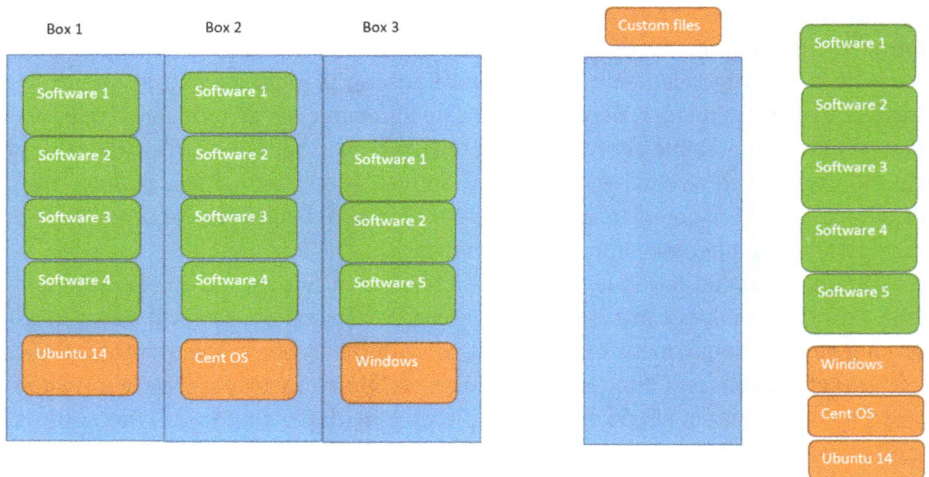

Figure 5.10: Space requirements of images and scripting are depicted. Imaging can become intensive for resources, with optimized resources managed by scripting.

Lower space requirement, flexibility, and adaptation make scripting far superior compared to imaging. The human-readable ability of the Ansible, Salt, and Chef scripting language makes them easier for automation. Pass/Fail reporting simplification and error testing is incorporated by the smart scripting processes. With the complexity of the tasks in parallel different images can be accomplished with the new age scripting. Instead of bash, scripting tools have libraries for a specific action in a frequent way.

5.8.2 Infrastructure delivery and application decoupling

The central controller is assumed for resource provisioning for combining the application control and policy arbitration in many of the previous work (Parekh 2001, Rad et al. 2015, Soundararajan et al. 2006, Urgaonkar et al. 2005). To fine-tune the cloud server controller parameters like email, classical control theory was explored by Parekh et al. (Parekh 2001). For the database server's dynamic provisioning, Soundararajan et al. (2006) proposed control policies. To model multitier application Urgoankar et al. (2005) proposed queueing theory, which also looked at applications' needs of resources. The Infiniband network was used for resource partitioning of the network by Paul et al. (Rad et al. 2015).

REST-full API services are provided to end users by platforms of cloud infrastructure. Knowledge of application performance, impact of its allocation, and placement choices of service quality for applications are all not available this makes APIs of cloud service providers out of state services. API services help separate service providers from the need to understand customers' applications and for customers, it provides a covering against knowing the physical resources involved.

Authors of this work propose to decouple application topology and configuration in application delivery model, from that of the deployment of infrastructure, and make it an external service as defined from service providers of the cloud. Cloud providers need to get rid of the playbooks of the application that needs to go into hand of customers for customization is the intent of the model. As the application needs to keep changing over time to avoid complexity in the cloud architecture, application control policies and the playbooks must be segregated.

Using the cloud providers' infrastructure APIs, the end users application is separated to provide customization possibilities for playbooks and policies to the customers and this is separated from the service providers using multi-cloud–application deployment controller (ADC). In the case of multi-cloud deployments, the innovation of control policies, and customization of playbooks for their applications, application developers will get potential from the principled layers offered.

Cloud interface translation plugin runs on multi-clouds that provide separation, which is the independence of playbooks functions from one another provided by the controller for application deployment. Resource reservation contracts from multi-cloud

ADC have terms expressed in property list exchange with negotiation. These terms for reservation provide a timeline for active resources for cloud servers.

Speed, minimal impact, consistency, reliability, and ease of learning are the main goals of these management tools for configuration. Software deployment to all required nodes and configuration explanation must be known to the users. Users also need to have information on any new installations and the failure points. Faster deployment is needed if a large enterprise package installation is needed; else if node-wise installation is done then it may run to many weeks. For the server's normal operations, resource need of the management tool should not increase after deployment of the software.

5.8.3 Software delivery tools

Ansible is an agentless architecture compared to that of SaltStack and Chef which depends on the architecture of agent based model. There are no agents at the node which are controlled in case of agentless setup.

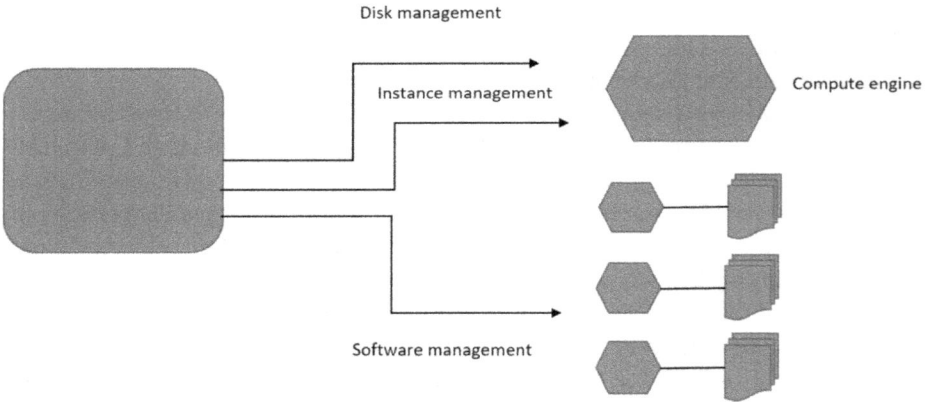

Figure 5.11: Agentless node management, which is standalone management, as shown in figure (Google Cloud Platform 2015).

This kind of setup helps as nodes do not have to check back if the process is not working, which avoids background processing and network traffic add-ons. SSH keys not needed, besides SSH, no other port opening is needed; root access is also not needed in case if they are needed, things can be configured with sudo, these aspects make Ansible different than SaltStack and Chef. There is no need for additional monitoring software (Ansible Inc 2014), which makes it a very relevant option in a case where security is critical with better stability and performance. Execution of the code over SSH from inventory files with required hosts is done by Ansible. SSH can slow down the process through other software and reasonable speed like ZeroMQ communication in Salt (SaltStack –

Github 2015). YAML file descriptive language makes it easy to learn with the execution of the code. Companies like EdX, Twitter, Evernote, and care.com have Ansible for deployment. Figure 5.12 depicts Chef's master/slave management.

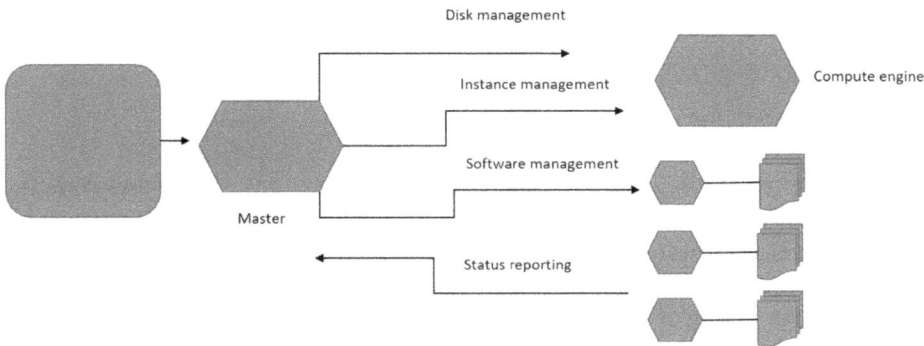

Figure 5.12: Depiction of Chef's master/slave management.

The chef will have a main master node or server as part of a master/slave configuration that is used for deployments to the client with recipes uploaded. A cookbook that provides the information to deploy and configure field-tested systems is a major use of Chef. Chefs create cookbooks at supermarket.chef.io. The administrator can view and search node activity to assign roles to nodes and cookbooks, these are Chef's website interface capabilities.

The cookbook can be imported or can be created one to use the client in Chef. Ruby programming skills will be handy as the cookbooks are written in that language and official documentation for the advanced functions also would get complex in some cases. Easy to use an administrator can use WebUI after importing into the server, it can either proceed with the command line, or by assigning the node's recipe, and executing the same. Using the certificates, communication and authentication are done with SSF by Chef. To manage ecosystems companies like Bloomberg, Ancestry.com, Etsy, and Bonobos use Chef (Chef Software, Inc 2015).

SaltStack case master node joining request is provided by clients; the client can be controlled once the master accepts. Resilience, speed optimality, and scalability are key advantages of the SaltStack. All these are possible with multitiers of the masters with the capacity to distribute the load and reduce redundancy and asynchronous file usage possibility on the servers. Masters can be asked questions from the minion, to get a complete understanding. For example a real-time lookup of the database can be done. This looked-up information will help minion configuration completion (SaltStack, Reactor System).

Out of various steps involved in SaltStack, assignation of the parts into servers by defining a state tree is important one. The current state is compared by minions with the state tree after the download of the state definition. Two are matched up by minions making necessary changes and reporting back to masters. Minion on the check from state tree about Apache installation can check and, if the installation is not done, then proceed with the installation and setup as defined by the tree. Node-specific fine graining from its coarse grain form for the tasks on the node is Salt's capability.

5.8.4 Summarization

Management tools discussed have the capability of handling simple and complex tasks. Agentless feature of the Ansible is an exception; otherwise, all these tools have comparative features. Ansible leads the configuration of the host machine for this purpose. System availability with IP address to host an SSH connection is all Ansible needs (Wong 2015). Agent installation on the host machine is needed for both Chef and SaltStack (Wong 2015).

SaltStack and Chef outperformed Ansible when looked at execution. Some of the limitations of the work reviewed here are to avoid bias due to the code; the optimization of code was not done but kept at as simple as possible. Also, as more additional nodes are added scale-out of code linearly may happen, as with limited bandwidth connections, nodes can have parallelized connections with these tools. Parallelization is not tested beyond two nodes. The installation of the software was also limited for the packages tested for deployment. To get an understanding of the true potential of the management tool, comprehensive test is needed for hundreds of nodes that are targeted for deploying of specific packages, this also should include configuration of software and start of the services.

Based on the use case of the company the products can be used. SaltStack will be a good choice if code simplicity is critical for the system. SaltStack uses half of the code used by Chef and four times less code than Ansible. Chef leads for speed with half of the time needed compared to SaltStack and Ansible comparison one-fifth of the speed. If the speed and size are not the concern, then Ansible makes an interesting choice with no agent scenario. Each host's time needed to install and configure agents is not accounted in time comparison; anyways this is not a significant part of the calculation. Ansible also provides the option to do a dry run to check if code works before the execution and this saves a lot of time in case of complex tasks particularly (Abhishek et al. 2019, Benson et al. 2016).

5.9 Source code modeling with deep transfer learning

Deployment is a prominent area of software development that needs attention. There is a need for advanced research in the area of one-shot learning and transfer learning, to facilitate some of the key challenges involved. This section focusses on understanding the deep transfer learning for addressing the problem involved, resulting in the need for machine learning models to evolve with these changes in software deployment scenarios, which cost time and money. As frequent training of the models for these rapid changes in the ecosystem becomes a need and adds complexity, one-shot learning and deep transfer learning may have answers.

Source code modeling and analysis have seen deep learning as a potential area. Deep learning goes well with excess data available and application will be problem-specific. One of the challenges is the need for training from the start even for the related areas. Transfer learning explored in this section helps the situation improve the performance significantly for source code models running with deep learning models. Unlike earlier methods transfer learning facilitates taking up learning in one area to another area. Recurrent neural network (RNN) models and gated recurrent units (GRU) are the base for transfer learning in source code modeling. Feature extraction is possible with pretrained RNN and GRU models. For various other downstream tasks, attention layers get these extracted features. Pretrained models learning are used by attention layers and fine-tuning the learning as needed for the tasks under operation. An event without a training model from scratch the performance is better than the state-of-the-art methods.

In the study source code suggestion and syntax, error fix is used as a reference, but the usage of the concepts in deployment context would be the takeaway that we will look for. Also, for deployment and debugging these features are of prominent help. Language models–based deep learning models have seen their usage in many of the software development tasks, (Akshay et al. 2017, Dhvaniet al 2018, Martin et al. 2015, Miltiadis et al. 2016, Ranca et al. 2017, Rahul et al. 2018, Srinivasan et al. 2016, Uri et al. 2019, Veseliner al 2014, Yasir et al. 2019).

Some of the works here on source code suggestion help to predict the next part of the source code. Syntax error detection and correction is another area of work. Source code files are corrected for error based on the feature of source code syntax that is learned. Source code summarization is another area of work that facilitated a better understanding of the code. Meaningful naming of the source code methods is another area. Source code generation using natural language is another area of work done.

As highlighted all these solutions have the challenge of learning from scratch based on the type of problem. Also, the need for a large amount of data for deep learning models makes it further challenging. Time taken on the deep learning models takes days together adding on to the complex situation. So, transfer learning will help gather the learning from the task and use it in similar downstream tasks.

The transfer learning approach also improves the deep learning models significantly, specifically for source code models. The source code–based language model is the primary area of exploration for deep learning models with the transfer learning approach. Source code language model's learned knowledge needs to be used in different problem contexts. Two variants of RNN, like GRU and normal RNN, are explored for transfer learning. These variants of learning are combined into the attention layer to be used in downstream-related tasks. Pretrained models use as their base the captured learning and then refine the learned features according to the downstream tasks, helping avoid the need for retraining.

5.9.1 Approach

The pretrained transfer learning and attention learner model that plays the role of combining the learned knowledge are the key components. The attention layer is featured at paying attention to those features that matter for the task under consideration. Figure 5.13 depicts Architecture of the proposed approach.

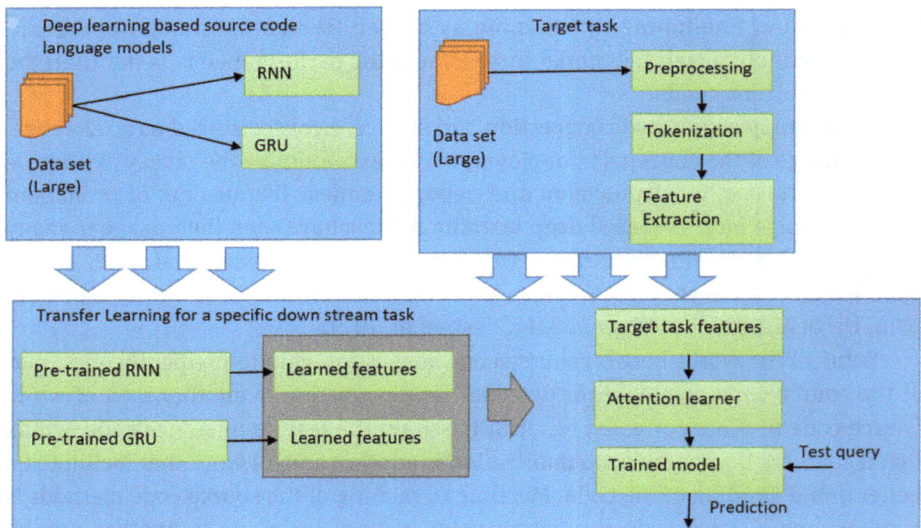

Figure 5.13: Architecture of the proposed approach is shown in the figure.

5.9.2 Transfer learning

Machine learning (Lixin et al. 2009), metric learning (Hu et al. 2015), and dimensionality reduction (Sinno et al. 2008) are the prominent areas for transfer learning. Image and text classification is another area of application.

In transfer learning knowledge gathered in one area of the domain is utilized for another domain. In the case of transfer learning, a pretrained model is required.

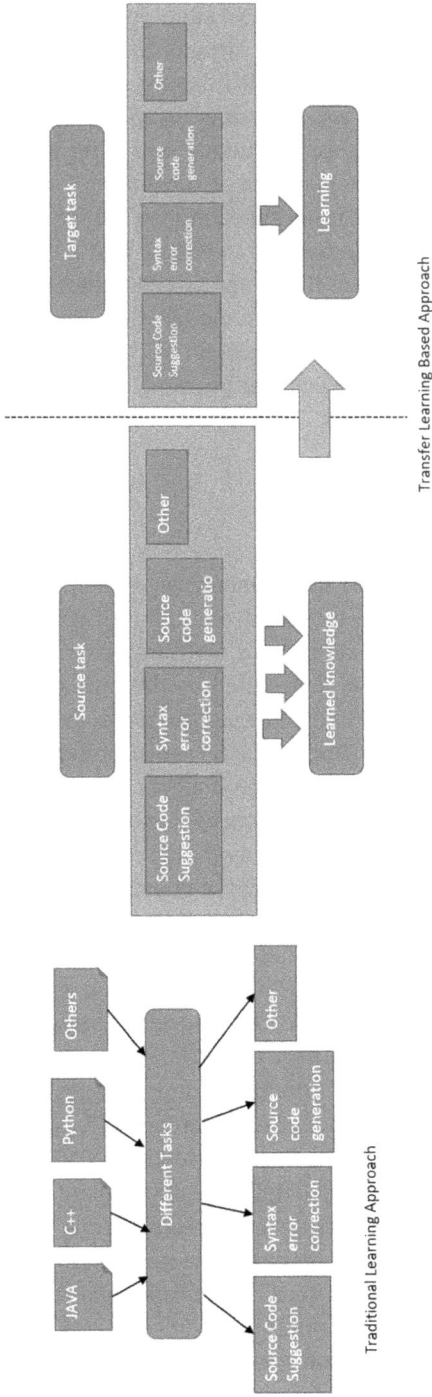

Figure 5.14: Transfer learning and conventional learning comparison for source code modeling.

Since the code strictly follows the rules set in language, its most suitable for statistical modeling. Here in this work, recurrent neural network variants like RNN and GRU are utilized for transfer learning purposes. RNN and GRU have demonstrated recent success and hence they are chosen for this task. Pretrained models are used as input. Datasets are preprocessed, to remove the less frequently occurring items and create a global vocabulary. Like word2vec dense mapping of vocabulary to continuous feature vector is done. Three hundred hidden units are configured with Adam optimizer used for model training. Early stopping is employed to ensure that the run is maintained until convergence. An important aspect of this approach is training at one time. Transfer learning can be done with these pretrained models that are publicly available. Preprocessing, tokenization, and feature extraction are the key tasks to be conducted (Yasir et al. 2019). Figure 5.14 depicts Transfer learning and conventional learning comparison for source code modeling.

5.10 Conclusion

As part of the discussion on the history of software deployment, we explored deployments in key fields (Dearle 2007). Missing common understanding of what is assembly, component, and a package or metadata associated with the system has caused setbacks to the field's progress and it is also contributed by the technology difference across operating systems, languages, and so on. The complexity of deployment is due to the danger associated with complex architecture, meta-architecture, language, tools, and other aspects. Virtualization provides the hope to address the complexity by putting everything into virtual images. Web service–based technologies offer language and machine-independent approaches in case of typed service interfaces. Ambiguity in the deployment lifecycle is removed by the standard naming convention provided by virtualization.

(Abhishek et al. 2019) Further extension of the guidelines provided in the review of "Guidelines and Benchmarks for Deployment of deep learning models on smartphones" can explore Open Neural Network Exchange (ONNX) to provide a computational data flow graph that is unified for the case of DNN. ONNX is Facebook and Microsoft's community project. Interchangeable usage of the ONNX supporting model would be possible with that sort of extension of the work. Caffe2, Pytorch, CNTK frameworks with TensorFlow, and converters for CoreML are the ones developed by ONNX.

(Benson et al. 2016) As discussed in the "Automated software deployment survey" section, automation of the deployment of tools is critical irrespective of the kind of tools used by the companies in the context of cloud computing system world, with the capability to bring the system up or down based on need. Significant benefits are seen in speed, reproduction, and mass deployment through the choice of the tool, which depends on the comfort of the company's users and policies.

Chapter 6
Enabling intelligent IDEs with probabilistic models

6.1 Introduction

Incorporating intelligence into software development has been the principal focus in recent times. In this work (Lin et al. 2017), the focus is on the intelligent development environment (IntelliDE) and software knowledge graph. IntelliDE system's objective is to leverage the capability of big data, gather them, analyze them, and facilitate smarter support for software development processes. Work here produces an architecture of this system and discusses the further research challenges. As part of software knowledge representation with a management framework, software knowledge graph contributes to the IntelliDE system (Kaiser et al. 1998).

The general idea of intelligent software development is to bring smart support for the software development process. Rapid progress in artificial intelligence (AI) and big data has brought a lot of hope for the advancement of software development processes to make them smarter. There is funding in big data mining, and analysis to understand better the hidden patterns that can help software development technologies grow smarter. Intelligent recommendations and intelligent questions and answering are the two prominent areas. The background of intelligent recommendations would be to guide software engineers based on their operating context. The proposal will be about the resources they need to employ and learned from the patterns associated with software big data. In the case of the question answering idea, answers are derived for the users, from the software resources that are available at the system's disposal. This aspect is more critical as the system needs better insights at the semantic level of the information hidden in big data and its associated knowledge sources (Robillard et al. 2010).

IntelliDE looks at three areas: aggregation of data, acquisition of knowledge, and smart assistance. So, the work corresponds to complete architecture with the ecosystem needed for software development and services for smart development. The software knowledge graph plays its role as an aggregator of the knowledge from big data that can be used by the smart development environment, also this connects the knowledge components, facilitating easier use of the knowledge.

6.2 IntelliDE

The modern trend is about open-source data on the web with software data of web-scale. Engineering practices that are widely spread within the enterprises also contribute large-scale data. These data's complexity is in its rich scale, deep knowledge, growth rate, and its diverse distribution. Growth in information extraction techniques

https://doi.org/10.1515/9783110703399-006

and machine learning approaches has facilitated better knowledge gathering. These boost software knowledge possibilities in software development and make it a good exploration area for researchers.

Generally, the smart capabilities of software development are focused on specific areas of software development like defects locating (Ye et al. 2014), prediction (Shepperd et al. 2014), and code completion (Raychev et al. 2014). These are also looked at as different tasks with different sources of data being explored, and approaches are worked out accordingly. Diverse tooling (Trautsch et al. 2016) is the name given for this approach. With this being the case, it is not possible to use the knowledge gathered for one focus area in other areas, including the applied resources. So, the work here intends to provide an integrated environment to optimize learning.

6.3 Aggregation of data and acquiring the knowledge

With the scale of data on the web, it becomes imperative to have the vast capability in the system as well to support such systems, which is the aim of IntelliDE. Challenges faced in this system are around collecting, combining, and updating of the data. Concerning data collection challenges IntelliDE will have to understand the data on the web perfectly so that the right approach can be taken to download them and organize. A variety of patterns will be hidden in software data that represent the complex nature of the software. This presents the need for understanding the various software sources and their relations well, so that data of multiple sources can be leveraged better. Another challenge is the fast-growing nature of the software data, keeping it up to date. This makes it important to be clear about the data update strategy.

Complex knowledge associated with the data is critical to be gathered from these data and make it available for intelligent systems consumption. This makes it important to focus on devising a knowledge base that would be used by the intelligent system. Acquisition of knowledge, extracting it, and appropriately representing it are the challenge areas. As part of knowledge extraction, the system needs to make use of algorithms related to the extraction. Algorithms would be natural language processing (NLP), machine learning, data mining, and so on. To handle software knowledge representation and management, knowledge graphs are explored here.

6.4 General review of data aggregation (approaches)

Clustering techniques are good ones to be explored in data aggregation as part of data mining. Since the objective here is to separate the data into groups based on their similarity, this closely relates to data aggregation challenge in an IntelliDE. It is important to note the objective of clustering the data as part of data aggregation.

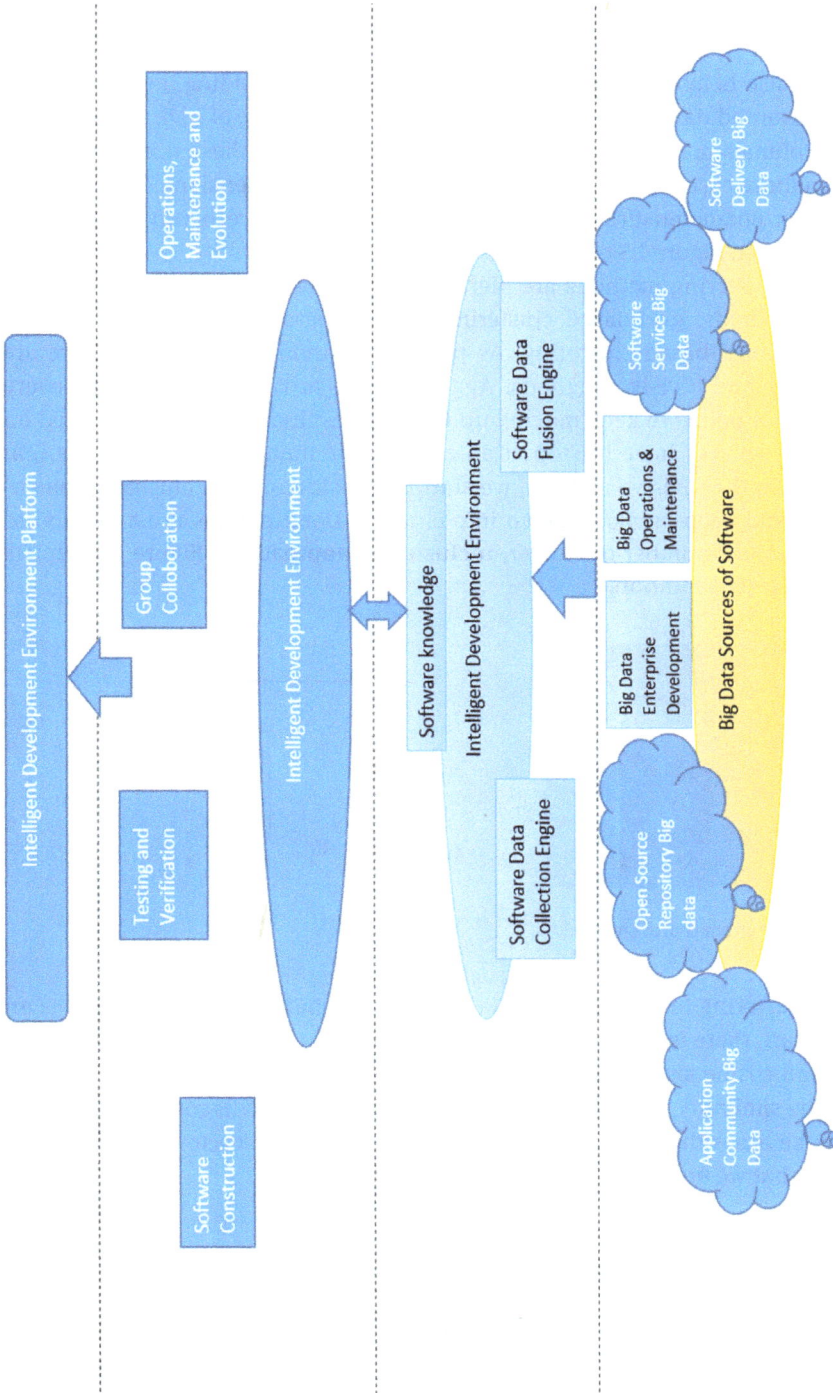

Figure 6.1: IntelliDE architecture.

We target the interpretability of the data, where we will explain the reasons for a specific pattern in data. This relates to the need for making sense out of data for IntelliDE. It also helps identify the important part of data, avoiding the need to process the unwanted data. Data also has characteristics of being of high dimensional, but their volumes may below. Or any other combination of these dimensions and volume. In those cases, clustering approaches will assist. Dimensionality here refers to features or characteristics of the data. As part of pattern recognition, this is an important aspect. Figure 1.5 depicts IntelliDE architecture.

Various clustering methods are hierarchical clustering methods, partitioning clustering methods, grid-based clustering, density-based clustering, constraint-based, and model-based clustering. As part of hierarchical clustering, there are agglomerative and divisive methods. Agglomerative involves starting from every individual data point we keep moving up until we reach one big cluster. Based on the use case we may have to stop at an appropriate number of the clusters that make sense. In the divisive method, we start by considering one cluster of the entire dataset and keep breaking down into clusters. Dendrograms are a method to compute the right number of cluster in clustering approaches. Figure 6.2 depicts Dendrogram representation.

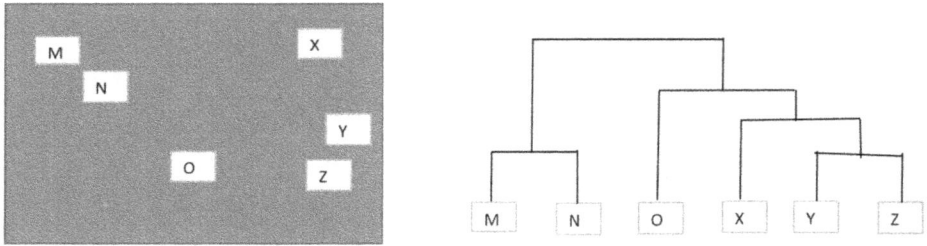

Figure 6.2: Dendrogram representation.

As part of clustering techniques for data aggregation similarity of the clusters is an important part, there is a need to devise means for similarity assessment. Euclidean distance and cosine similarity are the most commonly used methods that assist in deciding the similarity. Euclidean distance is based on the Pythagoras theorem. Cosine distance bases the similarity on the angle formed by the vectors that represent the entities that we are comparing.

$$d(a, b) = d(b, a) = \sqrt{(a_1 - b_1)^2 + (a_2 - b_2)^2 + \ldots + (a_n - b_n)^2} = \sqrt{\sum_{1=1}^{n} (a_i - b_i)^2} \quad (6.1)$$

$$\mathrm{Cos}(\theta) = \frac{a \cdot b}{||a|| \, ||b||} = \frac{\sum_{i}^{n} a_i b_i}{\sqrt{\sum_{i=1}^{n} (a_i^2)} \sqrt{\sum_{i=1}^{n} (b_i^2)}} \quad (6.2)$$

Minkowski distance and Mahalanobis distance are some of the other distances employed based on the scenario that we are handling.

6.5 Smart assistance

As part of clustering techniques for data aggregation similarity of the clusters is an important part, there is a need to devise means for similarity assessment. It will become a complex task to understand various aspects of the life cycle and devise the output accordingly. This makes it important for the system to devise the services based on the intelligence needed for the different tasks involved. In this work (Lin et al. 2017), the focus of knowledge gathering is looked at from the construction phase of software, testing phase, a collaboration of the users, operations, and maintenance.

For the construction phase, analysis of domain, program, and refactoring of the software is looked at. Services provided for these areas include visualization of the knowledge from software projects, retrieval of semantics from software, code synthesis, code completion, and software refactoring recommendations. It is also applied for for testing and verification, test case generation, code inspection, static analysis on warnings, software debugging, etc. For collaboration among the users with data around, skill prospect of the users, their ability to collaborate are to be gathered. These will be based on the historical knowledge of the collaboration and management of tasks. As part of the operations management area, a complete software lifecycle needs to be targeted. Areas separating development and maintenance process of software, changing requirements management, and other areas need to be focused upon. Smart services that must be devised in this area include anomaly analysis, smart load testing, and design-related ones.

6.6 Knowledge graph for software

Here exploration is done on how the knowledge graph concept is adopted for a knowledge representation framework in IntelliDE. Knowledge graphs are composed of nodes and edges used to represent domain knowledge (Liu et al. 2016). In the case of software, knowledge graphs, software systems, projects, and domains are represented. Nodes here represent the knowledge entities like class and objects of issues; edges that are directed represent the relation between the entities like traceability link and method invocation. The relation among nodes and edges characterizes patterns associated with the nodes and edges. For instance, the method entity will be represented with the return type, parameter list, description, etc.

Primitive and derivative are the two types of knowledge graphs. Data structure parsing is used to gather the primitive knowledge from the software. Code entities like methods and classes, abstract syntax tree (ABT) representation for source code files, and the association between them like method innovation would comprise knowledge graph components. Similarly, email archives can be parsed to derive mail components like mail users and emails and also they can be related to aspects like email sending, receiving, and replying.

Derivative knowledge is the knowledge gathered from software and represented in knowledge graphs. For instance, NLP of the sentences in software documents can yield business concepts of software. Also, the traceability can be established between document and code entities with this method. And it further helps to establish traceability of code entities and business concepts. Based on these definitions knowledge graphs can be derived from the big data of the software. Parsing of data and extraction of knowledge are the process involved. Data parsing is the process of deriving the knowledge from the software and adding on to the graphs, followed by knowledge extraction that adds on derivative knowledge.

Also, it is important to note that the software knowledge graphs need to have characteristics of continuous improvement. As the data extraction and knowledge extraction modules of these systems evolve better and various software knowledge graphs can evolve. Knowledge ecosystem can evolve for software in this way and IntelliDE will provide this framework for intelligent support for the software management process. Figure 6.3 depicts Knowledge graph representation.

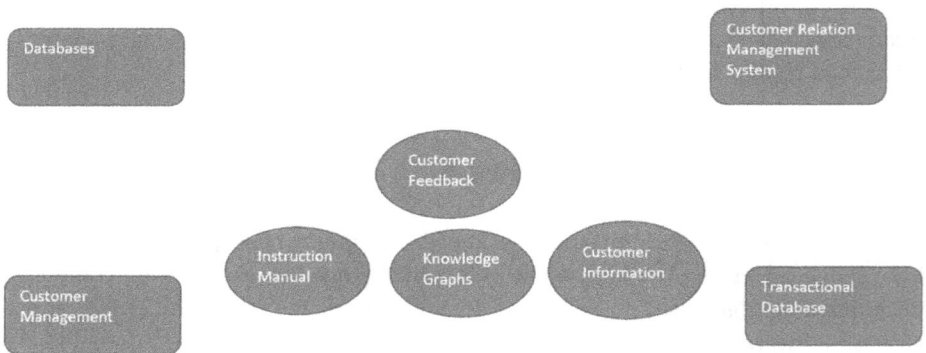

Figure 6.3: Knowledge graph representation.

6.7 Review of natural language processing (NLP) in software (approaches)

Software engineering and NLP are related fields of computer science. NLP involves the processing of the natural language with computational resources. In software engineering NLP can be looked up as a language that connects the human and machine. Software language is a way of human representing their knowledge to make it machine-understandable. Compared to natural language, software language is more systematic and pattern based.

Wherever the text information is associated with the software development lifecycle, NLP comes into the picture. Requirement documents use case detailing, and test case descriptions are some areas of natural language. Statistical NLP methods are explored in gathering the code corpora of complex size across GitHub and similar sources for facilitating the advancements in speech recognition kind of areas. There are explorations on speech-based programming; this can give an interesting dimension to the concept of an IntelliDE. This provides an opportunity for all walks of the software development community and removes the constraint of learning software programming. Domain knowledge experts can also get actively involved in the technical process.

Documentation being a key aspect of software engineering, NLP brings in capability. Bots coming into the scene for software development would be a bonus that will make use of NLP capabilities. The bot is envisioned in the context of the IntelliDE, which can manage, process, and provide knowledge based on the data it processes. If the bots' capability can interface the development capabilities with the customer interfacing, that would take the software development to the next level.

6.8 IntelliDE with the software knowledge graph

In the work of Lin et al. (2017), knowledge graph construction mechanism built-in IntelliDE is explored.

If the software engineer wants to create the knowledge graph with IntelliDE, related software data must be uploaded in the system from the front end of the web. These data sources can be of complex and varied sources. Mailing lists, version control system, source code files, Microsoft Office docs, issue tracking system inputs, and others are the data sources that can be plugged in. This is a knowledge construction platform with a data parsing framework. Since the platform provides a plugin capability for the data parsing, other plugins also can be experimented with.

Data parsing provides the primitive knowledge and later the platform extracts the derivative knowledge by making use of various algorithms. The knowledge extraction framework is another key component of the system. Even here there is an interface for knowledge extraction to introduce various knowledge extraction algorithm integration. Plugins provided in the platform for knowledge extraction are link recovery between issue and commit, traceability link extraction between document and code (Dagenais et al. 2012), estimation of lexical similarity between documents (Ye et al. 2016), latent topic modeling (Hua et al. 2014) for source code, API usage extraction (Wang et al. 2011), and so on. Figure 6.4 depicts Software knowledge graph construction in IntelliDE.

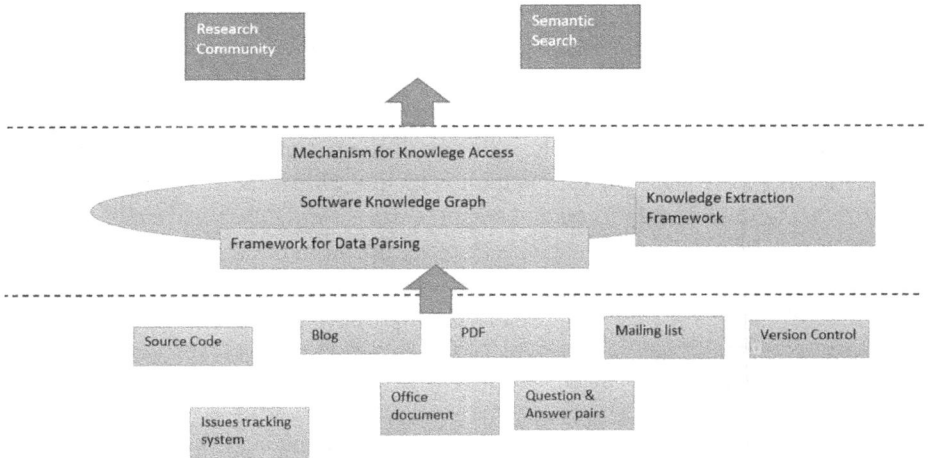

Figure 6.4: Software knowledge graph construction in IntelliDE.

The software knowledge graph is stored with Neo4j, a popular graph database. The software knowledge graph query is done with the Cypher, a query language for graph-based data. Java API of the Neo4j is also employed in addition to Cypher. Figure 6.5 depicts Cypher syntax reading representation.

6.9 Latent topic modeling of source code for IDE (approaches)

6.9.1 About latent topic modeling

As we discussed the complexity of the data that is associated in the software domain, which is supposed to be the input for the intelligent IDE. It is important to convert this data into information. Also, there is a need to derive a valuable approach for data mining. It is similar to exploring the possibilities of deriving the pattern from the data to make smarter decisions by integrated development environment (IDE). Topic modeling provides one such opportunity.

The topic modeling area focuses on gathering key topics from the corpus of documents. Latent Dirichlet allocation (LDA) is one such topic modeling approach. This approach focusses on mapping the text in the document to a related topic. This derives a

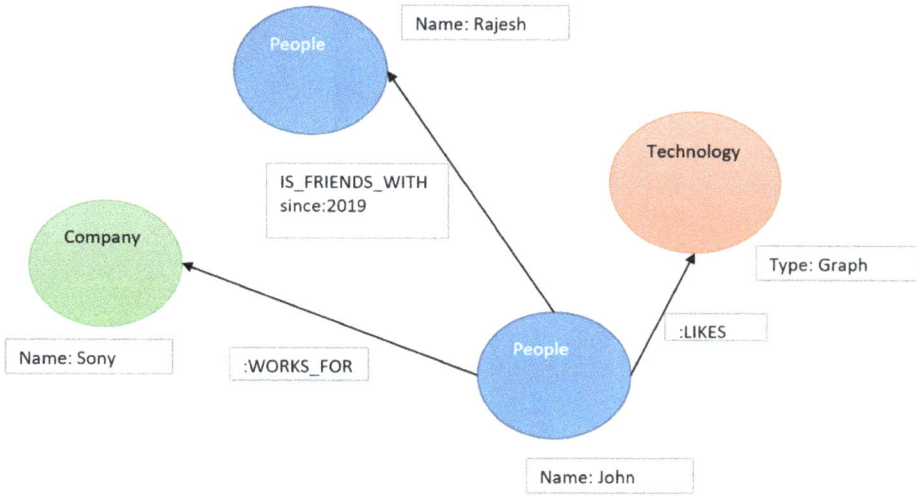

Figure 6.5: Cypher syntax reading representation.

model of a topic for documents and words for a specific topic model. Internally, the input text processing includes tokenization of the content and removal of words that are occurring in low frequency; here the decision of what is low is influenced by domain knowledge. Removal of stop words that are the words that do not provide any value for assessment of the text, such as "is" and "are", has to be done. Lemmatization of the words is to make them all into present tense and other transformation. Stemming of the words to bring them down to base word is enough for the algorithm to make sense. Like words stabilized, stability can be transformed into its base form of "stabil."

Topic modeling stands apart when compared to rule-based approaches, where regular expressions are used concerning the dictionary that helps look for a keyword in its search. It also relates to unsupervised learning of machine learning as it figures out the topics without having any labels, which again is a great fit in the context of data that is supposed to be fed into IDEs. Topics are generated based on the pattern recognized in the text corpus. Topic modeling will be good support to manage the challenges associated with data acquisition in building smart IDEs. Feature extraction, also being an important need of the smart IDE, will fit well. Feature here represents the key information from the data used as a reference in either the task of recommendations to the software engineer or for question-answering. Figure 6.6 depicts Topic modelling representation.

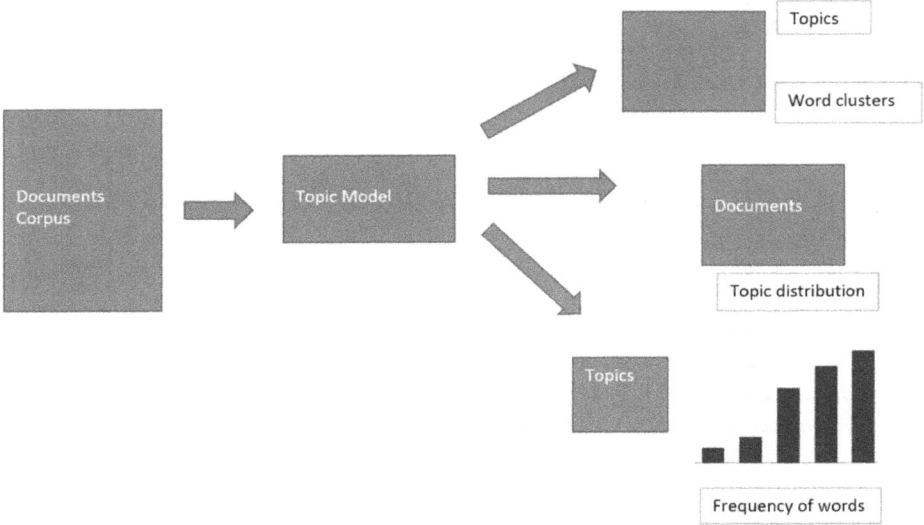

Figure 6.6: Topic modelling representation.

6.9.2 Latent semantic analysis approach

Terms frequency and inverse document frequency (TF-IDF) are other methods to derive the topics from text. One of the key assumptions made by this algorithm is that documents are produced from topics of a variety of mixtures. Based on the probabilities of the word the topic is represented by a distribution. Matrix factorization is involved in the process. The key concept here is to represent the document corpus in the vector space. This is the space in which the data entity that is converted as a vector is put together for inferencing purposes.

Say if we need "n" number of text documents, and "o" number of unique words. And we decide "l" number of topics to be extracted as specified by users. We need to represent this case in a document-term matrix of size $n \times 0$, with TF-IDF scores given. Figure 6.7 depicts Document term matrix representation.

		Terms				
		o_1	o_2	o_3	....	o_o
	n_1				....	
Documents	n_2				....	
	.....				....	
	n_n				....	

Figure 6.7: Document term matrix representation.

Singular value decomposition (SVD) method is employed to derive "*l*" number of topics from the above representation. SVD breaks the matrix into three components: "*V*," "*T*," and "*W^T*."

$$B = VTW^T \tag{6.3}$$

Document term matrix V_l in the figure of SVD representation has the rows that are representing the documents that are input. Length "*l*" represents these vectors which are the number of topics that we chose to derive. W_k is the term to topic representation matrix that takes in every terms or word in the document for representation. SVD generates vector representation for documents and words in them. The cosine similarity method is used to generate similar words and documents concerning the topics generated.

Compared to plain vector space method LSA is more effective. This is a linear model; in the case of complex nonlinear relations, this model may suffer. In the case of the intelligent IDEs data pipeline, we will have nonlinear data situations. SVD also poses the problem of computation intensity that needs to be accounted for in the system. Figure 6.8 depicts SVD representation.

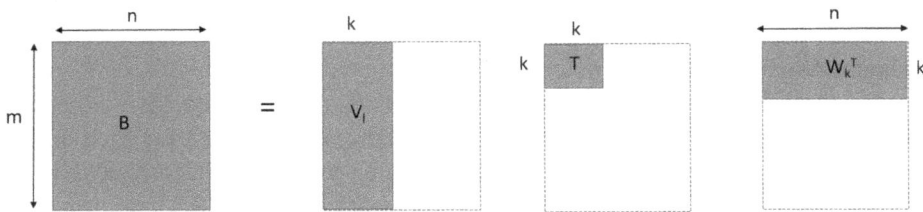

Figure 6.8: SVD representation.

6.10 Applications and case study of IntelliDE

As discussed IntelliDE aims at intelligent question answering and recommendation for the software development community. Further focus is on demonstrating the capability of the text semantic search which would be handy for the developers to extract knowledge from the knowledge graph.

This is the scenario of matching free text into the documents. As the text is the common mode of information representation in the software development space, the abstractions are represented in issue report, documentation, commits details, and user communications. But in the case of the keyword matching method of text search, there is a drawback of not understanding the free text's semantic meaning. This calls for knowing the graph-based text semantic search method. As the developer enters the search query, the list with rank for the items is listed from software knowledge graphs, based on its semantic association. Also, for each of the text items, there is a subgraph derived from the knowledge graph.

This points at the ability a well-structured knowledge of software source code brings when mining of the semantic relevance is the target. The approach taken is inclusive of code location, entity matching, and subgraph extraction.

In the case of code location relevant code entities are derived based on the code-free text provided in the search based on software knowledge graph; feature location technique (Marcus et al. 2004) is used for the purpose. Entity matching is done based on code entities' structural similarities using graph embedding techniques like TransR (Lin et al. 2015). Re-ranking of the search results of the text is facilitated by structural similarity. Query and related text items generated in search are validated with a subgraph generated for visualization. The minimum spanning tree algorithm (Schuhmacher et al. 2014) is employed for subgraph extraction.

In this work, the demonstration of the IntelliDE capability of knowledge graph extraction is done by applying the same to open-source software called Apache Lucene for the specific case of demonstration of the semantic search for text in software. Data sourced here include 800,000 lines of codes, emails of 244,000, 1.2 GB of version control logs, 500 web pages, 3,800 question-answer pairs, 5,200 issues reports, and so on. Extraction of relationships and software knowledge entities are done here. A software semantic text search engine is provisioned. A query goes in as input, with semantic search results as the outcome.

There will be a ranked list of items of text, which are related to query based on their semantics. The software knowledge subgraph is represented for each of the text items that are provided in the ranked list by a semantic search engine. For instance, if there is an item in the ranked list from stack overflow Q&A pair, user may choose this to visualize the corresponding subgraph, which represents nodes of query and corresponding keywords. Similarly, for the sections of the document, there is a node that represents Q&A pair and related keywords. This representation highlights how keywords are different in terms of the semantics as part of the software knowledge graphs. AtomicReader being superclass of Index-Reader class, there is a semantic relation between both as a Q&A pair, with IndexReader being in query and AtomicReader in the keyword. The knowledge graph can bridge the query and Q&A pair even in the context of common words existing between both, in the process to benefit the user.

This demonstrates the possibility of semantic understanding in the free text of software in the knowledge graph. To improve the software development process quality and productivity with the assistance of the intelligent integrated development environment, the software text search semantic engine is a great start.

6.11 Summary of work (Lin et al. 2017)

Key concepts of the IntelliDE and software knowledge graph were explored in this work. IntelliDE will be leading way to make the IDE smarter. Challenges that need focus and research are in the areas of aggregation of data, data acquisition, and smart assistance for development. Software knowledge graph is used as a management framework that provides the base for knowledge acquisition in IntelliDE. Apart from the primitive knowledge from source code and issue recording system, derivative knowledge from API usage cases is also integrated. To demonstrate the capability of IntelliDE on the semantic understanding of content during inferencing software text semantic search–based engine that uses the knowledge graph is demonstrated. This also provides hope for making question-answering in software development smarter. Figure 6.9 depicts Software text semantic search representation.

6.12 Review of probabilistic programming in the context of IDE

Probabilistic programming is a methodology in statistical modeling. Here the concepts and constructs of programming are utilized to apply for statistical model building. Based on the probabilities of the events that characterized the past, future events can be predicted. It also provides the ability to comprehend the complex interactions between the elements of the system. Since we are talking about complex data sources for IDE to make it smart, this helps. It works with the logic of simulating the data elements for future forecast and feeding back the inference for further refinement. With the pipeline of tasks and information sources associated around the IDE, this concept makes sense, so that we pick the pattern across time and feed it back to the prediction model and improvement model. Probabilistic program with the backdrop of AI has been said to change the landscape of understanding and deploying a probabilistic based system. Probabilistic programming language is a domain where probabilistic models inferencing capability is combined with programming language representational capability. Since the outcome gets modeled as probability distribution, it provides a good opportunity to visualize and change the unknowns associated with the outcomes.

   This is an appropriate fit for the IDE to use the knowledge hidden in a variety of data sources. It also helps in accounting for the growing complexities of the software engineering domain. Since software programming is more structured in its construct, a probabilistic approach is best suited. Also, the time-bound growth of complexities makes it a good fit.

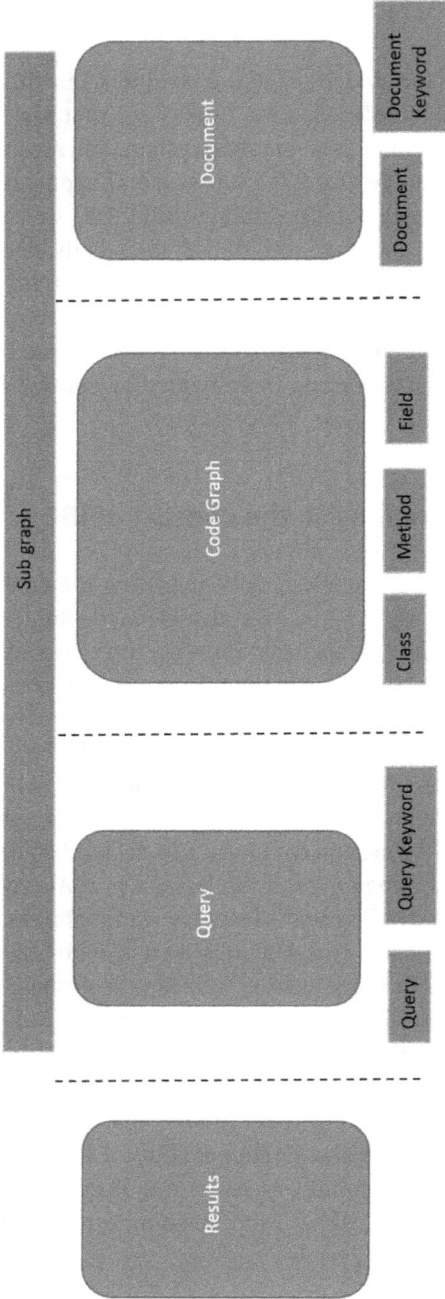

Figure 6.9: Software text semantic search representation.

6.13 Knowledge graph possibilities in software development

With advancements in the analysis of graphs, software evolution can be better understood using this thought process. Authors in their work in Pamela et al. (2012) have attempted to leverage those progress to use it to facilitate better software development and maintenance activities. We explore the study done on this topic and assess how we would leverage these thought processes in our primary objective of building an intelligent IDE. With software reliability as an important aspect, it is difficult to ensure the same even by taking care of the evolving, collaborative software development process. Graph mining techniques have been at the forefront in these cases for many areas that involve complex processes. The author's work herein (Pamela et al. 2012) focuses on a graph-based system that can represent software characteristics to account for its evolution and make sure ease of development process, extending this format for software development areas such as predicting the releases with possible defects, bug severity identification, refactoring aspect, and other areas.

The work's focus is around constructing the graph for the product at source code and module level and process level that covers the collaboration of developers. There is also an exploration of the metrics associated with the graphs that provide properties of the graph. Authors also have extended their study on open-source software over a decade of the period. Further, the work explores building predictors from the graph to handle bug severity, maintenance of key software components, and assessing the releases for their possible issues. The authors also strongly recommend that progress in these areas will facilitate research in many areas of software engineering.

Software development and maintenance is a cost factor and has a direct financial impact. A large part of the software cost is associated with the software (Koskinen 2015). Even with such high investment, the impact of bugs can be at a high level. That is exactly the reason for the area of model building to account for software evolution becomes significant.

In the recent past, the complex system's graph analysis has found its way back to the light, as network science or graph topology mining. This is due to the impact created by graph mining techniques in complex systems. For example, there has been considerable progress in areas of the topology of web and networks (Faloutsos et al. 1999), network traffic classification (Iliofotou et al. 2010), and so on. All these systems have a common thread of graph-based topology representation. The graph model provides opportunities to explore various pattern recognition frameworks, including the detection of outliers and abnormal behaviors. The author explores to draw a parallel in software engineering tasks and make use of them. We will have a close look at all the capabilities that will help our intelligent IDEs. Scoping down authors will look at focused components that will help address the issues and prioritize them for debugging, test, and refactor. Also, there is an extended lookout for predicting defects in upcoming releases.

Though the ideal solution will not be targeted by the graph-based models' capability to make progress in these areas are the focus. Authors have explicitly highlighted that their work would help look at multiple other areas like ours which is on intelligent IDEs.

The effort is put toward building graph-based systems that can represent an evolving software. Results seen are said to demonstrate graph metrics capability to highlight structural changes and help estimate bugs and facilitate predicting defective releases and help bring in priority to the debugging effort.

6.14 Focus areas of the work

Software process properties can be extracted by software graphs topological analysis. The software context is explored in two different levels of granularities. For software as a product, they are assessed at function-level and module-level interactions. For assessing the interaction of developers in bug fixing and new feature creation model level is a process. Developers' collaboration for handling bugs fixes within themselves is one of the construction methods and other one being commit logs analysis to assess the developers' collaboration beyond bug fixing.

With graph metrics having high potential to capture events of the software lifecycle, graph models of these processes are studied throughout since one to two decades. There is a highlight of fundamental similarities and structural differences. Among the various programs, some of the graph metrics show a high level of differences whereas few of the other metrics show consistent behavior of evolution for all the programs under examination. Observations have been made about the graph metrics revealing major changes in case of mid-level releases than their versions that are rounded numbers. There are instances where a major change in code structure is revealed with edit distances due to major bug fixes done in the software, these changes have not come out, while effective lines of code (eLOC) kind of metrics was observed.

These metrics of graphs brings out bug severity, maintenance effort, and defect-prone characteristics of releases. The work's prime contribution is toward suggestions, predictions, and inferencing that is possible in graph metrics explored in this work. The authors also propose the work done here will assist further research on identifying the key functions and modules and estimating maintenance effort and other key aspects by facilitating the construction of the evolution model with predictors.

Bug severity prediction with NodeRank metrics that is inline with PageRank is proposed. Modules that take the high cost of maintenance can be figured out with the modularity ratio metric. Releases with failure possibility can be explored with analysis of the developer's collaboration graph's edit distance. The value and deficiency of these metrics are analyzed in this study.

6.15 Possible extension of Smart IDE (approaches)

Topological data analysis (TDA) being a powerful tool to bring out the structure and shape in the complex data for answering the questions in the domain. Aspects of the domain are coded in the topological signatures There are topological representations that take care of connectedness of the entities with the management of continuum of the data, this can assist in managing the traceability of the various tasks of developers in software development. This can bridge the collaboration among developers for various activities, through the IDEs and make them smarter. TDA also can manage complex dimensionalities of the data but maintain the completeness of their structure within. It also helps assess the utilization of the features for various tasks, this gives opportunities to understand what is most important for the system. This characteristic will help us understand the components of the IDEs better and keep refining their components.

The primary driving factor for the TDA is the proximity of the data points. Euclidean distances are means for providing mathematical sense of the distance of the data points.

TDA approach provides a good intuitive explanation of the way the complex data structures are conceptualized by the machines, particularly in problems like that of image recognition, breakdown of the images into their underlying patterns, and then using them for prediction. Mapper concept in TDA build the interactive representation of the data, this interprets the complex model easier. This specific capability will make a great contribution to building smart systems like in the case of IDE. Also, the learning abilities of the models built will create the knowledge repositories for the smart IDE, which will assist the users of the system in this case developers as well as the researchers that are building the system. Since we also highlighted the fact of unbalanced data in cases of security issues management and similar such areas, this approach provides a better means of managing these adversarial situations or cases that are like outlier detection. Figure 6.10 depicts Representation of topological data analysis.

Since the recommendation is one of the key aspects that we have been talking about for smart IDEs, the concept of recommender systems with graph analysis can be a good fit. Unlike with the normal graph analysis system, in this context, we derive the pattern from the graphs specifically from the recommendations point of view. Here recommendation consumers will be software development communities.

Recommender systems in the context of graphs will look like users and item coordination map. In a smart IDE context, they will be software developers and software development artifacts such as source code, bug reports, code commits, and other information. The similarity matrix and metrics space representation together contribute to the strong approach here. Figure 6.11 depicts Basic recommender system.

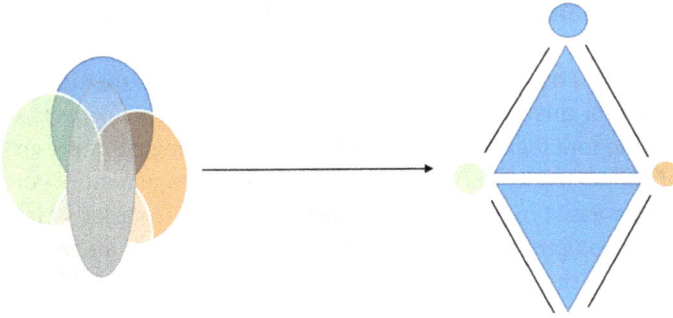

Figure 6.10: Representation of topological data analysis.

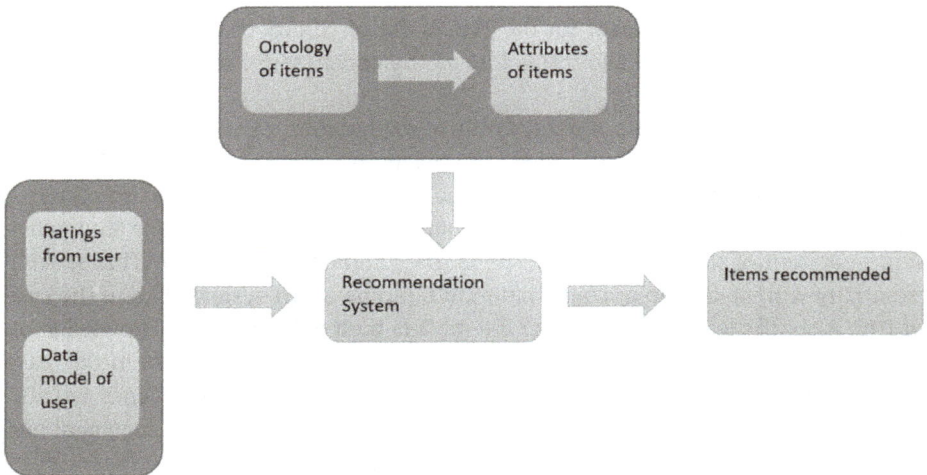

Figure 6.11: Basic recommender system.

When it comes to comparing the graph of the system components it is a complex task. This task is important to create graph outputs like the similarity matrix. General method is to compute the pairwise distance to assess the similarity. Common mechanisms are spectral distances and node affinities-based distances. There is a need to study a variety of these metrics that help to understand the graph characteristics. This study will help device the right metrics for graph understanding and will play a prominent role in smart IDEs. This idea of distance metrics and other metrics built for understanding data characteristics will align well with the thought of understanding if the data entities have originated from the same probability distributions.

We have also looked at the influence of time on software evolution, so exploration of a dynamic network of graph accounting for the time with time series aspects of the graph will be added advantage. So, idea is to explore the major changes that happen in time steps based on distance metrics. This is a positive hope toward accounting for the dynamic evolution of the software. It is also important to use the graphs with time accounting based on the specific need for tasks in hand. There may be software development tasks that may look at large scale features in graph or there may be scenarios where local structures of the graph for local aspects may be important. For instance, traceability of the requirements across phases of the life cycle may be complex, whereas defect localization may be related to local aspects.

Since the complex systems are in play, it becomes important to integrate the acquired knowledge. Ensembling methods need to be looked at to put together multiple sets of graphs derived for the data space. The primary difference in the knowledge graphs' thought process comes from their characteristics of being distributive rather than generative characteristics of other models. The general focus in traditional modeling is over simple focused models that are deterministic. Whereas to gain the advantage of these knowledge graph–based complex structure representation we may have to explore methods that can integrate the knowledge graphs.

6.16 Constructing the graphs

Continuing the review of work done by authors in Pamela et al. (2012), to understand the graphs' building, it involves graph construction phase, collection of required data, and computation of graph metrics. Figure 6.12 depicts Overview of the system proposed in work (Pamela et al. 2012).

Bug tracking system and source code repository are the primary sources of information for this system. Various aspects of the source code such as developers' interaction around code, commit logs, and patches done are the information gathered. The bug tracking system provides similar interactive data from a bug perspective. Graph edges combine this information and represent it as a system. As the group of graphs is generated in the process, the system will assist to put together the pattern that demonstrates how the evolution process works in software and leads to a prominent expectation of building influencing factors for the software metrics like bugs severity and maintenance.

With focused attention on graphs that may come from source code, function level call graph and module-level graphs can be used to demonstrate collaboration of source code-based graphs. Global variable usage and function call extraction is done with static code analysis tool CodeViz. The call of the function in the graph covers the basic relationship between the entities involved in the call transfer. Here the entities transferring the function calls will be represented as nodes and the actual call itself will be edge connecting them. Function call graphs play a significant role in the case of large programs where extraction of architecture is involved.

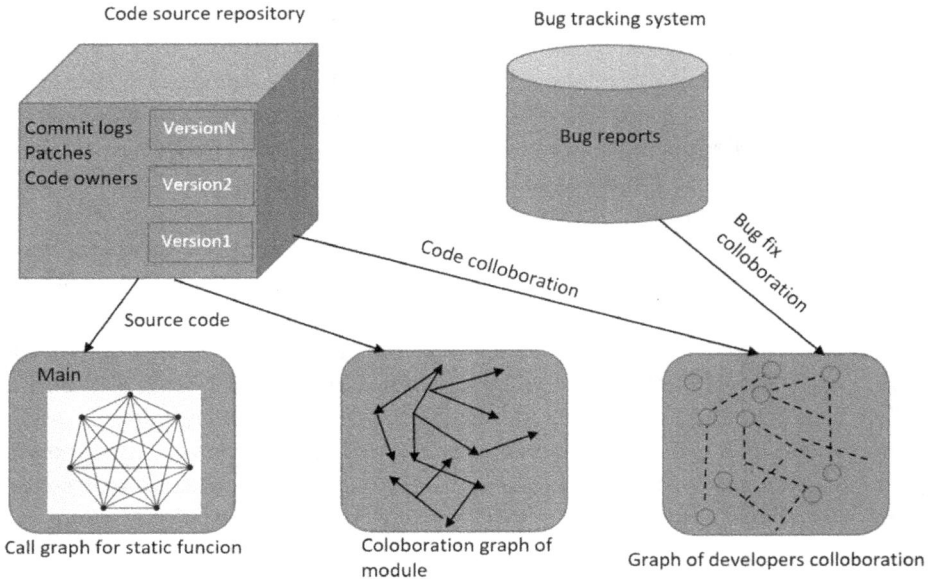

Figure 6.12: Overview of the system proposed in work (Pamela et al. 2012).

In the case of module collaboration graphs which are bigger than function call graphs which are finer, Here the function calls between functions across modules are looked at and representation would be of the involved modules that will be represented as nodes and directed connection between them as edges. Similarly, if the function of one module calls variable of other modules, similar connections are established. Modular collaboration graph representation is a good way to understand the interactions of the software components.

As it was highlighted developer-related factors are very much important in the overall context. Developer collaboration graphs are built with a specific interest in understanding the influence on the evolution of the software. Since collaboration revolves around commit management and bug management, developer's collaboration also can be explored around these. Bug tossing graphs proposed by the same authors in their work (Bhattacharya et al. 2010) are utilized to build developer collaboration graphs. A simple case is the directed edge used to connect two developers *A* and *B* when developer *A* could not resolve the bug and was moved to *B*.

Apart from the collaboration that happens over the bug, effort-based collaboration attempts to put together all other activities, where commit logs provide the base. Undirected edges are used to represent the work done by developers on the same file.

The study has revolved around popular open-source applications that are written mainly in C language and a combination of C and C++. While the applications are chosen careful consideration is made for choosing the applications that have a significant size in terms of modules or lines of code, extended-release history, a large user base that can contribute bug information and patches and large development users' community as well. These play a significant role in the case of deriving statistical, behavioral, and evolutionary observations. Collation of the data would include information on applications, the period of their active state, count of releases, languages used, size during releases, and so on. The entire approach provides a statistical means of representing the evolution of software applications. Source code official releases of the application are gathered with the version control systems. The bug tracking system from the application was referred to gather bug information for the application.

6.17 Collaboration graph and its contribution to IDE (approaches)

With the concept of the collaboration-based graphs, IDEs can get smarter with the possibilities of combining the multiple languages that are involved across the applications. This builds the software's capabilities to gather knowledge from across applications, domains, and languages involved. So, the recommendations become smarter and deeper. Since the intelligent IDEs are mostly built around open source, it would be meaningful to develop the knowledge and extend it to other software as well, which will create greater value. As the developer's community is involved in the digital transformation initiatives, knowledge graph collaboration can play a significant role in providing the right direction. As the industry is thinking about language-agnostic development platforms, there is a good pathway with collaboration-based graphs for functions and modules.

The people factor is said to have a major influence on the software development processes like any other domain. Developers' collaboration graphs open the avenue to explore the possibilities of quantifying the people factors that can help build smarter prediction systems. This system can extend to understand the skill sets of the people and their influence on the software's evolution. There is also a focus needed on metrics that also account for the roles of the developers. This can also be strengthened with the recommendations provided by IDEs which accounts for the developers' contributions as well.

There is a difference in social network kind of representation and slightly advanced in terms of information like in the case of hierarchy-based collaboration. Finer details are brought out when we get into granularity of the developer collaboration graphs, like restricting to only basic metrics like experience and bug closure rate, may mislead whereas granular details of the role, if the person is triage manager of the bug or the actual fixer of the patch, can add more value. Aspects like these bring in a merit-based system of tracking and make it more meaningful. Also, there is a need for expanding these roles to more granular to gather more in-depth information. This builds a comprehensive system that can also cover the recommendations on the weaker aspects of development. On the lines of collaboration built for bug-based and code-based attributes, even the developer collaboration can revolve around these aspects. It aligns well with the open-source system for development, which comprises source code and bug tracking mechanism.

Source code-based profiling will focus around contribution evolving from source code like lines of code added and changed. Attributes such as time spent on file, ownership on the file, and distribution across software will provide good insights. Another important aspect is to balance the amount of data with that of the complexity of the model involved. To strike a proper balance between a sophisticated model and interpretable one, Interpretability is crucial to make these systems smarter.

6.18 Metrics in graph-based analysis

Continuing the exploration of the work in Pamela et al. (2012), while the metrics are devised for these graphs it is important to account for the directed and undirected graphs; else there may be possibilities of these getting overlapped. In the case of a graph represented as G(V, E), where V is the set of nodes and E the edges, the following equation gives the average degree for the directed graph.

$$\bar{k} = \frac{2|E|}{|V|} \tag{6.4}$$

In the case of undirected graphs, clustering coefficients are explored. The clustering coefficient of the node represents the probability of two neighbors connecting or local connection possibility. The definition is the ratio of edges between neighbors over the maximum edges possible. The clustering coefficient can be represented by the equation given.

$$C(u) = \frac{2|\{e_{jk}\}|}{k_u(k_u - 1)} \tag{6.5}$$

C(u) is the clustering coefficient for node u or probability of neighbors of u being connected

$|\{e_{jk}\}|$ is number of edges between u neighbors
k_u is number of u's neighbors

All the nodes must be accounted for to assess the average clustering coefficient to obtain graph level metrics. Like PageRank (Brin et al. 1998), NodeRank is devised here, which provides weight for each of the nodes in the graph to arrive at the relative importance of the graphs concerning nodes in the software representation. In terms of the Markov chain, this represents stationary distribution. The recursive calculation method is used for PageRank computation. NodeRank can be represented as in the equation.

$$NR(u) = \sum\nolimits_{v \in IN_u} \frac{NR(v)}{OutDegree(v)} \tag{6.6}$$

u is the node
$NR(u)$ is the node rank
IN_u contains all the node v with outgoing edge u

First, equal NodeRank values are assigned, and with every iteration, the new NodeRank will be assigned. At the point of convergence of the NodeRank, the iterations conclude. Normalizing the values to get a sum of all values as zero is an important part of the convergence process. At the intuition level, for a vertex u, if the NodeRank is higher its importance in the program will be higher as there will be a higher dependency on this from other functions and modules, basically this would be most called one. On similar lines for developers, the NR(D) will be higher for a developer who has a better reputation in, a collaboration graph for developer D.

The longest and shortest path of any two vertices in the graph denotes undirected graph diameter. The correlation coefficient for the degree of the node on two ends of the edge is assortativity. Here the node preferences for its connection based on its similarity or differences are quantified. The node will tend to connect to another node of similar nature to the same degree, then its positive assortativity. Assortativity is an important measure in network science, for instance, those nodes that are highly connected will tend to have connections with other nodes of higher connection order (McPherson et al. 2001). Unusually in the case of biological networks nodes with a high degree tend to connect with nodes with a low degree, which shows disassortativity.

Metrics that were seen so far focus on one program, but if the focus is to assess the evolution of program concerning structure change over the time we need to account for changes in nodes and edges about two graphs, specifically between back to back releases. The edit distance graph for a directed graph can be represented as follows:

$$ED(G_1, G_2) = |V_1| + |V_2| - 2^*|V_1 \cap V_2| + |E_1| + |E_2| - 2^*|E_1 \cap E_2| \tag{6.7}$$

ED is edit distance between graph G_1 and G_2

Graph G_1 is represented as $G_1(V_1, E_1)$ with V_1 vertex and E_1 edge
Graph G_2 is represented as $G_2(V_2, E_2)$ with V_2 vertex and E_2 edge

Representation here is the G_1 and G_2 depicting the software structure model for release 1 and 2, a higher value of $ED(G_1, G_2)$ indicates large scale change in structure across the release cycles.

The modularity ratio for the directed graph is the next metric to explore. One of the software engineering practices' observations is the ease of software testing, understanding, and evolution contributed by its characteristics of software design being highly cohesive and exhibiting low coupling. Modularity is then defined as the ratio between cohesiveness value and coupling value.

$$\text{Modularity}(M) = \frac{\text{Cohesion}(M)}{\text{Coupling}(M)} \tag{6.8}$$

Cohesion(M) represents a total number of variables or calls with the module M. Coupling(M) will represents calls and variable counts across modules for module M.

Now for an area of defects and effort of software engineering practices, defect density provides information about the quality of the applications. To make sure the quality representation is accurate, information on bug database changelogs is referred. This helps bug to be related to the version of a release. A bug can be attributed to the previous release once we explore release tags, such as commit messages and bugs, which denotes the version in which bug was reported. For purpose of effort tracking associated with development and maintenance counting of several lines of codes that were added or changed must be taken with a count of commits, for every file in the release, as in the work of Fern´andez-Ramil et al. (2009) and Nagappan et al. (2005). Log files will help with gathering this information. Authors have done work on defect density and effort in their work (Bhattacharya et al. 2011).

6.19 Importance of ranking metrics in the context of smart IDE (approaches)

Ranking is an important aspect in information retrieval, data mining and NLP. With question-answering, recommendations being top areas of the exploration in our use case of smart IDEs, ranking plays an important role. The ranking also finds its place in machine translation, which can be explored to derive value for our thought process of making IDEs language-agnostic in software engineering. To take the analogy of information retrieval, input documents or data are assessed with the input query to rank the input data and present it sorted based on this ranking. Conventionally the conditional probability distribution is derived based on the query and documents presented the possibility of the output, denoted by $P(O|q,d)$. This is a conditional probability of the output "O" for the input of query "q" and input data documented, "d." Key aspect of model gets down to words occurring in the query and documents. Constructing these ranking models utilizing the benefits of the machine learning capabilities is a good area for exploration and strengthening this aspect. This is particularly important when we are looking at the complexity of the data space involved. Also, the model constructed out of this information will have to be flexible enough to learn frequently accounting for the dynamics of the software development processes. Focused research is happening in the area of web search with tons of data accumulating in logs. With the thought process of building a robust smart IDE, which can plug into multiple data sources, advances in this area will be a good contribution.

There are works on support vector machines (SVM) being used for information retrieval (Cao et al. 2006). Intuitively SVM seems to be a good fit in exploring the data points in the space of information retrieval, and more so in our case of smart IDEs. SVM is a classification-based machine learning technique that characterizes the classification by fixing a vector as a pivot that supports separation. Some of the key challenges that need to be accounted for in this area are the refined way of training data creation, a balanced approach between active learning and semi-supervised learning, making learning process efficient, and adapting to the domain areas, possibilities of the ensemble learning in the ranking.

6.20 Software structure evolution based on graph characterization

Some of the earlier work had been focusing on the cases of single release or shorter evolution span of longer evolution period for the single program (Louridas et al. 2008, Myers 2003, Valverde et al. 2007), here the focus was to study complete life span of the large projects and the progressive nature of the graphs. In the work Pamela et al. (2012), there is an attempt to look at answers for identifying the key moments in the program's evolution with graph metrics, where these moments are not the obvious ones. Also, to explore if there are any invariants, trends of evolution, and metrics values that can apply across all programs.

Further exploration is toward understanding how graph metrics evolve with time and their influence on software engineering aspects in the case of process and product. Evolution of nodes and edges are studied; it is observed that there are cases of programs that exhibit linear growth of nodes, and few others super linear growth over time. Similar observations are seen for edges. This is straightforward since the eLOC grows over the period.

An interesting observation is that, one of the programs did not have any increase in nodes or edges, even with the increase of eLOC, over time, which can be attributed to the program's maturity and, mostly code updates happening within functions rather than outside. Count of eLOC per node and per edge varied significantly across the programs. The values of these metrics decrease in some programs whereas an increase in a few others.

The intuition of average degree metrics is the popularity of the function or module. Coarseness quantification is done with average degree metrics of the graph and tight connection can be seen in graphs that show a high average degree (Mahadevan et al. 2007). The most increase in the programs' average degree makes the program the most popular one when increase is of a higher order.

Graph topology of Firefox significantly differed from that of other programs. Most of the cases of nodes exhibit a low degree with no connections between. Most of the high-degree nodes are connected to form a dense core of nodes in the graph.

Further low-degree nodes have a connection with these dense cores. A common library of Mozilla used by most of the products includes nodes in groups that exhibit high degree and connections. In cases of the projects where the average degree itself is low, most of the nodes will have a closer degree as that of the average degree of the graph itself. And high degree nodes which are few in numbers and rarely connected.

The clustering coefficient determines the connection capability of the neighboring nodes in the graph; bipartite graph type is seen in case of zero value. In this context, metric being of high value, it represents a tighter connection and misses to represent the hierarchy in its abstraction. This nature is due to horizontal and back edges being present. This also leads to software engineering practices that are good and make it a complex situation for understanding the program for its evolution and further difficulties in its testing as well. An interesting observation like significant intra-module calls being updated resulted in an increased clustering coefficient, though there was no change in function signatures compared to earlier releases in the process of a rewritten version of the code. Specific granular observations of this kind make it a fascinating study of the software program evolution patterns, leading to great input toward smart IDEs.

The analogy of the software engineering of looking for making everything work to make sure system works (Parnas 1972) holds good in the situation of cycles involved in the case of software structure. These cycles are ones like module graphs collaboration showing the cycle which is circular dependencies on the module. This will lead to difficulty in understanding and testing of the modules. An important aspect is these cycles hurt the quality of the software. The group of mutually recursive functions indicates the cycles in call graphs of functions resulting in difficulties in understanding the function and testing. This will then call for the condition of end-recursion that is well composed. Refactoring becomes a need in the case of nodes increasing across releases which means reduced software quality. Several nodes are seen to be increasing, in case of programs where functions were added that made a stronger connection among the graph components. It needs careful attention when there are cases of constant nodes over time as well.

Graph diameter helps in identifying ease of maintenance, large-diameter ones are tough to maintain. Debugging and understanding of the program are impacted in the case of the large diameter of the graphs. The diameter of the graphs indicates the distance between nodes. It is observed in most of the cases that the diameter tends to remain constant or maybe vary slightly.

Assortativity indicates connections among the nodes that are of a high degree in nature. Similarly, if the value is low, high-degree nodes connect with low-degree nodes (Newman 2002). Observations made in the experiments demonstrate the similarity among the software networks and biological networks. As the assortativity values are negative, there is higher tendency of higher degree nodes tending to connect with lower degree nodes.

Code reusability is also figured out in cases of program exhibiting low assortativity. In some of the functions that are used by many other nodes like in case of memory management kind of own functions seen in Mozilla, high-degree nodes tend to connect low-degree nodes. There is a possible further exploration area to assess the influence of assortativity on the code's reusability.

Edit distance metrics depict the dynamics of the graph structure changes. Here, the observation is on the extent of the change in the graph's topology with subsequent versions of the program. Similar kind of observations is seen in the programs on edit distance. There is a sharp rise in the edit distance at the beginning, which flattens later or shows a slight increase in trend. This goes well with the observation made in the work (Vasa et al. 2007), about software structures settling down in terms of its structural stability after a beginning spike period. Interestingly these sharp raise period validates the attainment of the mature period in the software program life cycle. Interesting observations can be made about the magnitude of changes happening over releases. This makes the release measurement more granular and meaningful. There was a measurement of the eLOC differences in the releases that highlight the fact that it would not be observed if we just stick to LOC measurement. Better modularity is possible with the increase in the ratio of cohesiveness to coupling over time. This comes out in modularity metrics and demonstrates good practices in software engineering.

All these indicate that there is a common theme of structural changes in the software program. This is concerning the quantitative observations made over range seen in metrics graph diameter, average degree, and clustering coefficient. So, there can be common lower and upper bounds of variations across programs. A similar evolutionary pattern can also be seen in programs through edit distance, modularity ratio, and clustering coefficient, with similar trends and patterns observed. Key moments of the software evolutions were found to influence significant variations in graph metrics over releases. These trends of evolution provide visual trend on how the program has evolved and would provide inputs to relate the key events of the software program evolution and help plan further evolution effectively.

6.21 Influence of machine translation approach on Smart IDE (approaches)

Machine translation (MT) is explored here to make IDE programming language agnostic contributing toward smart IDE. MT is the domain of changing the content from source language to its target language. It is mostly used in the case of text and speech being translated. The field has its history with a tremendous amount of effort spent over translation, which makes it a complex task with current data flood. Generic MT target the people where there is a need to do an ad hoc translations. Customizable MT focus on domain-specific knowledge. Adaptive one will keep learning as it processes. Among various approaches taken in MT, algorithms that focus on syntax and grammar are a basic one. The next approach is statistically based that learns the pattern in the content like any other machine learning techniques. Neural-based MT focuses on teaching the program for best results and involves high computing resources.

Sequence-to-sequence is one of the good approaches for MT that usually tops the list in the case of NLP use case. This brought in strength in NLP approaches, when there was lack in terms of handling the continuous pattern, over a long sequence of information, like in case of a large corpus of text. The basic structure is an encode and decoder setup with recurrent neural network (RNN) in its basic structure.

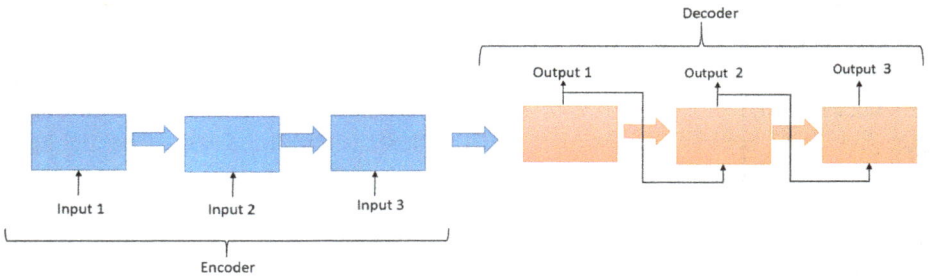

Figure 6.13: Sequence to sequence model outline.

This model has its influence on speech recognition technology, chatbot capabilities, and general question-answering systems.

MT must tackle various challenges of the flexibility of the language and ambiguity involved. Neural machine translation (NMT) is the latest state of the art that provides one model in place of multiple models operating in the pipeline for the same processing. In the case of NMT neural networks are involved in the statistical analysis. A basic multilayer perceptron would have done the job except for its limitation of the fixed input sequence that can produce the same output length. With RNNs improved the situation, and further encoder-decoder structure took it to the next level. A key aspect of this structure is its ability to form a context vector that represents internal fixed-length representation. After encoding, a variety of decoders would be used concerning the task on hand.

Encoder-decoder algorithms have its limitation when the sequence of the input grows. This is to do with the internal representation which is fixed and there is need to refer at word level for the decoding process. Improvement comes with attention mechanisms built into the architecture, which determines the input sequence portion that needs attention. So, it makes the encoder-decoder structure of RNN with attention possibility the state-of-the-art architecture. This is the core of the Google

Neural Machine Translation system. Even with such advanced structure, scalability is limited in cases of a large corpus of data.

This backdrop of the MT inspires the build of smarter components for IDEs, which forms the central platform to bring in the software development ecosystem's diversity and complexity. Interest in this exploration will bring together multiple programming languages and explore the possibilities of linking the common pattern among the programming languages.

The pattern of the work reviewed so far in Pamela et al. (2012) was around understanding the structural evolution of the program based on graph metrics; this provides further lead to figure out nonstructural attributes like effort and bug count including their severity.

6.22 Bug severity prediction

Work by the authors in Pamela et al. (2012) has a unique focus on using graph-based metrics on functions and modules that predict function or a module. The first part of data recorded when an issue or a bug is seen is to assess it and assign the severity, based on its impact on the program. For understanding the bug tracking basics, in Bugzilla bug tracking system details are as follows. Bug severity is blocker when it avoids the testing work in development and is ranked six. Critical if program crashes, lose data, or high leak of memory and will take rank five. Major issue if there is a significant loss of function and will be ranked four. Normal severity in case of issues that are regular and with some functionality loss. Minor if there is the loss of function of minor nature and is ranked two. The trivial case for the cosmetic issue and ranks one and enhancements will be ranked zero.

So, the software developers' focus will be to make sure that there are no critical issues in addition to the general focus of doing away with the issues or bugs. Also, the challenge would be to manage this objective with limited resources of testers and developers. This puts importance on any effort toward helping people with a tool to provide information on what is important. This helps optimize test and verification effort and, in turn, helps robustness and quality of the software. NodeRank will help identify the most important modules and functions by looking at how impactful the bugs will be in these entities. The node function's relative importance or a module in a call in a collaboration graph is the NodeRank. This helps users to identify the critical module and functions based on the NodeRank. Next author looks at putting all these into a formal hypothesis.

The first hypothesis is the higher-severity bugs are the possibility in the case of functions and modules of higher NodeRank. Bug severity in the bug reports of the program was the input. Further information on associated patches and the number of code changes that it has triggered is gathered. This provides information about the median value of bug severity and how buggy the functions are. The authors confirm that they would validate this hypothesis by correlating NodeRank with that of bug severity median value for the functions and modules. The correlation coefficient is assessed for programs across parameters of NodeRank, cyclomatic complexity, interface complexity, in degree, and out degree. The primary focus was on the top 1% of the NodeRank as there was a skewed distribution among this data in terms of most of the bugs influencing these top NodeRank. It is important to note that large program, top 1% itself will be a substantial amount of data of functions. NodeRank and bug severity demonstrated a high range of correlation indicating that the NodeRank would be a good indicator of the critical functions or modules concerning the bug severity. The study also was done on engineering quality–related metrics like cyclomatic complexity and bug severity. Interface complexity and cyclomatic complexity showed a low correlation coefficient, indicating that they are less likely to indicate the importance of functions and modules.

Now on to assess bug severity influencing node degree as a predictor. Node degree was looked at to see if it has a similar influence like that of NodeRank for bug severity. A node in and out-degree came out to be poor indicators of the bug severity with lower correlation values.

Apart from the top 1% of nodes, the rest of the 99% nodes also demonstrated a similar trend for NodeRank, pointing to lower values due to high scale noise associated with it which is also statistically significant. Though there is no correlation, a high-level trend exists between bug severity and NodeRank. The overall indication of the study is for cases where resources are constrained in testing and verification, high NodeRank nodes must be considered as a top priority. The noise that are talked about here is the cases like shared bugs between Firefox and Mozilla projects, where these bugs are rooted in shared functions and not so much part of shared modules. So, it will be tough to attribute these bugs to only Firefox and hamper the correlations and be a noise in the data.

In smart IDEs, bug severity would be a good part of smart recommendations, which will help the users prioritize their work. It will also provide a good resource for improving the system by providing inputs on the poor-quality system's root cause. Over time this information will build the maturity of the smart IDEs.

6.23 Effort prediction

A major part of software cost for maintenance is to do with the complexity of changing the source code for refactoring and the addition of new functionality. Modularity ratio is proposed as a metric that will help identify the modules that are complex in terms of changing them. Basics are the modularity ratio of a module represented as cohesion/coupling of a module that indicates ease of changing the module. The maintenance effort involved for each release in a module can be assessed with metrics number of commits divided by eLOC churned out. For effort estimation, this is one of the widely used metrics.

So, the hypothesis to be validated is modules demonstrate lower maintenance costs when they have a higher modularity ratio. Results of the effort data validation indicate that as the modularity ratio increases there is a related decrease in the effort involved for maintenance; this signifies that the structure of the software improves. Granger causality test is run in the experiment for graph data. This test is a statistical hypothesis-based test to determine the influence of one-time series data on another. The correlation test to be replaced with a causality test is due to the time lag involved. Hypothesis here was to check influence of modularity ratio in one release on the change in the effort of a future release. F-prob and lag values are the measures in the Granger causality test. For some of the programs lag value indicated that change of effort needed in a release is determined by change in the modularity ratio of the previous release.

This approach would contribute to smart IDE for a recommendation on future releases that will assist the software development teams for their release planning. Since release planning is an important part of business, the product's time to market is a key business metric. Effective release planning can help the business to manage the competitive state in the market. So, this recommendation for future release can play a handy role. Information around the maintenance effort would be a great benefit, as that is one of the unknown factors of software development, where the team base their estimations on their experience and not so much with data.

6.24 Defect count prediction

Logically, the outcome of the well-connected team will be of higher quality than a team with high turnover and disconnected. Authors here have explored the influence of structured and stable teams resulting in higher-quality software with a collaboration of high level being demonstrated. Developer collaboration graphs provide a good indicator of team stability and its characteristics feature. Edit distance metrics can help us indicate how much the graph has matured over time. So, it is evident from the authors' tests that the period in which the team involved was stable resulted in a period when the defect count was low. In the bug developer collaboration graph, this was validated by the hypothesis; increase of defect count can be figured out with the increase in edit distance.

Input data for this hypothesis was the bug report for a few programs over the period which was used to build graphs for developer collaboration. With the data, bug-oriented developer collaboration graph was built but at year level rather than release level, as bug counts in case of some releases were low. Graph edit distance was computed for year on year with correlation analysis done with edit distance for a year data with a defect at the end of the year. A stronger correlation is seen between these two. So, there is an indication of team stability influencing the bug injection and indicating that when developers work with the team that they have been working with, there is a favorable situation than working with new team members. Begel et al. (2010) report similar observations of productivity going up in the case of working with a known team. Interestingly with open-source software not being governed by a central team and no social structure for the same, but this will also demonstrate collaboration of team having its influence on the quality of the software.

In comparison to commit-based collaboration, bug tossing collaboration is a better predictor for the defect. The same results are confirmed with the developer's collaboration graph based on commits. Though there was no reported correlation between edit graph and effort, which indicates the bug-tossing graphs to be better ones in comparison to that of commit based graphs, this will be a prospect in the area of the understanding relationship of developers, in open-source projects.

6.25 Conclusion of work done on knowledge graphs

Works done by authors in Pamela et al. (2012) have focused on the analysis of the range of project in C and C++ or both. With multiple releases of the project studied, it provided the trend of release patterns helping to study the graphs' evolution. Software quality being influenced by the developers' collaboration, is demonstrated with the study of developer collaboration graph in large open-source projects.

In the context of software network failure prediction, this study emphasizes analysis of the topology of the graph representation over multiple releases for a project, bringing out the key spots of the software with metrics like NodeRank. The modularity ratio metric brings out maintenance effort. NodeRank's ability to work at the module and function level makes it appropriate to predict the bug even before its report. Developers' collaboration is brought out by graph method to observe the yearly transition of them. Also, a positive correlation was demonstrated with defects and developer graphs sequence concerning edit distance.

Overall, the work proposes the graph structure with related metrics to observe software outcomes' evolution in terms of process and products. Source code graph metrics have shown granular changes in structure across programs and highlight the key event in the process that would get missed out in other metrics. Defect-prone releases, bug severity, and maintenance effort have been demonstrated with graph metrics for predicting over releases.

6.26 Overall conclusion

Statistically modeling of software source code for enabling the intelligent IDE is done so far. We derive the inspiration from the work that is done so far in the space, to make the IDEs smart. There is an exploration of the overall landscape for understanding the ecosystem and underlying challenges involved. This lays a good platform to carve out the future focus areas. The knowledge graph is a prominent area in recent advancement, this will nicely fit it to challenge of handling vast data, and it dynamically changes the landscape. Areas of NLP will contribute to some of the prospective areas involved in the challenge. There are key areas of application discussed, which will motivate us to think beyond for more such use cases. Understanding the underlying principles of the graph constructions and building metrics around the same is a great way to relate to the smart IDE thought process's challenges. In the pursuit of making the smart IDE, it also covers the chances of understanding the software development product's inefficiencies and process, thus helping with the overall improvement of the domain.

Chapter 7
Natural language processing–based deep learning in the context of statistical modeling for software source code

7.1 Introduction

Statistical modeling for software source code work is targeted at all the issues faced by the software development processes. The applications' primary areas are focusing on addressing the challenge of low confidence that the community is facing today on the quality of the deliverables across the life cycle of software development. To build in transparency and trust using the knowledge hidden in the software code is the focus. Lack of a better understanding of the business needs contributes to the challenge. Deficiency in the domain and technical knowledge for needed analysis contributes toward a disconnect in understanding customer needs. This points toward maintaining the traceability of the software requirements across the life cycle.

A big picture of the final delivery is missing, and the focus is more on the structure of the code than the holistic design approach. The gap between real-world user needs and the technical descriptions of the system is evident. Business domain understanding becomes critical here. There is an effort toward making software development a community experience customer point of view. Though there is an effort toward the same, the results are not satisfying.

Though responsible, Business system analyst were not able to take care of this, due to the lack of a robust end-to-end system. This aspect hints at building an intelligent system that is traceable end to end and helps integrate all the knowledge across and feed in to, as needed, to make meaningful decisions. Uncommon focus on unit testing is another important part of building a robust construction phase. Though the practices exist, they are not vital, for the case of continuous integration and continuous deployment (CICD), and quality assurance focus takes back seat. Defects are handled as new requirements, making people lose visibility on work's quality progress. Maturity of the CICD practice is a concern leading to the question of the people's skill during the transition from traditional methods to CICD way. First Time Right is focused, but the real sense is missing in the operations, resulting in suboptimal deliverables across the life cycle phases.

The lack of initial analysis extends itself into nonfunctional aspects of the requirements, leading to security vulnerabilities costing organizations on reputation and money. Lack of knowledge about the possible security vulnerabilities also contributes towards security lapses.

With these areas as the focus, statistical modeling for software source code tries to derive the knowledge hidden in various software artifacts, including code and

https://doi.org/10.1515/9783110703399-007

text, and build a robust and intelligent system. Natural language processing (NLP) and its application using the non-code content in the software domain are an important aspect. This information complements the code-related information and pattern associated with it.

In this chapter, we explore the deep learning methods and NLP areas. Distributed representation of language is explored, including models like word2vec and character embeddings. We explore the convolutional neural network (CNN) and its variant. Discussion includes recurrent neural network (RNN) and its contribution towards moving advances in deep learning applications and its variants like Recursive Neural Network. Then we wrap up by relating the areas of deep learning with NLP into the approaches with software source code modeling. Then we look at all prospective progress happening in deep NLP (Young et al. 2018).

7.2 Deep learning–based NLP review

Deep learning uses a layered approach to identify the data patterns in sequential format and has resulted in the state of art outcomes. There have been extensive experiments and introduction of design and approaches in the NLP in the recent past. In this review, the authors provide insight into deep learning models in NLP tasks. All the methods are summarized, and a review of the evolution of these methods is discussed.

Automatic analysis of human language and their representation is the key motive behind NLP, and a variety of computation techniques are involved. NLP finds its root in the era for batch processing of the punch cards that took several minutes to analyze the sentence to today's capability to process millions of web documents in second. Machine translation, dialogue system, named entity recognition (NERs), and tagging based on parts of speech (POS) are some of the capabilities that NLP brings into the table.

Pattern recognition and the computer vision world are significantly influencing the mark of deep learning methods. Recent researches have increasingly trended on deep learning methods for NLP.

Association of Computational Linguistics (ACL), Empirical Methods in natural language processing (EMNLP), European Chapter of ACL (EACL), North American Chapter of ACL.

A shallow model like support vector machines (SVM) and logistic regression have dominated the research's earlier years. With these models' focus was on sparse features and high dimensional ones. Trends have shifted toward dense representation in the network for vectors that have produced superior performance. Word embedding (Mikolov et al. 2010, 2013) and deep learning methods (Socher et al. 2013) have fueled these trends. Automated feature representation learning at multiple levels is the significance of deep learning. Handmade features were a major dependency on conventional formats. These handmade features generally were not complete and consumed a lot of time.

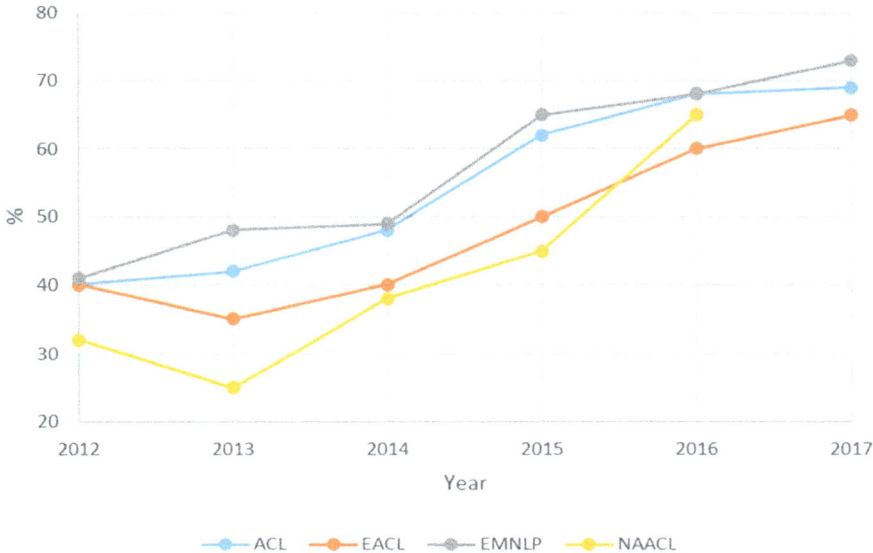

Figure 7.1: Percentage of deep learning papers across ACL, EACL, EMNLP, NAACL (detailed papers).

When compared to the state of the art method of NLP like NER, a simple version of deep learning methods can overcome their performance. As the trends continued a whole lot of deep learning methods are proposed for NLP tasks. Major advancements in the field like recursive neural network, CNN, and RNN will be reviewed here. Attention mechanisms that focus on memory augmentation strategies, workings of deep generative models, reinforcement learning (RL), and their applications for language-related tasks are discussed.

Tutorial-based work of Goldberg (curse of dimensionality) covered only basic neural network application principles for NLP. Review here covers the work comprehensively and also covers the area of distributed representation for the deep learning methods, then followed by a discussion of popular models and their application to NLP tasks.

7.3 Distributed representations

NLP tasks of complex nature fit well into statistical-based NLP. Curse of dimensionality was a major roadblock earlier for this, particularly in language model with joint probability functions. This was a major lead to focus on distributed learning of words in lower space dimensions.

7.3.1 Word embeddings

Distributional hypothesis for words having similar meaning coming together in the vector space is the key theme of word embeddings. Figure 7.2 depicts Distributional vectors representation.

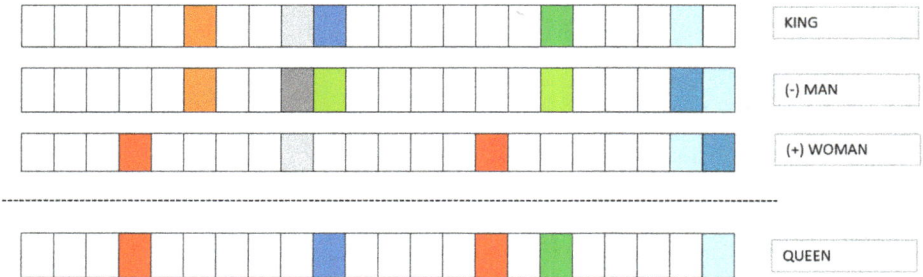

Figure 7.2: Distributional vectors representation.

Vector representation captures the characteristics of the neighbor words. Distribution vectors have a major advantage in capturing coexisting words. Cosine similarity is the metric that helps to measure the similarity between the words. Word embeddings are the entry points for deep learning models in terms of being the data processing layers. For a large corpus of data where the labeling is not done, pretraining the word embedding happens by setting a supplementary objective and targeting to optimize it. An example is to predict the word; looking at its context (Mikolov et al. 2013) (Autoencoder), general syntax and semantics of the word embeddings can be learned. Since the dimensions are manageable, time consumed is quite under control; also their efficiency is appreciable for capturing the analogies and similarity of the context.

Representation by these embeddings has been extended to words, characters, and phrases. These latest models take over conventional models that were based on word counts. Sentiment analysis accounting for the domain used stacked denoising autoencoders (Autoencoder) in combination with embeddings (Glenberg et al. 2000).

Context is the key to distributed representation learning. Distributional semantics was fueled by research work that is done extensively in the 1990s (Collobert et al. 2008). Topic model latent Dirichlet allocation (LDA) (Blei et al. 2003) and language models (LM) (Mikolov et al. 2013) derive their inputs from the work that is done in these works (Dumais et al. 2004, Glenberg et al. 2000). Representation learning got inspiration from all these works. For words, distributional representation learned from the neural language model was inspired from work by Bengio et al. (2003).

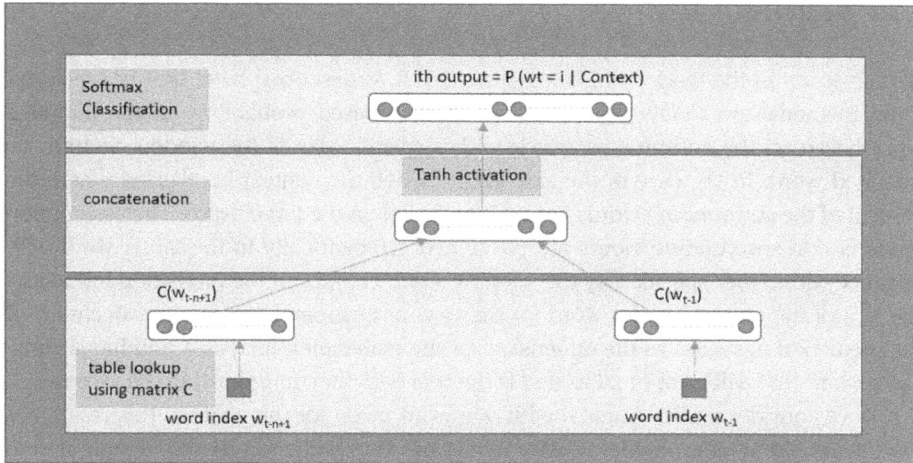

Figure 7.3: Neural-based language model representation by Bengio et al. ith word embedding is consider.

Work here proposed that combining the word embeddings with joint probability for the sequence of those words resulted in a neighboring sentence with semantic similarity. Generalization was a key advantage as the words with similar context had a higher confidence level even if they have not appeared in the training corpus. Figure 7.3 depicts Neural-based language model representation by Bengio et al. where ith word embedding is considered.

Pretrained word embeddings utility was first explored in the work of Collobert and Weston (2008). Neural network architecture from this work forms the input for many works. Work also shows that word embeddings are a great way for NLP tasks. Skip-gram models and Continuous Bag of Words (CBOW) coming from the work (Mikolov et al. 2013) popularized the word embeddings in the context of constructing distributed word vectors that are of high quality. Compositionality is another major extension of these works where two-word embedding combined to provide combined semantics representation of individual words, like "king" – "man" = "queen." An assumption like the need for words being distributed uniformly in the word vector space is an important condition for this characteristic, as Gitens et al. (2017) explained.

Count-based famous model from Pennington (Johnson et al. 2015) is another important one. In this approach co-occurrence count matrix is normalized in preprocessing and log-smoothing applied for the word. Further lower-dimensional representation is derived based on reconstruction loss minimization.

7.3.2 Word2vec

Word2vec, as in the work by Mikolov et al. (2013), is described here, that houses skip-gram methods and CBOW. CBOW uses condition-based probability on the intended word based on the surrounding words with a specific size of the window around the intended word. In the case of the skip-gram model, the context is reversed where the context of the surrounding words is predicted based on the target word. The assumption made is that surrounding words are positioned symmetrically to the target word with distance within that specified by the window size. Accuracy of the prediction influences the size of the embedding for word for the case of unsupervised learning. Accuracy of the prediction increases as the dimension for the embedding increases and hits a stabilizing point that will be optimal as that is the best without compromising the accuracy.

If we consider a CBOW model with one word taken for the context, this resembles the case of the bi-gram model. CBOW model has one hidden layer in a simple neural network that is fully connected. One-hot vector input for context word with V count of neurons in the first layer, which is the input layer, and hidden layer with N count of neurons. Softmax summarizes words sitting in the output layer. Weight-based matrix W RVxN and join layers $W' \in R_{HxV}$ for input and output. Two learned vectors come from the words in vocabulary represented as v_c and v_w, which are context and target words respectively. So, for kth word representation of the vocabulary will be as follows:

$$v_c = W_{(k,.)} \text{ and } v_w = W'_{(.,k)} \tag{7.1}$$

Generally, for given word w_i for a context c as input,

$$P\left(\frac{w_i}{c}\right) = yi = e^{\mu_i} / \sum_{(i=1)}^{v} e^{(u_i)} \tag{7.2}$$

Where, $u_i = v_{wi}^T . V_c$

Based on defined objective function in the form of log-likelihood and for gradient denoted as:

$$I(\theta) = \Sigma_{w \in vocabulary} \log\left(P\left(\frac{w_i}{c}\right)\right) \tag{7.3}$$

$$\partial i(\theta)/\partial(V_w) = Vc(1 - P(w/c)) \tag{7.4}$$

Context of all hot vectors for context words are considered together for CBOW model,

$$h = W^T(x_1 + x_2 + \ldots + x_c) \tag{7.5}$$

Inability to represent phrases with a combination of words like "hot potato" or "Boston Globe" where name entity is involved is a limitation of word embeddings. Here the meaning of the combination of words cannot be represented. The solution of looking at the co-occurrence of these words and then use it to train the embeddings separately was proposed by Mikolov et al. (2013). Learning n-gram embeddings directly on unlabeled data has been some other recent approach (Johnson et al. 2015).

In semantic analysis, there is a limitation of fixed window size being used in the embedding process; words like *good* and *bad* many times have similar embeddings in this context. These are cases of opposite polarity words but have a semantic similarity. This results in poor performance for downstream models where the polarities that are opposite are not recognized. Sentiment-specific word embeddings is a good format to tackle this problem as proposed by Tang et al. (2014). As part of the loss function here in this approach, sentence polarity from supervised learning for text is accounted for while learning embeddings.

Word embeddings are largely specific to the application they are involved in, which is a key drawback. Aligning the word embedding to task space by retaining them in the task space is an approach from Labutov and Lipson (2013). This holds the key as training the embedding from scratch will consume a huge amount of time and resources. Negative sampling was attempted to address this issue by Mikolov et al. Here negative terms are sampled based on the frequency and used in word2vec model training. Each word gets a unique vector in word embedding representation traditionally. So, polysemy does not get accounted for. Multilingual parallel data was used in their work by Upadhyay et al. (2017); this helps to learn multisense word embeddings. An example can be the French meaning of the English word "bank" can be "banc" and "banque" meaning finance and geographical. Polysemy accounting was possible with this multilingual representation.

7.3.3 Character embedding

Syntactic and semantic information is captured in word embeddings intra-word morphology and shapes information will play a good role for cases of NER and POS tagging. Character-level natural language understanding system work has encouraged large research work. Certain NLP tasks show good performance in morphologically rich languages. Neural language model with only character-based embedding showed good results in case of work from Kim et al. For NER, character trigram-based word embeddings with others were used to consider hierarchical information for learning label embedding.

Out of vocabulary (OOV) is the issue in the case of large vocabulary languages. Since individual letters are considered as the composition of the word in case of character embeddings they work well. In languages where the characters compose the text and not words like Chinese, character embedding will be the primary choice; this will help prevent word segmentation (Chen et al. 2015). Instead of word vectors, character embedding fits well for deep learning applications (Zheng et al. 2013). Sentiment classification performance will help significantly with radical-level processing as per work from Peng et al. (2017). For Chinese, radical-based hierarchical embeddings were proposed, which incorporated sentiment information with character-level semantics. For morphologically rich languages character-level information representation was used for improvisation by Bojanowski et al. (2016). Bag of character n-grams was used for representation in the skip-gram method. Thus, this method took care of the issues seen in embeddings with the utilization of the skip-gram method in an effective way. Large corpus training was possible as the method was fast. When looking at speed and scalability, FastText stands out compared to other work.

To address the OOV issue apart from character embedding Herbelot and Baroni (2017) proposed unknown words to be initiated taking the sum of context word and then with high learning rate refining them. The approach must be explored on NLP tasks. Pinter et al. (2017) attempted to recreate pretrained embeddings using the character-based model. Thus, the OOV issue was tackled by composing character to word embeddings.

Distributional vectors have been making sound but there has been discussion on their long-term reliability. Like conceptual meaning provided by the word, vectors are being assessed in Lucy and Gauthier (2014). Perceptions involved in the words as a concept present limitations as found by this work, which cannot be made up by distributional semantics. Grounded learning is making its mark and can handle some of these issues.

7.4 Convolutional neural network (CNN)

Word representation in distributed space by word embeddings paved the way for extracting high-level features making use of the n-grams in feature function. NLP tasks like machine translation, summarization, and question answering will benefit from these extracted features. Computer vision has a greater contribution from CNN so that automatically takes preference.

Collobert and Weston work mark the initial traces for work on sentence modeling with CNN (Collobert et al. 2008). For NLP tasks like NER, semantic similar words, and LM, multiple predictions being done with multi-task learning were used in this work. The user-defined dimension vector representation was done for each word with a lookup table. Series of vectors like $x_{i1}, x_{i2}, \ldots, x_{in}$ achieved with input sequence like $i1, i2, \ldots, in$, using a words lookup table, like shown in the figure.

This can be looked at as weights being learned during network training like in the case of conventional word embedding methods. The performance of CNN was reliable as CNNs had already proven the capabilities in computer vision tasks.

For sentences, latent semantic representation was able to be created by CNN's extracting n-gram features from input from the downstream tasks. These were outcomes of the work done by Collobert et al. (2011), Kalchbrenner et al. (2014), and Kim (2014), which propelled CNN-based frameworks. Figure 7.4 depicts Wordwise class prediction with CNN framework as proposed by Collobert and Weston.

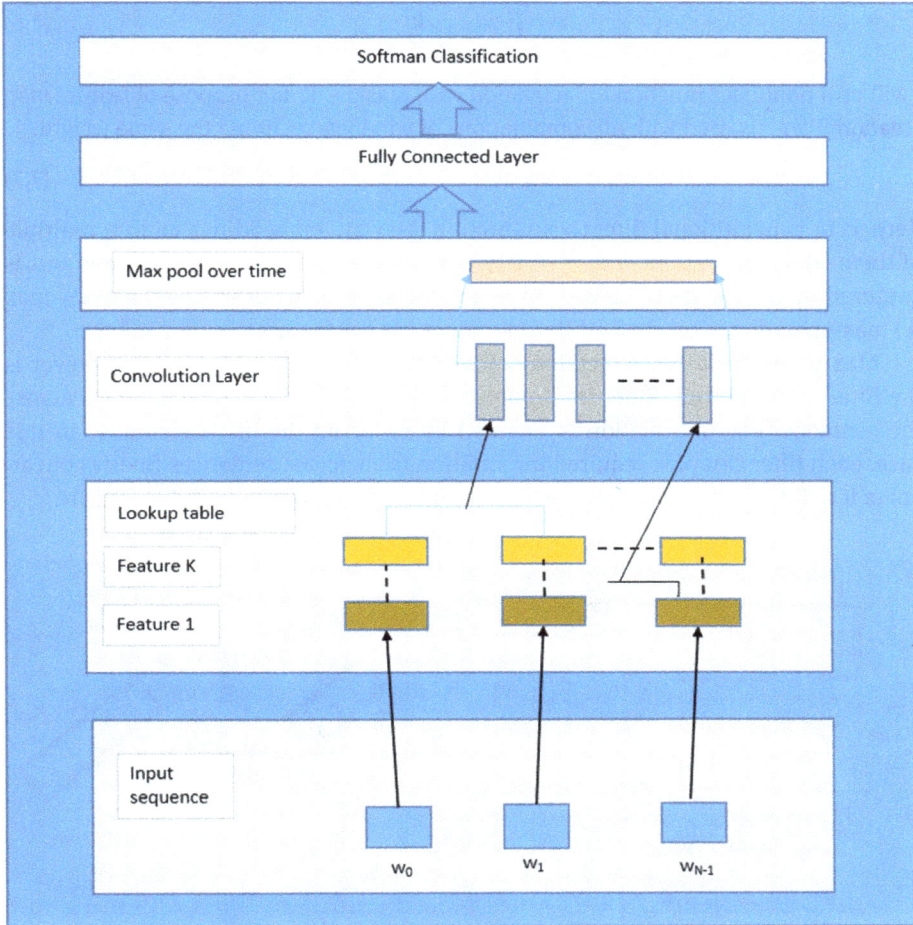

Figure 7.4: Wordwise class prediction with CNN framework as proposed by Collobert and Weston.

7.4.1 Basic CNN

Sentence modeling: For d dimension word embedding, for i^{th} word, each sentence can be represented as $w_i \in R^d$. Sentence representation becomes $W \in R^{nXd}$ for n-words sentence. A sentence from here will be input to the CNN framework as shown in the figure.

For vectors w_i, w_{i+1}, ..., w_j, concatenation can be represented as $w_{i:i+j}$. New feature production, with a window of size h words, the filter of $k \in R^{hd}$ involved. The window of words $w_{i:i+h-1}$ is used to generate feature c_i and can be represented by:

$$ci = f\left(w_{i:i+h-1}.k^{T}{}_{+b}\right) \qquad (7.6)$$

f will be a nonlinear function for activation with bias $b \in R$. For purpose of feature map creation filter k is used with all combinations of windows retaining the same weight.

$$c =[c_1,c_2,.....,c_{n-h+1}] \qquad (7.7)$$

Kernels or convolutional filters scan through the word embeddings matrix; multiple of them would be present with a variety of width. The specific pattern of n-grams is captured by each of these kernels. Max pooling strategy would be next part $\hat{c} = \max$ {c} max operation is used on a filter to sample out the features.

Max poling helps classification tasks with fixed-length output, so whatever is the filter's size, output dimension representation will be standardized. From across the sentence, key information is retained by reducing the dimensionality. In this case, each filter can pick required information from across sentences finally concatenate it to final representation. Figure 7.5 depicts Text model applied with CNN.

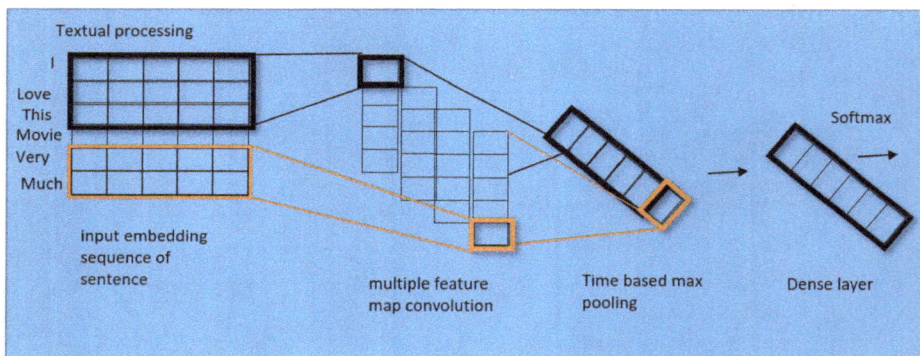

Figure 7.5: Text model applied with CNN.

Large unlabeled corpora can be used for pretraining or random initialization of word embedding. The former method is better when there is only limited labeled data (Kim 2014). Deep CNN networks are formed with a combination of layers of convolution and max pool. This model of filtering the data across layers will help abstract the information and derive semantic information from it. Deeper the convolution layers, filters or kernels will cover a large part of the sentence and create summarization for sentence features.

Window approach: The approach discussed above helps to generate sentence representation from a sentence. But for cases of many NLP tasks like POS and NER, they need word base predictions. So, in those cases window approach is used where the word tag is dependent on neighboring words. A fixed-size window is taken to assess word with its words around it, to gather a range of subsentence within a window range. Further separate CNN is applied to this, and predictions are done based on the word that appears in the window's center.

Assigning a group of labels to an entire sentence is the primary objective of word-level classification. Conditional random field (CRF) kind of prediction methods are used to derive association among class labels close by and then arrive at an overall label sequence to generate a score for a complete sentence.

Time delay neural network (TDNN) (Waibel et al. 1989) with windows approach can capture larger contextual information. Across all windows in the sequence convolutions are performed. Kernel width definition will be a limitation for convolutions. TDNN has capabilities to consider all the words in all windows at the point of time unlike window approach only considering words in the window. To focus on globally available features in a higher layer and picking lower-level features in lower layers, TDNN also takes the form of CNN architecture.

Applications: Above architecture application was explored by Kim (2014) for sentence classification tasks, with effective results. This was the architecture of choice for researchers for its effectiveness and simplicity. Randomly generated filters after being trained on the intended task, behaved like feature detectors similar to n-gram. Among the many issues in this network, not able to model long-distance dependency was an important one. Semantic modeling of sentences related to prominent work from Kalchbrenner et al. (2014), with dynamic convolutional neural network (DCNN), addressed the issue. Here k most active features were picked from sequence p with dynamic k-max-pooling technique. Positions were not identified through the order of features that were captured as in the figure. Figure 7.6 depicts Dynamic pooling and filter for small width at a higher level help assess input sentences for their far apart phrases.

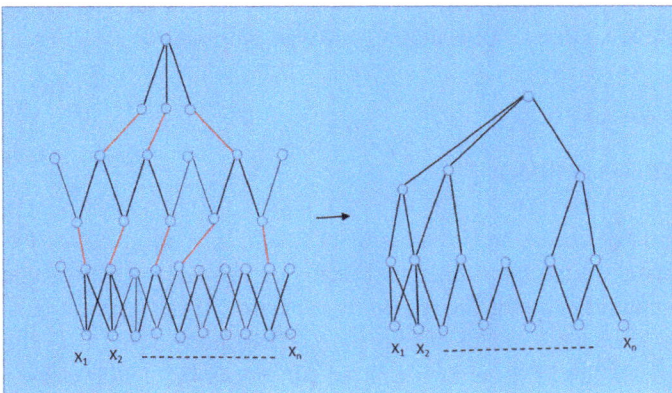

Figure 7.6: Dynamic pooling and filter for small width at a higher level help assess input sentences for their far apart phrases. This is DCNN subgraph representation as proposed by Kalchbrenner et al.

The *k*-max strategy was applied with pooling for sentence model creation based on the concept of TDNN. Using small width filters was possible with this approach to get the information from across sentences with covering long-range within the target sentence. Significantly variable ranges were demonstrated by higher-dimensional features that ranged from short focused to long and global. Good results were seen by applying these for question type classification. Work focused on a range of filters with context as a lookout for deriving contextual semantics that way to make sure they reach out to a larger span. Sentiment polarities and aspects were key need of sentiment analysis tasks (Mukherjee et al. 2012). To get good results at the input of the CNN vector representing aspect was added on. Based on the length of the text approach for CNN model differs. Variation is the performance on longer and shorter texts were seen in the works. To perform well external knowledge is needed for CNN working on micro texts of NLP. This came out in work that was focusing on sarcasm detection in Twitter texts (Poria et al. 2016). The add facilitated this on the network that was trained on emotion and sentiment.

To assess the document's meaning based on the summarized document, DCNN was applied by Denil et al. (2014). High-level semantics are derived from lexical features with sequence-based learning on hierarchy looking at sentence and document using convolution filters in DCNN. This also includes visualization of the learned representation in the learning process, and automatic summarization of the texts.

For fixed dimension semantic space, representation for query and documents, are done using cosine similarity to rank the document based on a query with CNN. For word sequence contextual window was considered, intending to gather rich context-based features using this model. Context-based feature from n-gram level was considered. The overall sentence vector was derived from convolutional and max-pooling layers.

Based on the knowledge base, questions' similarity is derived based on semantics, to provide supporting information to derive answer (Yih et al. 2014). For understanding the questions from multiple aspects, multicolumn CNN was proposed by Dong et al. (2015).

7.4.2 Recurrent Neural Network (RNN)

Sequential-based information processing is the theme of RNN (Elman et al. 1990). Repeated computation happens on the same token and there is a dependency of the steps on its earlier steps of computation that is the reason for the name. On a recurrent unit, token is fed one by one to obtain a vector of fixed size. RNNs are featured with the memory from previous steps, and the same is utilized for processing of the current steps. This fits well for NLP tasks like machine translation and speech recognition.

For various NLP tasks, the characteristics of RNN that made its best fit are the next part of the discussion. The overall theme of processing sequentially is useful for capturing the inherent language characteristics, which is also a sequential combination of character, word, and sentence. Earlier words influence the semantic meaning of future words. An example of the difference between "hotdog" and "dog" can be looked at for understanding this difference. Language's context dependencies modeling and similar sequence task models make the RNN choose researchers over CNN.

RNNs ability to process longer length sequences of variable length and similar documents makes RNN popular (Tang et al. 2015). The flexibility of the calculation and ability to capture contexts that are not bounded score RNN over CNN. Whole sentence-level semantic modeling is the need for many NLP tasks. For this, in a vec-tor space, a fixed dimension representation of the sentence is needed. Summarizing a sentence into the fixed vector and comparing the same to variable-length target is a key aspect in machine translation (Cho et al. 2014), and RNN provides this ability.

Joint processing based on time is the network support provided by RNN. POS tagging are tasks that are based on sequence labeling, and are covered in this do-main. Subjectivity identification and multi-label categorization of the text are some of the applications. Though the above points are highlighting the superiority of the RNN over CNN, the works have proved the CNNs ability even in RNN kind of scenarios; also other deep networks have proved useful (Dauphin et al 2016). While sentence mod-eled in CNN and RNN have a different lookout, CNN extracts most key n-grams and RNN look for unbounded context for a long sentence of varying length and create a composition from them.

7.4.3 Variations of RNN models

Simple RNN: Elman networks (Elman et al. 1990), which are three-layered, are the basis for RNNs in NLP scenarios. Figure 7.7 depicts General RNN representing the whole sequence with unfolding across the timeline.

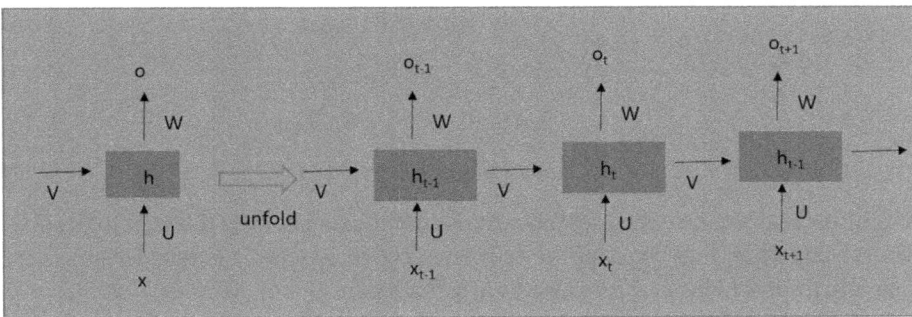

Figure 7.7: General RNN representing the whole sequence with unfolding across the timeline.

For time step t, x_t represents the input to the network and s_t hidden state at the same time point. s_t calculated as $\text{pers}_t = f(Ux_t + Ws_{t-1})$

Previous time steps hidden input and current step input is used to compute s_t. U, V, W are the weight across timeline whereas f represents rectified linear unit (ReLU) and the kind of nonlinear transformations. One hot encoding or embeddings are composed in it for NLP scenarios.

Long short-term memory (LSTM): Compared to simple RNN-based LSTM has additional "forget" gates; this facilitates the error backpropagation across time steps. A combination of input forget and output gates are considered to derive hidden states.

$$x = [h_{t-1}, x_t] \tag{7.8}$$

$$f_t = \sigma(W_f \cdot x + b_f) \tag{7.9}$$

$$i_t = \sigma(W_i \cdot x + b_i) \tag{7.10}$$

$$o_t = \sigma(W_o \cdot x + b_o) \tag{7.11}$$

$$C_t = f_t \Theta \, C_{t-1} + i_t \, \Theta \tanh (W_{c.x} + b_c) \tag{7.12}$$

$$h_t = o_t \Theta \tanh (c_t) \tag{7.13}$$

7.4.4 Gated recurrent units (GRU)

To simplify the RNN and to maintain the performance GRU was introduced. GRU is specialized by reset gate and a gate for the update to handle information flow, without memory like LSTM. So here without any control over hidden content, it gets exposed. GRU performs more efficiently compared to the LSTM version of RNN. GRU representation is as follows:

$$z = \sigma (U_z \cdot x_t + W_z.h_{ti-1}) \tag{7.14}$$

$$r = \sigma (U_r \cdot x_t + W_r.h_{t-1}) \tag{7.15}$$

$$s_t = \tanh (U_z \cdot x_t + W_s.(h_{t-1} \Theta r)) \tag{7.16}$$

$$h_t = (1 - z)\Theta \, s_t + z \, \Theta \, h_{t-1} \tag{7.17}$$

Attention mechanism: encoding the information that is not fully relevant to the context is one of the drawbacks of the encode-decoder network. Sentences being very long and information-intensive also pose a challenge.

Sequence-to-sequence learning problem example is text summarization where the input of the original text is converted into a reduced version. Representing the long text into a fixed-size vector is intuitively not possible. Some alignment between input and output text like token generated for output is highly related to input text; these are seen in machine translation and text summarization tasks. This background

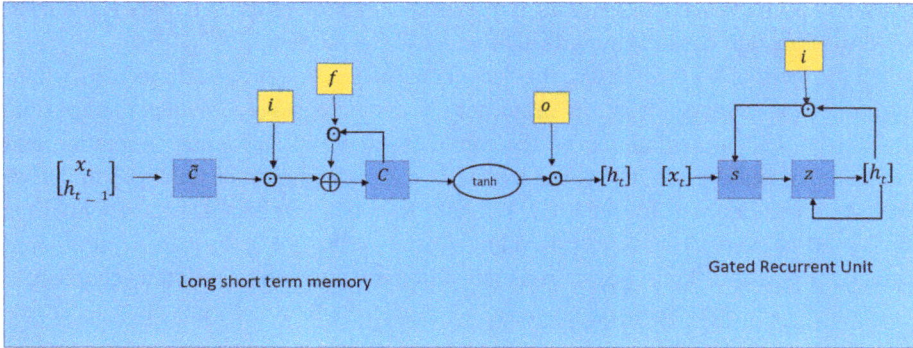

Figure 7.8: LSTM and GRU gate representation.

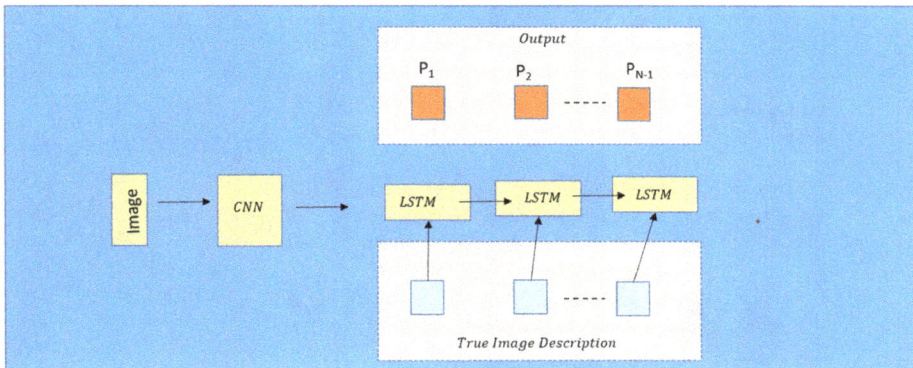

Figure 7.9: CNN image embedding with LSTM decoder in case of image captioning. The architecture proposed by Vinyals et al. (2015).

fuels the attention mechanism. The decoder can access the previous sequence to create this capability. Last hidden state of the decoder was the key part of the work, in multi-layer perceptron, with attention signal on hidden state sequence of input. Attention mechanism applied on long sequence improves performance for machine translation in the work of Bahdanau et al. (2014). Last hidden state of the decoder in multilayer perceptron, attention signal on hidden state sequence of input in this setup was a key part of this work. The source and target's alignment was possible by looking at the input sequence and their signal of attention at every decoder step in the demonstration. Figure 7.8 depicts LSTM and GRU gate representation.

Every output word was conditioned with the input-based sequence using attention mechanism as per the work of Rush et al. (2015), in-text summarization task, which is a similar approach as mentioned above. With just minimum linguistic feedback scaling

up of the approach is possible for abstract based extraction of the summary when compared to extraction-based summarization which is conventional.

In the work of Xu et al. (2015), the decoder of the LSTM was conditioned on a different part of the sequence from the input, for the decoding stage. Previous hidden state
and other CNN features influenced the signal of attention. For dialog generation and
text summarization cases, there was a need to copy a sequence of the input to output
with some conditions (Vinyals 2015). Whether to copy or generate the sequence was
considered at every time step in decoding (Paulus et al. 2017). To provide additional
support for classification aspect-based embedding was fed into attention mechanism.

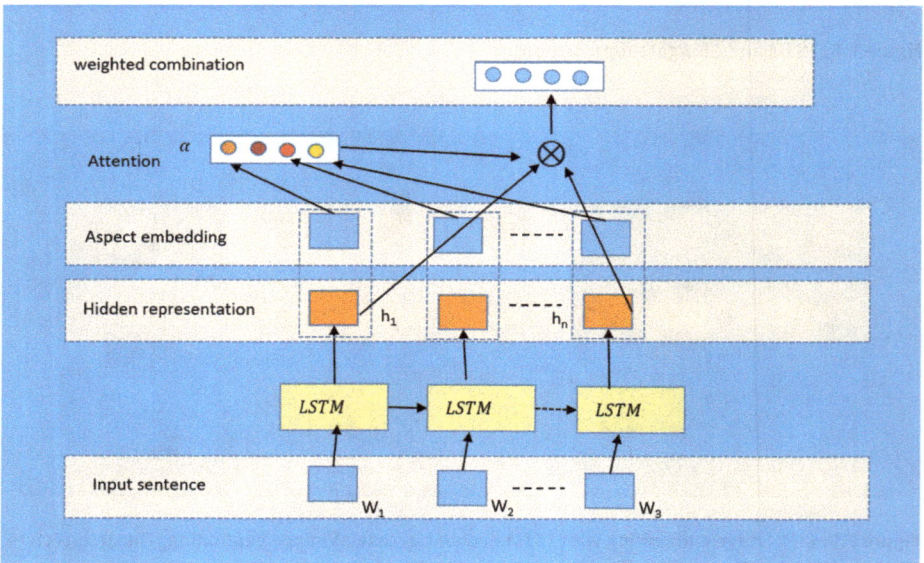

Figure 7.10: Attention-based model from Wang et al. considering aspect classification.

The selected region of the sentence was considered in attention mechanism based on
the influence on the aspect that was to be classified. Sentence level and target level
attention, based on a hierarchy of attention are used to bring out the common-sense
information on the aspects, which is targeted for sentiment analysis. Intuition-based
application of the attention networks makes them an active area of exploration for the
researchers. Figure 7.10 depicts Attention-based model from Wang et al. considering
aspect classification.

7.4.5 Applications of RNN

Word-level classification with RNN: Word-level classification is a strong area for RNN application proving the state of art results. NER with bidirectional LSTM was proposed by Lample et al. (2016). This took care of the limitation of the fixed window issue, taking into account the long length of the inputs, taking into account context around the target word. Output considered the fully connected layer that is built on two fixed-size vector representations. Finally, entity tagging is done by the CRF layer.

RNNs also have been used to model for LM compared to traditional models based on the count's statistics. Graves et al. (2013) introduced multiple hidden layers in RNN to improve the model and counter the complex sequence accounting for context over a long range. This work would be extended beyond NLP. For predicting the words, replacing the feedforward network with RNN was checked by the work of Sundermeyer et al. (2015). Their work demonstrated the sequential improvement in the count-based tradition method's approaches, then feedforward networks, followed by RNN and LSTMs. Extension of their work on statistical machine translation was an important aspect.

7.5 Recursive neural networks

Natural recursive structure hidden within the language makes RNN a good fit here. Words are combined in a hierarchical format to form the representation that is depicted with a parse tree structure. Sentence structure syntactic representation is thus done with tree-based models. Child representations for parsing tree's nonterminal node are devised in case of recursive neural net.

Network *g* is composed of words (a, b, c) in too high-level phrases (p1 or p2). All nodes are represented in the same form. Parsing is the key area of application for this. The score is done on top of high-level phrase representation to arrive at the likelihood of the phrase. Best trees are figured out with beam search. Node-level sentiment tagging was possible with phrase-based sentiment analysis by Socher et al. (2013). Acting upon the gradient vanishing problem is the LSTM application on tree structures (Tai et al. 2015). Improvement in sentence relation and sentiment was demonstrated in this work compared to the linear model of LSTMs. Figure 7.11 depicts Lower-level representation formed into high-level representation in an iterative fashion using recursive neural net.

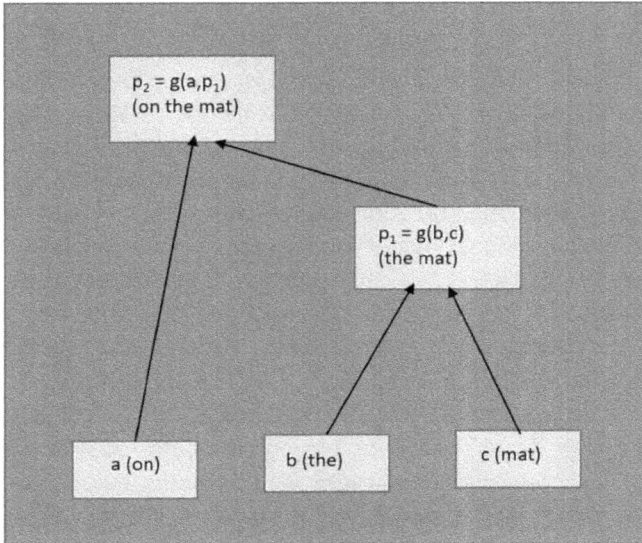

Figure 7.11: Lower-level representation formed into high-level representation in an iterative fashion using recursive neural net.

7.6 Deep learning for NLP with statistical source code modeling context

Machine translation context is used for the software source code which must be migrated from one language to another. Code completion for the developer in software development in the integrated development environment (IDE) context is explored. Text summarization of the code and code convention are the key areas. Software engineering applications' accuracy can be improved based on the application of DNN-based LM, treating the code parts as a text for the processing purpose.

One of the major possibilities is the hidden structure that is seen in the source code and its regularity compared to natural language context. As part of searching the code snippets from the vast amount of code sitting in a public repository, natural language content associated with those code content provides good input. Extension of this can be seen if the content must be used for education purposes. Software source code modeling looks at overcoming the dependencies on the software metrics needed to build the features as data collection on these metrics may be a burden.

Software source code bug summarization is a key area to help in the optimization of software quality. Approaches focusing on the utilization of only the source code part only, another approach of using only natural language part with source code, and the third part where a combination of the both are looked at. Automatic documentation of the source code, for purpose of end user support is a focus as well.

Source code presents the challenge of vocabulary getting refined due to the extensive change in the parameter names over the period. Exploration in the open vocabulary neural language model can help, where the size of the vocabulary need not be fixed. Exploring training corpus from similar projects and learning from different domain projects has helped extend the key knowledge from one area to the other, and transfer learning has its part to play in these contexts.

Utilizing the open-source code for building tools for software development is a great area. Software code naturalness is another exploration area that paves the way to draw similarities between source code and the natural language. This study also helps to look at statistical language modeling for the purpose. Software requirements traceability is manageable across the life cycle using language models that depend on the sequence to sequence embedding.

Recommender system -based models are used for the facilitation of the maintenance of the code by developers. The growth of the vocabulary is a higher rate compared to that of natural language. Peer review is a key practice in software development for ensuring code quality; this presents the opportunity as there is a similarity to the large portion of code review feedback being in the area of code style and documentation, to name few. This kind of trend makes the process less focused to look at all-around issues, and hence automating this area is an important one. Possibility of deep learning to make use of the historical code review information will be key.

In the contexts of rule-based system scenarios, it is helpful to use decision tree algorithms as a starting point of optimization. Variable naming areas of the software domain have found the application of deep learning graphs that can be extended to the identification of issues.

7.7 Key areas of application of deep learning with NLP

Deep learning brings the hope of reducing the expert support needed from linguists, managing with a good amount of data. Machine translation and speech recognition are the areas that have a groundbreaking demonstration of the capabilities of deep learning in NLP and living up to some of the hype that is created across it.

Possibility of integrating the deep learning techniques as a component plugin into the NLP scenarios are area of exploration. Sequence-to-sequence prediction is great hope for a wide variety of NLP challenges. Feature engineering automation would be effective rather than depending on the handmade features from the expert. Models have demonstrated the continued refinement and improvement over each other. Possibility of building the end-to-end solution in the landscape of NLP, would take it to next level.

The exploitation of the nonlinear relationship is a major contribution. Long-range sequence learning by the power of the deep learning methods are to be explored. Models refining to make use of the Marconian principles have a greater contribution. In the context of automated learning, it wins compared to a conventional handmade feature as it creates the possibility of continued learning to keep improving over time and avoid the limitation of oversimplification that happens in case of hand made features.

Compared to building the special model's integrated architecture, deep NLP provides the model that has end-to-end capabilities that make the development and management of the system easier. Neural machine translation systems build robust learning capabilities based on the sequence to sequence modeling format, rather than looking at a piece by piece model work. Even with the end-to-end models, the composition of the various neural networks brings in the challenge of various errors in each of the networks that point toward building one-shot learning of the entire model landscape.

Sentiment analysis of the review on Amazon and movie review on Netflix is a high-profile use case. Text-to-speech and speech-to-text translation finds its core area as well. Speech recognition of the acoustic signal is processed into a corresponding sequence of readable words by a human. Text captioning of the movie and TV shows are a prominent area. Image content captioning finds another prospective area; this comes in handy and is related to object detection from the images. Question-answering systems are under extensive work with online support for the customers needing attention. Bots have been taking over, but the need for perfection in the system is prevalent.

In the area where data availability is the challenge data augmentation can play a role. Here, the existing training data is modified by retaining the original labels' characteristics. In the case of computer vision images are cropped, flipped, and rotated to synthesize the data. These work in the case of images but not on text.

Generative adversarial networks (GANs) have been making rounds in the artificial intelligence world. This will include two parts of network one that generates the output based on labeled data; the other one is adversarial that tries to figure out if the generated output is coming from true training corpus or not. In the process of refinement, each part of the network tries to get better than each other. In the context of challenges involved in data collection, this may be a handy area for synthesizing the data. Deep fake videos and the authenticity of the information that floats online have seen the ongoing battle between the group that is trying to create a bad impact on society with the fakes and the other group trying to counter it.

Transfer learning is finding its place again to optimize the resources involved in training, as the knowledge gained in one area can prominently be used in the other area with minor refinement to be done on the new area as needed. This contributes to the area of feature extraction as that is an expensive part of machine learning.

Pretrained word vectors have opened the field with Word2Vec and FastText making a mark for having embedding layers readily trained for large corpus data in an unsupervised way that can be applied further on the domain of our choice with data from the new domain. These play a key role in reducing the parameter involved in the model as they capture the pattern from a large corpus of data, which helps prevent the overfitting.

Model interpretability is a big area of exploration; many approaches have been proposed, including the one-off using the output of the black box model as input for the interpretation model to generalize the interpretation. Medical and banking applications have seen these drawbacks being impactful. Methods of using the sensitivity of the inputs on the model output are explored. Also breaking up the final decision in conjunction with the input parameters is explored. This is also important to build trust among the community that invests in the deployment of the model.

The self-driving car is another major talk of the town that involves RL. What was science fiction in the past is soon going to become a reality. RNN, CNN, and RL have been composing the architecture in this area. Planning the path, perception of driving scene, behavior simulation, and controlling motion are the key aspects of these algorithms. For the driving scene, simulation key areas are objected detector based on bounding box, segmentation by semantic, instance, and localization policies. The safety of these self-driving systems has been a larger concern in the community. Pedestrian detection with CNN with no common definition of safety in deep learning scenarios is a key area to look at.

Chatbots are replacing human conversations and have made an impact on programs like Siri, Google Assistant, and Alexa. For profitability increase and cost optimization retail and telecom have been largely interested in chatbot-enabled systems. NLP plays a bridging role between the conversations in which the bots are getting involved to automate the human intervention. Overall chatbot architecture will consist of intent and entity classification to gather the input message's intent and entities. Response generation based is on context identified and shortlisting the response to be output. Stability, scalability, and flexibility are the key areas that need to be focused on ensuring that the chatbots can effectively be built for complex scenarios.

Chapter 8
Impact of machine learning in cognitive computing

8.1 Introduction

In case of ambiguity involved in complex situations, computerized models simulate human-like thought in cognitive computing. Cognitive computing has a closer association with that of machine learning (ML)and artificial intelligence (AI) and shares common grounds. Cognitive computing will help the systems gather information from various sources and weigh them based on multiple factors. Cognitive learning includes data mining–related self-learning, natural language processing, and pattern recognition capability to mimic the human brain. Cognitive computing needs the ability to be adaptive, interactive, iterative, stateful, and context based.

In this chapter, exploration covers the possibilities of AI and ML for cognitive computing. It starts by exploring cognitive computing for ML architecture and also exploring challenges presented by ML. The computing stack of cognitive systems explored includes ML capabilities. Learn the cognitive radio and its importance by digging into the learning capabilities of CR (Cognitive Radio). From ML perspectives, exploration happens in supervised and unsupervised areas. The cognitive radios decision-making process and various modes are studied.

Cognitive computing is the miniature of human thinking in computing language. The main objective of these systems is to create self-managing computerized systems that can solve complex problems. In simple terms, these are systems that use existing knowledge to create new knowledge. Cognitive computing is making its presence in chatbots, sentiment analysis, face detection, risk management, fraud detection, and host of other applications. The natural language landscape will drive the future of cognitive computing and its new progress. Natural Language Processing (NLP) creates opportunities for contextual analytics, managing sensor data, and visualization based on cognitive capabilities.

8.2 Architecture for cognitive computing for machine learning

Model organization in a heterogeneous computing environment for ML is explored. For ML to be at the scale, it is essential to have a structured knowledge representation in place. Work here explores a combination of distributed computational graph knowledge with other knowledge areas resulting in accelerated ML applications. Use cases like recommender systems are attempted for resolution by extending them to a distributed cognitive system of learning where real-time ML issues get tackled.

Business intelligence was the start of information technology, which evolved later to analytics and further into real-time insights.

https://doi.org/10.1515/9783110703399-008

During the crises, the business had derived key insights. These come from live data analytics keeping the context in purview as per knowledge from the historical data. Extract transform load (ETL) job managed processes are now handled with inference engines that are based on deep learning (DL) and ML. Knowledge base systems are advancing at rapid speed. The ecosystem is getting universal acceptability in the light of open-source frameworks and algorithms being explored for commercial infrastructure. An area like autonomous driving, personalized medicine, and other areas need scaling. Figure 8.1 depicts Information technology evolution.

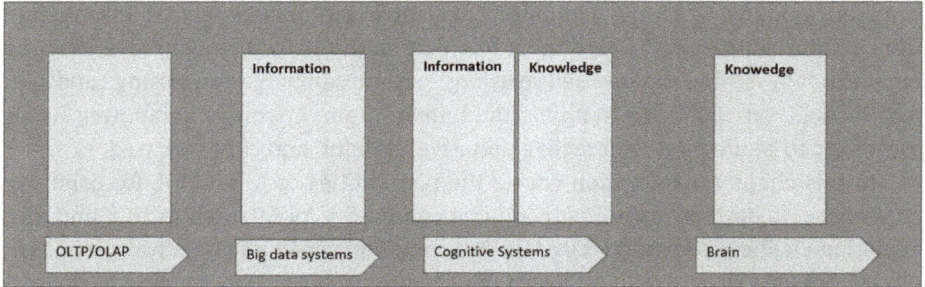

Figure 8.1: Information technology evolution.

Machine translation is closing up toward human accuracy, the large-scale image classification problems overtake the human. Figure 8.2 depicts Language translation capability of DL and ML. Figure 8.3 depicts Image classification process automated.

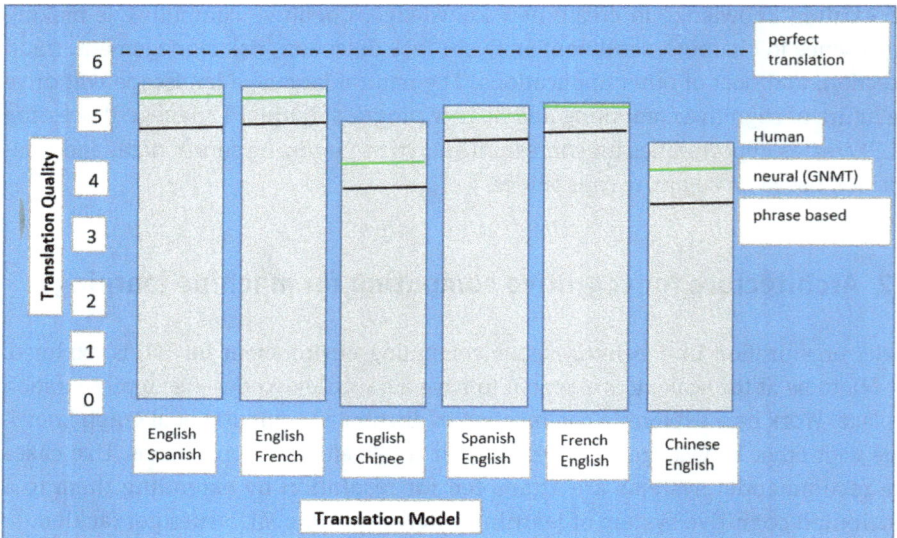

Figure 8.2: Language translation capability of DL and ML.

Image Net Image Classification Task

Top 5 Classification Error %

Error rate reduction
due to deep CNN

Hand crafted
feature
design

Deep CNN based design

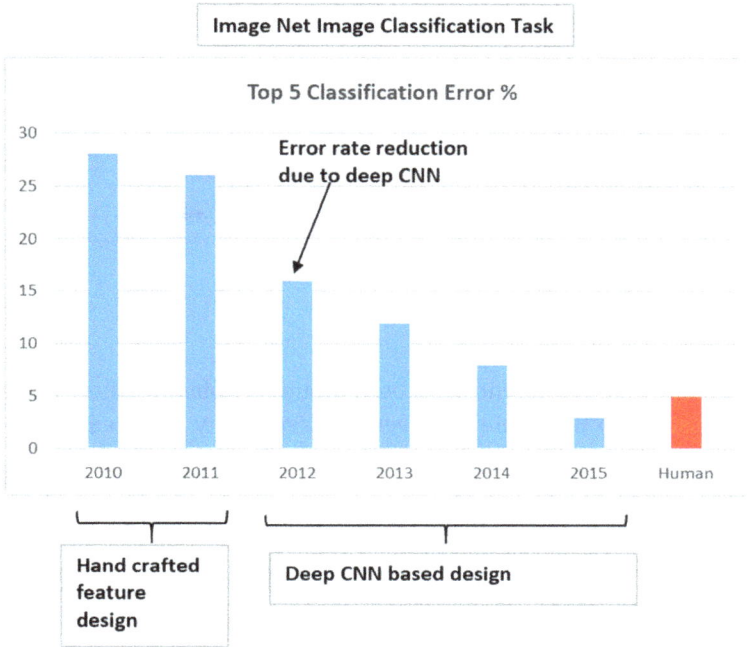

Figure 8.3: Image classification process automated.

8.2.1 Machine learning challenges

Thousands of engineers must get together to build ML solutions that can build successful user stories. As the model gets larger and consumes more data, it gets better accuracy. With complex use cases, computation needed also grows nonlinearly.

All these throw up challenges around the model's complexity, time consumed during experiments, and scalable deployments. ML and DL involve complex data pipelines with computational complexities. Codes of ML and DL are said to cover only 10% of real-world systems of ML (Sculley et al. 2015). Infrastructure management and data management take a major part in the effort (Palkar et al. 2017). Figure 8.4 depicts Scaled learning and its computation needs.

Petabytes of data are created in models computing with exaflops, which make ML and DL systems compute-intensive. Experiments of medium size with well-known benchmarks run for days, with the data scientists' productivity traded off. Just after a hand full of nodes, distributed scaling gets impacted by messaging, data locality, and synchronization issues. Economic returns have largely impacted data growth, beating Moore's law. Due to model partitioning, messaging, synchronization, and locking issues crop up. Challenging storage and poor data locality issues are results of computational complexity. These are the key challenges of infrastructure.

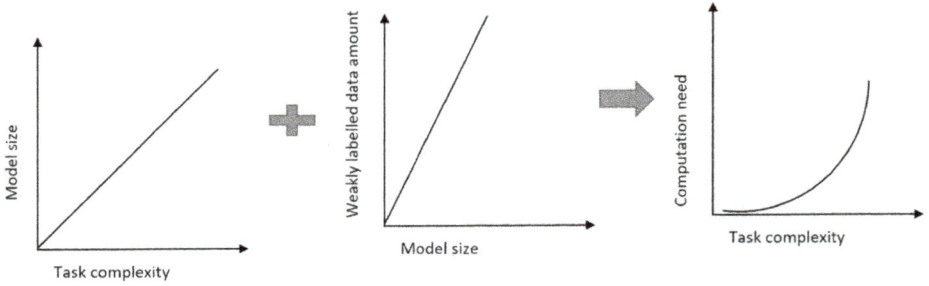

Figure 8.4: Scaled learning and its computation needs.

A major focus of the research recently has been around automating these pipelines as much as possible, so that there is universal acceptability to get experts to start focusing on the quality of the production systems. Figure 1.5 depicts Machine learning systems are hidden technical debt.

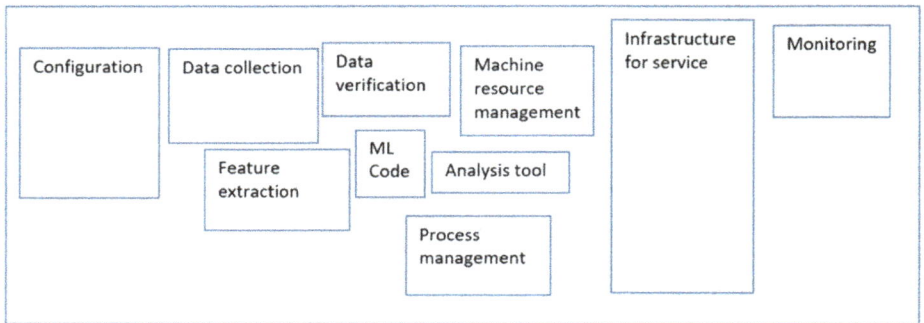

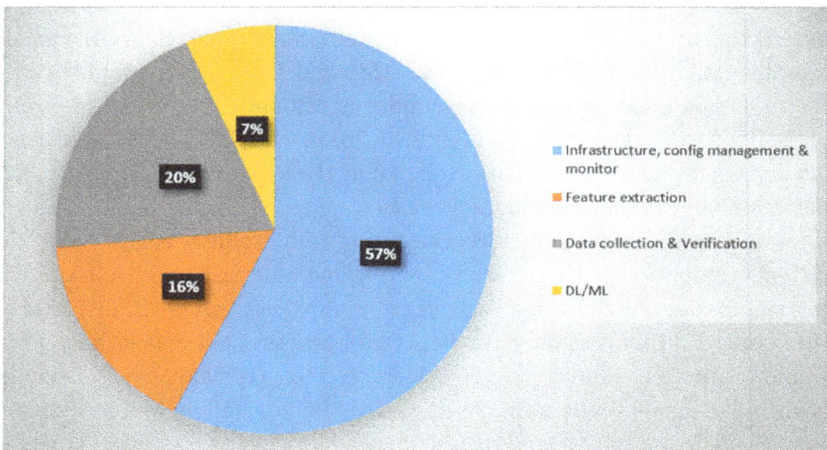

Figure 8.5: Machine learning systems are hidden technical debt.

8.2.2 General review so far

Designing a recommender system for IoT that is networked on multiple devices is a challenging area of exploration. Cognitive capabilities come in when the systems can learn from their known environment and improve their performance. A context-based recommender system is a key part of research in autonomous driving. The complexity comes from the fact that the systems in this IoT network are not just limited to consume the context-based information but are expected to execute a wide range of processing for ML. The general framework lacking in the recommender system is a challenging aspect. Cognitive systems can handle this challenge to manage a common framework.

The general cognitive framework needs to focus on the part that focuses on behavior goals, then the part that analyzes the current proceedings of the network and takes decisions and monitors the environment. A recommender system is a content-based, ratings-based, demography-based system to name a few. Based on these thoughts IoT systems with cognitive recommender systems can be focused on.

Various Machine translation areas are, statistical ML, rule-based ML, hybrid ML and neural ML. Neural machine translation (NMT) is a new trend that is inspired by AI. In these systems, multiple layers focus on refining the data and getting the accuracy high. Systems have similarities with the way human neurons operate. NMT excels in similarities capturied between words, which compete for sentence analysis. One of the key challenges is, the system is a black box, due to the complexity of calculations involved, gradients computed and attention mechanism in play, though the outcomes are extremely exciting, they are not explainable.

Statistical machine translation (SMT) was the approach until Google came out with NMT. It is effective for a specific subject area; it needs less space and operates with multiple existing platforms. But when it comes to the complexity of the language's syntactic structures, it does not do well. Rule-based machine translation is the first of approaches in the area that works with grammatical, lexical, and software program rules. Translations are based on the pattern similarity between the languages. The goodness of these systems is its ability to explore semantic and syntactic levels of the language. Many rules may at times be its weakness to manage the dependencies between the rules.

Hybrid machine translations are a combination of approaches like the statistical and rule-based combination. Machine translations are not intended to replace human translators but to get the speed and precision up. These would be handy in cases where massive data translation, proofreading, and reviews is to be done.

ML finds its challenge in the security of the data, infrastructure needed for experiments, business models on different formats, and investment possibilities from the enterprises. Data collection and managing is the key concern to start with challenges on infrastructure side popping up with much less progress that has happened on the understanding in the industry around the benefits of deployed ML solutions. There is a greater need to provide the lab setup for experimenting with the optimal approaches and then roll the best production options. Collaboration is needed in the industry to share the experimentation of the approaches and try the best one, with enough room for personalization that will enable innovation.

Having an adaptable business framework is a need for ML to thrive, there should be a good ecosystem to change over once there is a clue of low performances. The organization being ready to invest in the costly resources of data science is also a key thing, but to trade off this investment with the approaches taken to establish ML team will help to make better decisions.

Challenges related to ML infrastructure: Data constraints due to silos, experienceed by the data from different sources is a concern. Reliability and capacity is a question with a centralized database. Scalability of the systems and ethical principles should be accounted for even if the automation gets more sophisticated. The integration of the ML systems into existing organization systems do not have the best frameworks. To get confidence in the ML system, it is important to build a pipeline that can efficiently handle the training, testing, and deployment with scalable capability; this is as important as getting confidence in the model outputs.

Internal tool development also holds key; here smaller companies suffer, whereas the larger setup can afford to invest. For effective ML implementation, some of the key inputs needed are managing large datasets, on-demand resources for computation, in case of cloud platforms, required accelerators and potential to manage with rapid advancement in technologies. In the overall ecosystem of ML, only a small portion of the system is ML code, surrounded by supporting infrastructure. These supporting systems are about testing, debugging, server management, monitoring, and so on.

To handle these infrastructures it is important to look at the DevOps principles in the light of MLOps (ML operation). Like DevOps focus on optimizing the development cycle, improve the deployment pace, with a focus on continuous integration and delivery, a similar line can be taken up for ML systems.

8.2.3 Computing stack for cognitive

To address the key issues of ML that were discussed in an earlier section of work under review (Samir et al. 2017), self-aware and self-managing computing capabilities are explored (Mikkilineni et al. 2012). With application performance and data security being the focus, in line with the application's objective, enabling automated management of the computing infrastructure is the focus. Application performance is governed by rules for computation, and the definition of workload and policies. Figure 8.6 depicts Dr. Rao Mikkilineni's computing model for cognitive distribution.

The objective mentioned above is promising to assist us in segregation of the core application logic from the concerns associated, which takes complexity to backend and

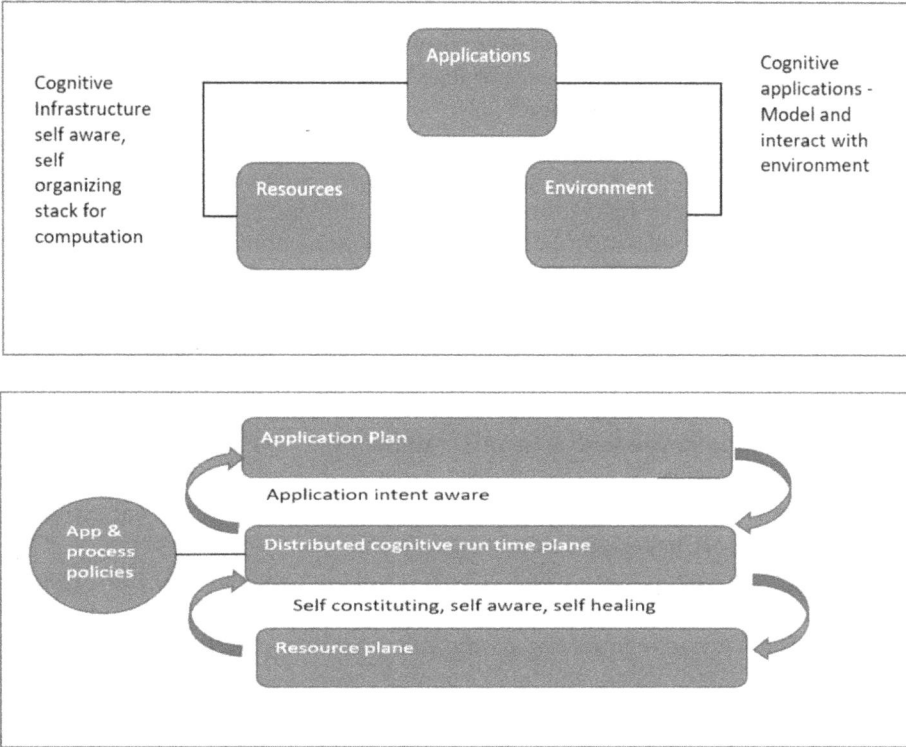

Figure 8.6: Dr. Rao Mikkilineni's computing model for cognitive distribution.

helps experts focus on ML and DL breakthroughs. Develop the capability to do scaling on the economic scale and achieve greater performance with less engineering from a human. With a focus on reliability and data movement, which are key concernsof ML and DL system design, the best of all the system components are made use of. Handling complexity with scaling capability, to achieve the goal with the maintenance of power and performance, the system will be able to leverage on the knowledge accumulated and make use of it. Data, applications, software and hardware systems' resilience can be managed, and the outcomes will be dependable (Samir et al. 2017).

8.3 Machine learning technique in cognitive radio

Cognitive radio (CR) and its learning problems are explored here, and looking at the role, AI can play to get the real-time cognitive communication systems operating. Decision-making and feature classification areas of learning problems for cognitive radio context are explored. Different observation model is derived from feature classification and policies, and decision rules are derived from the feature for decision-making. Unsupervised learning and supervised learning algorithms are the contexts. The discussion will revolve under decentralized network and non-markovian environments situational learning challenges.

Cognitive radio is a radio device that can understand and adapt to the environment (Mitola et al. 1999, 2000). Cognoscere (to know) from Latin is the meaning of cognition, and it stands for gaining knowledge, thinking, judging, and problem-solving (Giupponi et al. 2010). Self-learning or autonomous programming is the mark of CRs (Costlow 2003, Jayaweera et al. 2011).

CR needs to be knowing its radio frequency (RF) environment, sensing all types of RF activities. So, spectrum sensing has a greater significance, and multiple explorations have happened in a variety of sensing techniques (Letaief et al. 2009, Wang 2009, Yucek et al. 2009). Learning and reasoning is another key ability to be demonstrated by the CRs (Jayaweera et al. 2011, Mitola et al. 1999, 2000, Mitola 2009). These will be part of the core engine with cognitive capabilities. Cognitive engines will coordinate with the actions of CR with ML-based algorithms. Recent focus has been on applying ML to CRs (Bkassiny et al. 2011).

Learning takes importance when there is a situation where we do not know the impact of input on the output (Clancy et al. 2007). When putting it another way when we are not aware of input and output functions, the learning process needs to figure out functions that will facilitate designing the appropriate inputs. When taking the example of wireless communication, uncertainty can be seen due to not ideal wireless communications, so learning rates can be applied to reduce error probability over wireless by reducing the coding rate and getting it to an appropriate level (Clancy et al. 2007). Channel estimation is straightforward to solve. But when the degree of wireless network freedom increases, it is highly reconfigurable, especially in software-defined ones. This brings in many parameters that are to be configured like coding scheme, communication protocol, and sensing policy. This case brings in a need for stronger computation setup beyond simple formula. The primary contribution is from the complex interaction of these parameters within themselves and on the environment.

Learning algorithms then come into picture to put it all together even without understanding these parameters, facilitating easy adaptation of CRs to their environment (Sutton et al. 1998). Threshold-related algorithms took up reconfiguration of the sensing processes with uncertain conditions for CRs (Bkassiny et al. 2012, Gong et al. 2009). The situation gets more complicated when the CRs have to adapt within the network and across CRs. CRs must pick up appropriate action just by scanning only a few nodes. The dynamic programming of the Markovian decision process (MDP) (Puterman 1994) will not suffice in the case of CRs operating in an unknown RF environment. Reinforcement learning type of special algorithms enables optimal solution assessment for MDP even without probabilities in the transition. Thus, learning algorithms assist to adapt to the environment and coordinate with per radios. Focus also would be to have less complex learning algorithms to have fewer complex systems.

Work done on this shows that supervised and unsupervised approaches have been explored in the works of Dong et al. (2010), Huang et al. (2009), and Ramon et al. (2009) where neural networks based on support vector machines (SVMs) with supervised learning are explored for CR application. Reinforcement learning (RL) is leveraged for dynamic spectrum applications (Hamdaoui et al. 2009, Yau et al. 2010). CR applications have shown good results in the case of using the Q-learning algorithm (Galindo-Serrano et al. 2010). Weight-driven explorations have improved the efficiency of RL (Jiang et al. 2011). Signal classification (Bkassiny et al. 2012) used a Dirichlet process (DP) for learning nonparametric Bayesian unsupervised learning (Han et al. 2011). For, non-Markovian multi-agent systems through Q-learning-based RL provides a good framework for unsupervised automation learning, but performance is not up to mark (Baxter et al. 2001, Claus et al. 1998, Croon et al. 2005, Sutton et al. 2001). Among learning mechanisms, policy gradient methods, imitation learning, instruction learning, and evolutionary leanings have done better than RL. For optimal policies search from the environment policy gradient approach does well (Baxter et al. 2001, Sutton et al. 2001).

For designing learning policies for CR networks, multi-agent environment learning has been explored. Individual and group behavior was tracked like in a human society with a cognitive network (Xing et al. 2008), also a strategic learning framework of cognitive networks is proposed (van der Schaar et al. 2009). In a evolutionary game framework, during the users' strategic interaction, based on cognitive, adaptive learning as targetted (Wang et al. 2010). To avoid selfish learning at the node level and to have coordination across CR network, a cooperative scheme-based optimal learning method can be devised. This will be important considering the fact of the distributed nature of CRs.

Coordination of action is the main challenge in CR networks for learning (Claus et al. 1998). For the whole network to have optimal joint actions, centralized policies for CR networks will be a good approach. But these may not work in all scenarios. Cognitive nodes will help apply decentralized policies in a distributed network to avoid communication overhead among nodes and result in optimal behavior or close to it. Docitive network proposed in Palkar et al. (2017) which is knowledge transfer based on the wireless medium, focused on increasing the learning rate and generating reliable decisions. Here, there is interlinked teaching between radios which may take additional overhead during this knowledge transfer but will make up by achieving improvement of policies due to cooperative behavior (Palkar et al. 2017). The idea is for nodes to learn from more intelligent nodes, not only the results but also the same path (Bkassiny et al. 2013).

8.3.1 General review of the section

Looking at the cognitive system's abilities, the judgment needed and the available immediate memory is a key aspect. So, it becomes a need to break up the information into manageable parts to avoid these issues. Recoding is another important aspect of human intelligence; it is linguistic-based recoding that plays an important role in thought processes. These systems have a limitation on the risk analysis, need for a systematic training process, and need to augment intelligence rather than AI. A miss in risk analysis is primarily in unstructured data. Risks of accounting various dimensions for decisions like political, environmental, and people to name few. That calls for people involved to finalize the output of decision-making. The need for a large amount of data is anyways a common problem not excluding the cognitive services as well. Current capabilities of the cognitive system are more of an assistant based and not intelligence based. Cognitive services have started setting benchmarks for the next level computing but are limited by the rapid changes and uncertain situations. As the target of data increases, complexity builds. Cognitive systems need to be able to handle multiple technologies so that it is possible to get into more detailed insights.

Markovian models are based on the Markov process that has a variable of the probability distribution at any moment based on the conditional dependencies on many events at different points of time in the past which is equivalent to probability distribution value at the most immediate point in past. To study the Markov model's possible limitation in the area of credit risk rating, we need to explore that possibility of rating the next instance as going down will be based on the trend of last instance. Markov chains assume exponential distribution whereas the credit rating scenario does not fit due to variations when the rating is finalized by the rating authority from the point of assignment of the case. Also, the contributor is the rating and its dynamics change over the different point of time is influenced by the economy; there is no guideline which is time agnostic. These throw light on some dimensions of the Markov model that would be learning in other scenarios also.

Learning techniques may have a wide variety of advantages but will also have limits while we implement. Neural networks need a lot of data that are labeled, and they may not generalize well also. SVM also needs labeled data with previous knowledge of the system, and it gets complicated with the complexity of the problems. Genetic algorithms lack a feasible fitness function and get complex for the complex problem as well. Reinforcement learning (RL) needs policies supporting the training phase. Fuzzy logic needs derived rules and outcome is heavily dependent on those rules, decision trees may overfit and so on.

While we explore reinforcement possibilities it is important to know their drawbacks as well. They should not be used for simpler problems; it is operating with a base of Markovian principles that may not be appropriate in all situations. Curse of dimensionality impacts RL as the law highlights the nonexistence of the entities in the lower dimensions that are represented in higher dimensions. Practical issues like the consistency of data supplied, to take example of data collected from robots suffer the deterioration of the system over time, and the cost in installation of those systems. With all these concerns it may be suitable to use these systems in conjunction with the other systems like DL.

In general, DL has been backfooted due to their inability to be interpretable. Outcome quality heavily dependent on the input and no strong mechanism to validate the quality of input makes it a concern too. There will be a huge disconnect to make the outcomes and actions of these AI systems legal. Challenges in complying with the General Data Protection Regulation (GDPR) will be another area. One example is the need for a clear justification of AI systems being free of bias due to race and gender.

RL including the lights of Q-learning have challenges in terms of difficulty involved in designing their reward function, difficulty in making up over the local optima even with good reward function, the problem of fitting into the weird pattern from the environment, even if they tend to perform well but are hampered by unstable results and difficulty in reproducibility. Some of the life-saving points are local optima, which still can be a good point, better computational capability also helps to build scalability into the system to address data-volume-related concerns; RL can be a good refinement tool, exploring the option of learning reward function with ML with possibilities of imitation learning. Transfer learning is also a good area to explore and improve some of the RL implementation challenges' efficiencies.

DPs are utilized in the case of nonparametric Bayesian models, where mixture models of the DP are used. Sample picked from the DP is a distribution itself, that is, it is a distribution from a distribution. Finite-dimensional marginal distribution based on Dirichlet is the composition so the name.

8.3.2 Purpose

Work that we are reviewing here (Bkassiny et al. 2013) intends to explore the role learning plays in CRs and stresses on the importance of autonomous learning for achieving the objectives of the CR device. There is also and exploration of the best in class ML approaches for CR. Work in He et al. (2010) reviews the hidden Markov models and artificial neural nnetwork (ANN), metaheuristic algorithms, case-based reasoning, and so on. An example of ANN employed for simple lab testing equipment to build quicker prototype provides the confidence on this network to help CR optimize the parameter for channel state and requirement. Ultimately building on this approach to develop process flow-based, case-based reasoning framework for the cognitive engine.

Machine perception, knowledge representation, and ML are a variety of fields in AI. This work focuses on the challenges of using ML for CR with the importance of the learning highlighted earlier. Supervised and unsupervised areas have been explored. Centralization and decentralization are explored and challenged specifically in the case of Markovian environments.

8.3.3 Importance of learning in cognitive radio

Perception, learning, and reasoning are the key aspects of intelligence. Figure 8.7 depicts Learning the knowledge from information collected is possible with smart design.

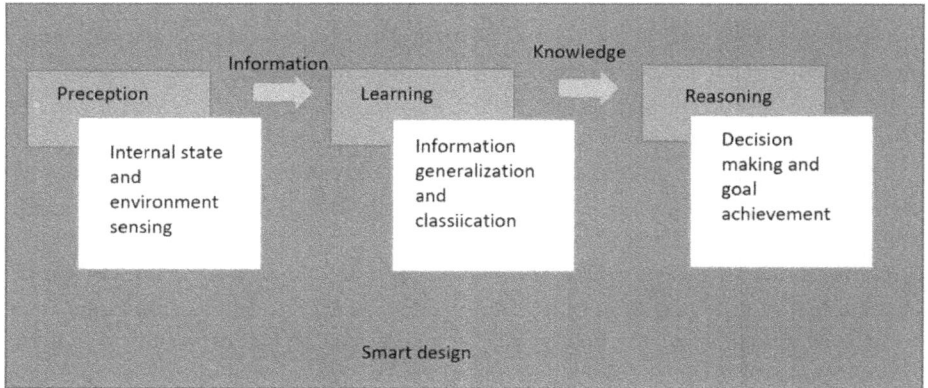

Figure 8.7: Learning the knowledge from information collected is possible with smart design.

For information gathering sensing the surrounding environment and the states of the system is the perception. Classification and generalization by hypothesis methods used to gather knowledge is learning. The goal achieved through reasoning is knowledge. CRs and any other intelligent systems will have learning at their core. That is the base for CR to gather the knowledge from the data.

CRs need to have knowledge, perception, and learning built into them. The sensing mechanism builds perception on how the ongoing activities of the environment get identified. Once these signals are gathered, learning happens to get suitable knowledge by organizing and classifying the needed information. Reasoning helps to associate the knowledge gathered to be used to achieve the required goals.

8.3.4 Cognitive radio learning problems characteristics

Cognitive radio is generally defined as radios that can adapt by sensing from its environment (Clancy et al. 2007, Jayaweera et al. 2011, Mitola et al. 1999, 2000). Learning problems are largely around CRs due to the way they operate and the environment of radio frequency (RF). Partial observations of the state variable only are gathered due to noise in data and sensor errors. So, for learning purpose only a partial environment is available. Figure 8.8 depicts Observations and learned knowledge are used for actions that are driven from decisions.

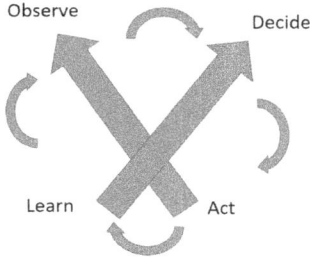

Figure 8.8: Observations and learned knowledge are used for actions that are driven from decisions. New learning happens by understanding the impact the action makes on the system. The cognitive cycle of autonomous cognitive radio is represented (Jayaweera et al. 2011).

CRs in the network will learn in a collaboration that makes it a multi-agent learning process. Cooperative or non-cooperative schemes may be used in learning policies and partial knowledge of other system users. So, in this context CR may apply algorithms that are learning to guess the actions of the other part of the network earlier to make its own decision (Claus et al. 1998). Autonomous learning in the RF environment is key for CRs to learn. Operating at any time and location will be key for the CR system apart from wireless systems (Jayaweera et al. 2011). This helps CR that need not have any earlier information from RF environments like noise distribution or user traffic. Instead, the algorithm will look for the underlying pattern of the environment in the autonomous learning setup. This makes it unsupervised learning for CR systems.

To put it all together partial environment learning, distributed CR networks, multi-agent learning and in case of an unknown environment, autonomous learning are focus areas. All these aspects need to be considered as part of the design of CR.

8.3.5 Supervised vs unsupervised learning perspectives

In case if the CRs operating environment is unknown then unsupervised learning makes sense (Jayaweera et al. 2011). For understanding the environment, and also to adapt to the same, unsupervised learning from autonomous will be useful, and this any operate without any knowledge (Bkassiny et al. 2012, Jayaweera et al. 2011). Supervised techniques can be used if CR has prior knowledge. Training algorithms will efficiently help CR to get the signals in case if CR knows certain characteristics of the signal forms.

Learning by instruction and learning by reinforcement are the forms of learning identified for supervised and unsupervised learning, respectively, in work (Dandurand et al. 2009). Learning by imitation is another format where the learning happens looking at other agents in action (Dandurand et al. 2009). This work highlights that the learner's capability depends on the management of the learner and its environment. So, to make it an efficient process, learner needs to choose a perfect learning option by imitation, instruction, or reinforcement (Dandurand et al. 2009). It will also depend on the circumstances, of which kind works well for the situation. One of the situations would be that learning by the instruction may not work out in the instructor's absence then other options have to be looked at. A good architecture for the CR would be one that can adapt accordingly to environmental needs and characteristics to choose the type of approach. Figure 8.9 depicts cognitive radio supervised and unsupervised learning techniques.

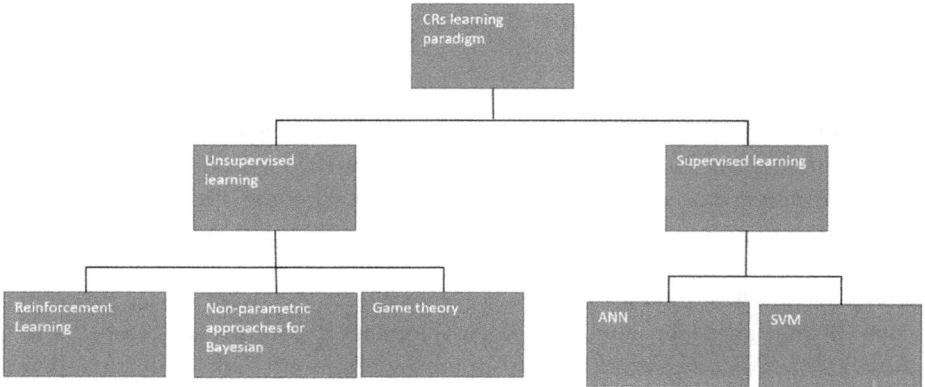

Figure 8.9: Cognitive radio supervised and unsupervised learning techniques.

8.3.6 Issues associated with learning in cognitive radio

To get a good insight into the learning algorithm's functions and understand the similarities, itis important to characterize them into buckets and explore.

Decision-making and feature classification are the areas that are depicted in Figure 8.10 for the problem faced by CRs. These are generalized to cover a wide variety of CR problems. For the determination of spectrum sensing policies, in case of adaptive modulation decision-making problems, classification problem for actual spectrum sensing is applied.

Classification problems are supervised and unsupervised. ANN and SVM algorithms are supervised with the need for labeled data. For risk-minimization algorithms like ANN, earlier knowledge holds the key, which comes from the observed process distribution in comparison to another set that depends on structural models (Hu et al. 2008, Tumuluru et al. 2010, Vapnik 1995). In smaller datasets, structural models like SVM have greater accuracy particularly for the smaller dataset, with their ability to handle overfitting (Hu et al. 2008, Vapnik 1995). Figure 8.10 depicts Learning algorithms for the problems in cognitive radio.

Assuming the dataset for training represented by $\{(x_1,y_1), \ldots, (x_N, y_N)\}$ with condition $x_i \in X$, $y_i \in Y$, $\forall_i \in \{1, \ldots, N\}$, supervised learning algorithms will maximize the given score function by finding an optimal function $g: X \to Y$ (Vapnik 1995). For

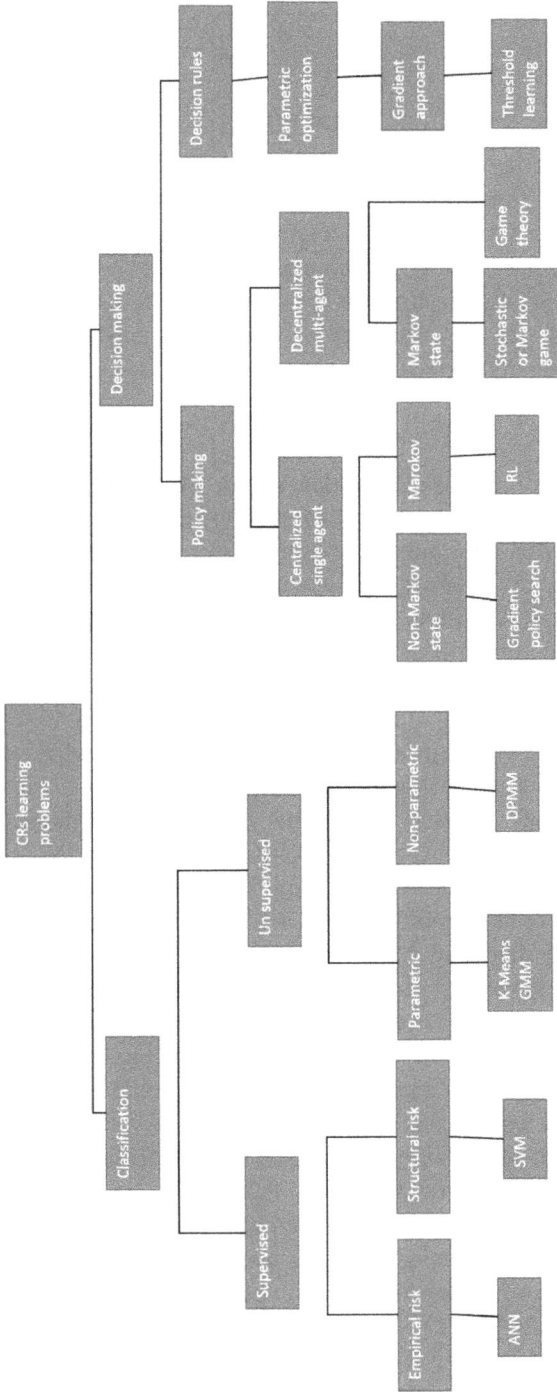

Figure 8.10: Learning algorithms for the problems in cognitive radio. Depiction of learning algorithms hierarchy and its connections.

the case of ANN, g is a function that minimizes risk represented by the following equation:

$$R(g) = R_{\text{emp}}(g) = 1/N \sum_{i=1}^{N} L(y_i, g(x_i)) \tag{8.1}$$

For $L: Y \times Y \rightarrow R^+$ which is the loss function. So, ANN finds optimal function given by g to fit data well. In the case the training set is not enough and the function includes a large number of features, overfitting may happen due to poor generalization in the process of empirical risk minimization. Structural risk minimization with the regularization penalty added to the optimization process will help avoid overfitting (Vapnik 1995). For this risk function to be minimized,

$$R(g) = R_{\text{emp}}(g) + \lambda C(g) \tag{8.2}$$

where C is the penalty function λ control bias and variance trade-off (Vapnik 1995).

Unsupervised learning does not need labeling like supervised approaches so they can be parametric or nonparametric ones. Gaussian mixture models (GMM) and K-means unsupervised algorithms need prior knowledge on the count of classes or the clusters; these are parametric classifiers. Prior knowledge is not required for nonparametric classifiers which are unsupervised. Based on observed data, several clusters can be figured out with the Dirichlet processes mixture model (DPMM) (Bkassiny et al. 2012, Tehetal 2006).

Decision-making is a key aspect of CR applications. Policymaking and rules for decision are two areas of decision-making, where the policymaking can be a centralized or decentralized one. The agent determines optimal policy for a specific period deciding on specific actions for policymaking decisions; this is also called an optimal strategy in game theory. Without earlier knowledge of transition probabilities, RL algorithms can provide an optimal solution for MDP for Markov state (Huang et al. 2009, Sutton et al. 1998). Also direct policy search is possible for the gradient, in a non-Markov Environment in the policy space for achieving a direct solution. The game theory fits in for interaction among the users and the distributed nature of the environment involved in multi-agent scenarios. Conventional game state can be used in normal scenarios but for those that can be defined in Markov state, Markov game or stochastic game model can be applied. To achieve equilibrium under uncertain conditions no-regret-learning of game theory can be used to apply learning algorithms (Han et al. 2007, Latifa et al. 2012, Zhu et al. 2010).

For certain models of observations, hypothesis-testing problems can be formulated which will be a class of decision-making problems. Learning tools will help implement decision rules for cases of observation model showing uncertainty. So that was about the problems classified under two heads and conditions needed for algorithm and its application. SVM finds that its application for labeled training data and for classification problems DPMM algorithm is applicable.

For the CR's uncertain conditions, learning algorithms looked at here will serve to optimize the learning agent's behavior. For a certain type of game Nash equilibrium created by game theory (Akkarajitsakul et al. 2011) utilizes optimal MDP policies provided by RL (Sutton et al. 1998). For empirical risk, local minima are targeted by ANN, whereas SVM finds global minima with structural risk optimization (Hu et al. 2008, Tumuluru et al. 2010). For nonparametric classification, DPMM comes in handy for Markov chain to converge to stationary probability distribution for Markov-chain Monte Carlo (MCMC) gibs sampling procedure (Escobar 1994, Tehetal 2006).

8.3.7 General review of the section

Choosing an appropriate optimization algorithm is a key aspect of neural networks. Gradient descent, stochastic descent, and Adam optimizer are few. Optimization algorithms are targeting to optimize the objective function that is building a relationship between inputs and outputs; the focus here is to work on learnable parameters of the model. The idea would be to minimize the loss by optimizing the objective like bias, and weights are the parameters to tune in the neural network case. Gradient descent is a first-order optimization algorithm that takes a tangential line at a point of optimization. Gradient represents the vector for the rate of change of output to input; gradient comes in handy if the computation of variation of one parameter concerning multiple parameters is to be looked at.

There are compute expensive second-order optimization algorithms, but this accounts for the curvature of the surface and does well in case of stepwise function. Newton's second-order optimization also does well compared to first-order ones as they can do well around the saddle points in optimization function space where gradient descent may get stuck. Gradient descent operates with back-propagation algorithms. Batch gradient descent works on making one update only after the complete dataset is skimmed into, causing memory fitting in, kind of problem.

In the stochastic gradient case, a descent update happens on every data point and is faster. But this frequent update does not help in converging into optimal point and may jump off the clip every time. Convergence instability and variance can be handled using mini-batch gradient descent. Here the appropriate size of the batch is decided and used for processing. These employ matric optimization formats used in DL techniques.

Choosing gradient descent and its variants bring in the challenge of appropriately selecting the learning rates, and the constraint of common learning rate for all parameters may not work well if we have scenarios where the learning rates need to depend on the data characteristics. Saddle points is another critical point during optimization where there is slopping up for one feature and slopping down for another feature. A momentum is a good approach that helps control the optimization by controlling the acceleration in only optimal direction and the rest of the directions are deaccelerated. Adagrad also is an approach that makes learning rates vary as per the parameters. Here, the learning rate also is decided by the time step, based on earlier instances. But the concern here is the depreciating learning rate.

GMMs represent the mix of clusters with several Gaussian, each having multiple clusters of datasets. These clusters have their mean width defined by covariance and size of Gaussian function or probability of the mix. Expectation-maximization (EM) is an approach used in case complexities of the GMM models. It includes initialization, expectation setting, and maximization step. Data clustering has a wide implementation with GMM. Scenarios, where distance-based clustering is done in K-means clustering, are not effective; distribution-based clustering like in GMM works well.

One of the important aspect of GMM is the predefined parameters for the model, like several clusters, where the DP can apply and help generate the clusters and lead to a DP-based phenomenon. Also, to experiment with various combinations of cluster numbers Gibbs sampling is utilized. DP being nonparametric, EM cannot apply for latent variables estimation for keeping the identity of cluster formation. Nonparametric hypothesis tests use the DP classical nonparametric type of hypothesis test which is Bayesian nonparametric versions. Another application of the DP is understanding word distribution in a text document. With the reference vocabulary of size n, then the given document will be represented by its probability function of reference vocabulary size.

For a given range of values, maxima or minima will be the smallest or largest value for a function respectively that is local. whereas in the case of global minima and maxima will be the smallest or largest value across the whole domain. In ML, this is associated with a cost function where the value is larger or smaller compared to its neighborhood. This points to ML as an optimization problem that is trying to achieve the objective with gradient descent.

8.3.8 Decision-making in cognitive radio

RL with centralized decision-making on policy with Markov states: Modifying the behavior based on interaction with the environment by an agent is what happens in RL (Sutton et al. 1998). This is what gives agents their capability to learn autonomously. So, after-action-execution agent gets feedback from the environment, which is the only source of input. Delayed reward and trial and error are characteristic features of RL. Without any earlier information, randomly acting in the environment is the trial and error. For every action taken, the feedback that the agent gets is the delayed reward. Rewards will be good or bad based on the action taken. The agent will look at maximizing the reward making the best moves in the environment.

Figure 8.11 is a depiction of the interaction between CR and its RF environment. For an observation o_t, at time t, for state s_t, delayed reward as another input given by $r_t(s_{t-1}, a_{t-1})$ for time instance t which is the outcome of the action taken a_{t-1} at a

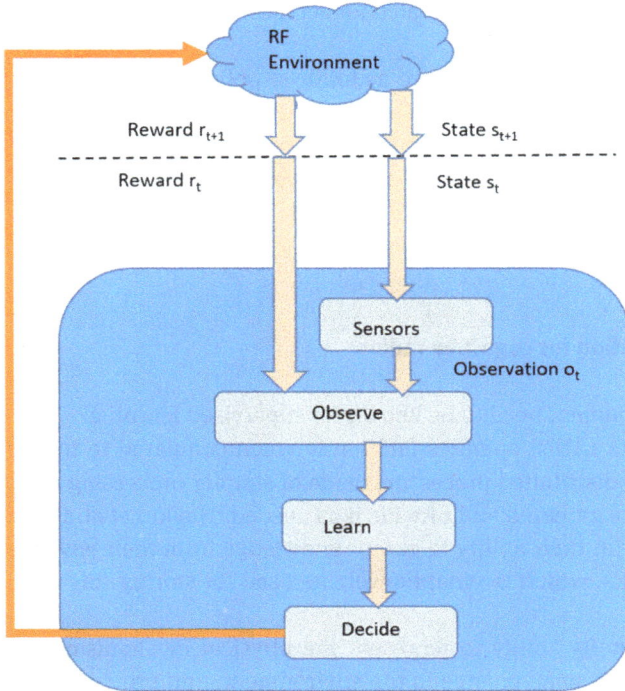

Figure 8.11: Learning cycle of RL: Rewards and observations received by agents in the beginning, policy updates based on these, which becomes the basis for decision-making for an agent. RL-based cognition cycle for CR (Yau et al. 2010).

state of s_{t-1} for time $t - 1$. At a time t, action is computed for observation o_t with delayed reward $r_t(s_{t-1}, a_{t-1})$. State transitioning from s_t to s_{t+1} for action given by a_t for a delayed reward of $r_{t+1}(s_t, a_t)$. The agent here is not passively influenced by the environment but also with its action influencing the environment to move the states to an optimal level with a lookout for the highest rewards.

For interference control at aggregation with RL: Agent environment creating MDP-based interaction is the assumption of RL algorithms. MDP is characterized by a set of decision epochs for different decision-making times and time intervals between decision epochs. The specific set of states for the agent and set of actions, agents can take at different states are important. Another important point is non-negative system state for an action chosen by the decision-making.

Centralized making of the policy for states which are non-Markovian with search based on Gradient policy: For MDP problems, RL can provide optimal policymaking but their performance in non-Markovian situations is not convincing. So, policy-search methods are alternate for these cases (Baxter et al. 2001, Croon et al. 2005, Sutton et al. 2001). Optimal policies are looked for in the policy space, without even looking at the system's accrual states. The policy vector is updated for achieving optimal policy levels in the non-Markovian environment.

Limitations in value-iteration approach are as follows, policies with deterministic action may get selected or not, based on a small change in estimated values of action. This would result in losing out on optimal actions as their values are not estimated appropriately. Gradient-policy approaches have done much better with lesser parameters needed in the learning process and apply in a model-free situation even without much knowledge needed from the system.

8.3.9 Supervised classification for cognitive radios

Characteristics of the environment need to be known for supervised learning.

Artificial neural network (ANN) operates differently when compared to that of computers. Large parallelly distributed processors made of simpler processing units that can store the experience gathered and provide it as needed (Haykin et al. 1999). ANN resembles the brain with their ability to gather knowledge from their environment and connect their units, which is synaptic weights used for storing this gathered knowledge.

ANN is characterized by its ability to represent the physical mechanism with the nonlinear fit, adapt to any small change in the environment, and provide confidence about the decision made. ANN has a limitation as it heavily depends on the training in the environment and is influenced greatly by the selected parameters.

CRs performance evaluation function is synthesized using a multilayered feedforward neural network (MFNN) (Baldo et al. 2008). This is useful as it operates as a black box for CR's measurements, being its function. And these characteristics can be refined with regular updates based on gathered knowledge. Environment measurement and status of the network and its effect on different channels from an ANN-based cognition engine are proposed (Baldo et al. 2009).

To understand how ANN operates in CR context as per work of Dong et al. (2010), multilayer perceptron (MLP) is in operation here where inputs are mapped to their appropriate outputs. Directed graphs connect the nodes in multiple layers in an MLP representing the fully connected network. Each node in MLP is a neuron except for input node, here each of these neurons has activation function of nonlinear nature that is functioning to compute the weighted summation of inputs from its previous layers. Sigmoid is one such common function for ANN:

$$f(a) = 1 / (1 + e^{-a}) \tag{8.3}$$

The architecture in Dong et al. (2010) comprises input, output, and intermittent hidden layer. The hidden layer's number will influence the ANN capability of establishing nonlinear relation between the variables, which are not linearly separable. Also, to be noted that more layers make network complex and time consuming for its computation.

In an MLP network of output y_j^l in jth neuron of lth layer, W_{ji}^l being the weight occurring between jth neuron of lth layer and for the neuron in ith position in $l-1$ layer, output y_j^l can be shown as

$$y_j^l = 1/(e^{-a}) \tag{8.4}$$

$$a = \sum_i w_{ji}^l y_i^{(l-1)} \tag{8.5}$$

Mean squared error (MSE) is used for the computation of error between output obtained and target:

$$\text{MSE} = 1/K \sum_{k=1}^{k} (t_k - o_k)^2 \tag{8.6}$$

where t_k is the target value, o_k is the output value, and k is the count of output nodes.

The update of the function will continue up to MSE reaching a value that is set as the threshold. Authors intended to implement a learner for a cognitive engine with MLP-based neural network. This study showed that the convergence speed reduces though the MSE goes down as the hidden layers count is increased. But more hidden layers call for more data. Accuracy at which the convergence happens and the pace of the same is a matter of trade-off for a given set of training data.

SVM: In pattern recognition and object detection, SVMs developed by Vapnik et al. (1998) are used. SVM specializes with anumber of support vectors, which are not dependent on the dimensions. In this case the margin is acted upon for control and local minima are not existent, and the solution is sparse. Robustness against noise and capability to generalize makes SVM a superior choice for many real-world problems (Atwood 2009).

Mapping the input vector to their higher dimensional space to get a linearly separable state is the feature of the SVMs. Kernel functions come in to play for the transformation of vectors from lower to higher dimensions. The type of kernel function also will depend on the area of application. Polynomial kernel natured Gaussian kernel for the infinite degree is a general choice for classification problems. The hyperplane is figured out that can facilitate large generalization for high-dimension space, which is also called maximal margin classifier (Boser et al. 1992). The distance of the hyperplane and data point closest to it is the margin. Figure 8.12 depicts SVM representation.

Among multiple hyperplanes between classes of data, one with maximum margin will get qualified. Hyperplane with maximum margin is the optimal one for separation and the closest points to hyperplane are the support vectors. The objective

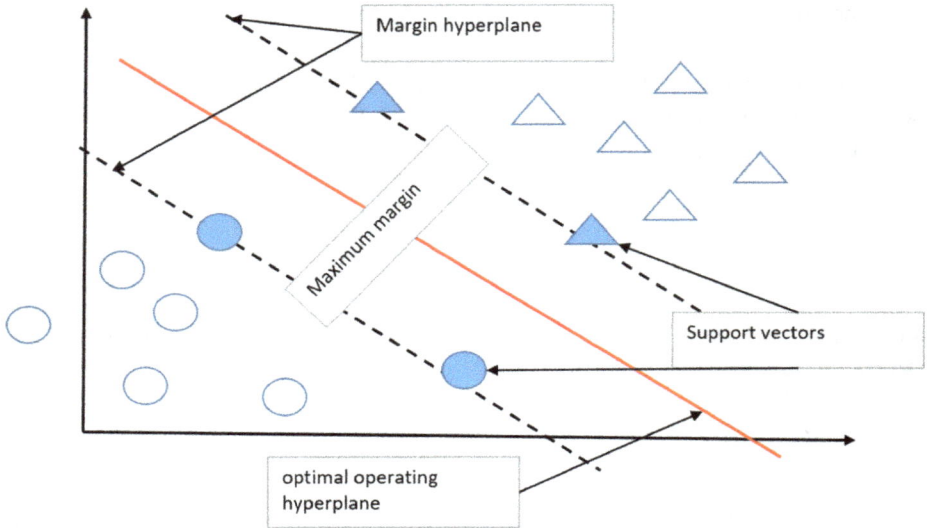

Figure 8.12: SVM representation: Red line depicting the optimal plan separating the data, dotted lines are marginal hyperplanes for binary classification problems, and bold items are support vectors.

is to make sure that the hyperplane separates the classes of data such that the distance of the support vectors to hyperplane increases. SVM also works well for cases of multi-class classification. Like in Petrova et al. (2010), automatic modulation classification was done for studying multi-class signal classification using SVMs. Seven modulation schemes with five digital and two analogs were recognized by training SVM signal–based classifier. The realistic carrier was used for signal generation; results were promising enough to categorize these signals.

8.3.10 Cognitive radio-centralized and decentralized learning

Dependable spectrum detection is not possible in cases where signal to noise ratio (SNR) is small and under certain acceptable range which is a limitation of spectrum sensing due to uncertainties in the noise. When the primary receiver is within the transmission range of secondary users and the primary transmitter not being detected by secondary users, it results in the hidden terminal problem, hampering primary users' transmission (Cabric et al. 2008, Jha et al. 2011, Sun et al. 2011, Zan et al. 2009). Cooperative spectrum sensing improves the spectrum sensing dependability to utilize the idle spectrum using multiple independent fading channels making use of its diversity.

Generally, there is a trade-off between centralized and decentralized control in radio networks in the case of CR networks also. Signaling and processing overhead bother the centralized scheme though efficient spectrum resource management is done. A decentralized scheme reduces the decision-making complexity of the cognitive network. In the case of this decentralized model, the radio may behave selfishly focusing on their utilities instead of network optimization. So, a solution was proposed for CR networks of heterogeneous nature, the network helps wireless users' decision by aggregated information being broadcasted to users in the hybrid approach setup (Haddad et al. 2011).

8.3.11 Review of the section so far

CR plays a key role in enabling users without a license to use the underutilized spectrum from the licensed part without disturbing the licensed users. RL has helped unlicensed users to take actions optimally so that the performance gets better with different approaches for CR, specifically in activities such as channel sensing and dynamic channel selection. Q-learning has been widely used in RL for CR applications, particularly, for cross-layer resource allocation using deep Q-learning.

The non-Markovian process does not demonstrate the Markov property of memory reduction. These are processes where historical events influence the current or future state. Also, in case of demand, if the future product depends on factors different from the factors which current demand depends on then it is Markovian process.

Genetic algorithm approaches are also explored in CR space. Friedberg's initial work tried to mutate small programs of FORTRAN, with an attempt to generate smaller machine code program series, this provides a program that does good performance on the simpler task. These algorithms look in search space to optimize the fitness function by targeting to develop a group of solutions, called chromosomes in this context, which strings binary digits. As parameter count increases, the string size also increases. Multi-objective problems are explored to optimize parameters associated with CR with the target of improving system performance with maximization of throughput also as the target.

The activation function decides the firing of the neuron or not in a neural network. Activation functions are also supposed to be efficient in computation as they are involved in the large computation for neurons across layers. Activation functions in the neural network are nonlinear to assess the complex pattern associated with the data. Activation functions are of a binary step function and linear activation function class. Nonlinear activation functions are more sophisticated and take care of all issues seen in the linear version. Some of the commonly used nonlinear activation functions are sigmoid, TanH, ReLU, Softmax, and Swish.

Kernels are part of the kernel trick in ML where the nonlinear problem is solved with a linear classifier. Where the data that are not linearly separable are transformed into linearly separable ones. Kernel function does transformation on every data point to transform them from lower space to higher space. Kernel helps inefficient memory management.

8.4 Conclusion

Cognitive computing is the miniature of human thinking in computing language. The main objective of these systems is to create self-managing computerized systems that can solve complex problems. In simple terms, these are systems that use existing knowledge to create new knowledge. Cognitive computing is making its presence in chatbots, sentiment analysis, face detection, risk management, fraud detection, and other applications. The future of cognitive computing will be driven by the natural language landscape. Opportunities will be extended into the field of contextual analytics, managing sensor data, visualization based on cognitive.

Chapter 9
Predicting rainfall with a regression model

9.1 Introduction

Machine learning is transforming all the fields, including healthcare, education, and transport. These machine learning algorithms learn every step in our lives to provide us back with more significant life experience. Intelligent gaming, self-driving cars, and even soon it is going to help us cover up the weakness that we humans have physically and mentally. It is going to improve life by taking over the jobs that include high risk and repetition and create fatigue. Environment safeguarding, being such a vital need today, can be rescued by these advancements. Building emotions to these data-based systems are at its peak of experiments. Healthcare will be at its best in history; banking would see the revolution, and digital media will get integrated into human lives. Smart homes and secured homes get targeted, logistics would get enhanced, andcustomization of the human needs and demands met.

Agriculture is the backbone of many developing economies; exploration of these advancements in these areas would make a lot of sense. Damages caused by adverse weather conditions are another central area of focus to be taken care. In this chapter, we explore dimensions of the machine learning and its advancements and relate to the work that is happening for the rainfall prediction to enable farmers.

In exploring the use of machine learning model for rainfall prediction, we will look at regression and time series approach with ensemble deep learning methods, exploring various forecasting methods in the backdrop of machine learning and deep learning models. Specialization of recurrent neural network (RNN), deep belief network (DBN), restricted Boltzmann machine (RBM), and convolutional neural network (CNN) explored. Ensembling of the models and their influence on effective modeling is discussed. In the specific context of rainfall prediction, artificial neural network (ANN) and various other versions of it unfolded for understanding. Exploration of all the work is done so far by multiple authors and their key features.

9.2 Regression and time series forecasting with ensemble deep learning

9.2.1 Context setting

The authors have proposed DBN ensembles for time series regression forecasting in this work. A notable critical feature of the work includes the using support vector regression (SVR) model to aggregate multiple DBN outcomes. Authors have explored the electricity providers as the reference for the case study. The industry is moving

https://doi.org/10.1515/9783110703399-009

toward the competitive arena; frequent power-load demand prediction is a crucial need. Load forecasting for electricity gets complex since the climatic conditions add due to the complex dependencies. These factors also are nonlinear in their influence (Hong et al. 2012, Sousa et al. 2014).

In the case of time series forecasting problem, linear statistical forecasting methods are explored for many decades, auto regressive moving average (ARMA) (Box et al. 2013) being a key one. Exploration of the classification, regression, and time series method (Aung et al. 2012, Al-Jamimi et al. 2011) with a variety of the ANN and various machine learning solutions are explored in recent times. SVR has made its mark with strong predictors compared to the weak predictors of ANN, where ANN get stuck in local minima SVR can hit optimal point. Least square support vector machine (LSSVM) is applied for forecasting the peak load with multivariate (Aung et al. 2012). Online learning methods are employed to do away with trend component of the time series. The support vector machines (SVM) and probabilistic neural network (PNN) fault prediction model is experimented by Al-Jamimi (2011). This work demonstrates PNN's capability on the large dataset, where they used NASA's PROMISE dataset repository.

Until kernel-based methods like SVR came in, ANNs were in full flow but were left behind. A further deep neural network with deep layers was introduced by Hinton et al. (2006) to bring back the interest in ANNs. DBNs with RBMs combination was proposed by Takashi Kuremoto (2012) for time series prediction. CATS benchmark data as used, with Particle warm Optimization PSO) algorithm. DBS proved its upper hand here compared to multilayer perceptron (MLP) and autoregressive integrated moving average (ARIMA). For simulation, deep learning methods are tried on the shallow neural networks by Busseti et al. (2012). This demonstrates the deep learning methods' ability to handle electricity load prediction areas.

For a problem with fault prediction, pattern recognition, and time series method, decision tree and neural networks will have ensembles on top of them, this will help in strengthening the unstable predictors (Breiman et al. 1996). The mining machine reliability forecasting model is proposed by Chatterjee (2014), which was ensemble-based. Genetic algorithm (GA) based parameter optimization is used on the LSSVM model. SVM combo was used to generalize the outputs of the models. A performance evaluation was conducted with a turbocharger dataset. Reliability forecasting and fault prediction saw the good performance with these ensemble models.

The output parameter is associated with all its dependent input parameters in regression problems. This was first explored for a biological setup by Francis Galton calling regression toward mean (Galton 1989). To manage variance in the regression redictor bagging methods various application are proposed by Breiman (1996). Bias variance and strong correlation are other perspectives on which the ensemble models have been demonstrated to work well (Geurts et al. 2006).

Authors here confirm that there have not been any other approaches proposed with ensemble deep learning methods used for time series forecasting. DBN and

SVR are used in the proposed model of deep learning methods. DBNs output rolling up is done by SVRs (Qiu et al. 2014).

9.2.2 General review

Time series problems present the case of studying not just the state but also the associated nature. The challenge here is to balance the short-term capability of the models with the long-term capabilities, as many times these come as a separate capability in the model. Transformation methods have been employed to manage the nonlinear and linear data to fit the appropriate models. In the quest of nonlinear and linear models, there are approaches to combine both effectively to improve overall models' predictive power. MLP as a nonlinear version and ARIMA as linear versions are integrated for the case.

Neural networks are looked at for time series data due to some of the lack of core time series methods like inability to handle missing data, assumption of linearity in the data spread, and focus mostly on single-variable cases. But the neural network's expectation to provide a fixed number of inputs to get specific output will be a concern. CNN's ability to auto feature engineers will make a great contribution.

ANNs ability to generate multiple outputs gives it an edge over the SVR which looks at the single output. The batch-training capability of the ANN also stands out compared to SVR's sequential training. SVRs operate in isolation, whereas ANN operates as a single architecture. One of the key advantages that LSSVM brings to the table is its computational efficiency in comparison to its counterpart SVM, specifically when the approach is about constraint optimization. Large-scale problems have shown considerable advantage using LSSVM versions.

In comparison to online learning other statistical learning techniques assume that data are coming from an unknown distribution. Online learning features, with no assumptions on the distribution and learning happening with individual observation at a time, are created in an adversarial way. The key objective is to gauge the cumulative loss suffered in comparing the performance against the expert baseline.

9.2.3 Forecasting models

SVR: Statistical learning theory is based on SVM as an approach for regression was roposed by Drucker (Drucker et al. 1997). The method works on the logic of structural risk minimization. SVM as an approach for regression was proposed in (Drucker et al. 1997). Beyond the time series problem, SVR finds its application in fault prediction (Sousa et al. 2014) also.

If the power system time series data can be shown as below:

$$D = \{(X_i, y_i)\}, \quad 1 \leq I \leq N \tag{9.1}$$

For m data points, at the time given by i with X_i and y_i representing input and output, respectively, regression function will be depicted as follows:

$$f(X_1) \; WT \, \phi(X_i) + b \tag{9.2}$$

(X_i) maps input vector X to feature space at a higher dimension with b as bias and W as a weight vector. W and b are obtained by solving the optimization problem as follows:

$$\text{Min} \tfrac{1}{2} \|W\|^2 + C \sum_{i=1}^{N} (\epsilon_i + \epsilon_i^*) \tag{9.3}$$

With condition:

$$y_i - W^T(\varphi(x)) - b \leq \xi + \in I \tag{9.4}$$

$$W^T(\varphi(x)) + b - y_i \leq \xi + \in i^* \tag{9.5}$$

$$\in I, \; \in i^* \geq 1$$

For trading off between model generalization and simplicity, parameter C is used with cost errors being measured by $\in i$ and $\in i^*$ which are slack variables.

In the case of nonlinear data involved, kernels can be utilized to represent such data in higher dimensional space to facilitate the usage of the linear model. So overall representation of function for SVR can be depicted as

$$y_i = f(X_i) = \sum_{i=1}^{N} ((\alpha_i - \alpha_i^*)) K(X_i, X_j)) + b \tag{9.6}$$

Lagrange multipliers α_i and α_i^* are involved. Gaussian radial basis function (RBF) is a commonly used kernel.

$$K(X_i, X_j) = \exp\left(-\|X_i - X_j\|^2\right) / (2\sigma^2) \tag{9.7}$$

ANN: ANNs derive their inspiration from the biological neuron.

Input vectors size is decided on the count of neurons in the input layer. The second layer of hidden neurons has a nonlinear activation function. Weighted summation of intermittent layers is done in output layers. Based on the input's influence on output, there is a weighted connection between each of the neurons.

Output representation of ANN is as shown below:

$$y_i = f\left(\sum_{i=1}^{N} (w_i x_i + b_i)\right) \tag{9.8}$$

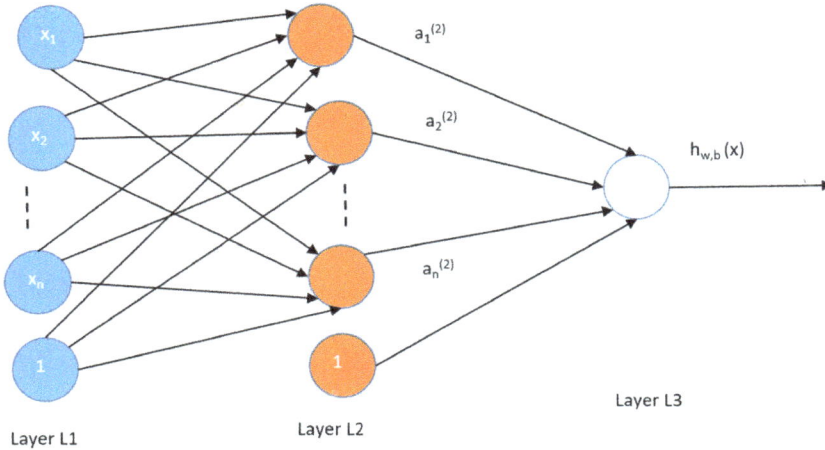

Figure 9.1: Three-layer ANN representation.

The nonlinear functions f holds together the weight w_i, the input x_i, and bias factor b_i to represent the output y_i.

RNN and feedforward backpropagation neural network (FNN) are the applicable ANNs for power load forecasting. Figure 9.1 depicts Three-layer ANN representation. Figure 9.2 depicts Schematic representation of the deep belief network with three layers.

FNN: For basic working of FNN required inputs are provided to the network with the output information mapped. If the network cannot match its prediction with output, then the weights are modified accordingly to reduce the error rate. Error is the difference of what the network predicts in comparison to the actual label. In the process, the network gets more intelligent and gets closer to the actual label. Error terms are represented by mean square error, with overall objective being, training the data points with the objective of reduction to the optimal point as much as possible.

RNN: In the case of RNN, there are additional state capturing neurons in the input layer that will capture the feedback. The context layer plays their role in taking the feedback from the previous layer and passes it on to the layer which is hidden with other inputs with intermittent processing being done on the inputs. The simple recurrent network is the name provided for the Jordan RNN and Elman RNN which are the most frequently used ones (Graves et al. 2009).

Deep learning algorithms: Distributional representation is used in the case of deep learning algorithms. Based on structures associated with nonlinear transformations, the deep learning methods learn high-level features. DBN and CNN are the most used ones.

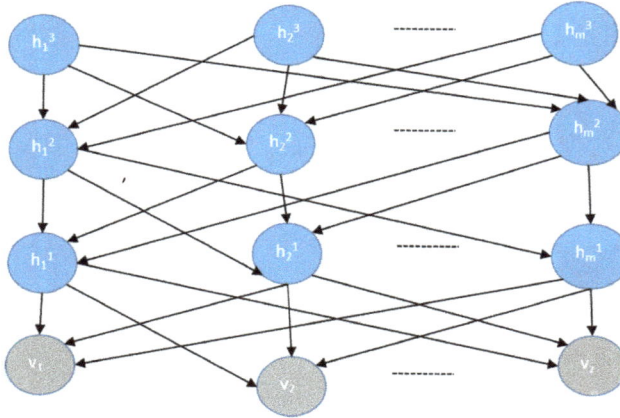

Figure 9.2: Schematic representation of the deep belief network with three layers.

DBN: Multiple hidden layers are characteristics in the case of DBN. Units in each layer are not connected (Hinton et al. 2006).

Unsupervised format of feature extraction is possible with DBN. Further topped up by the supervised component of softmax or SVR.

RBM: RBM is configured to learn distribution based on the input data probabilities. Figure 9.3 depicts Schematic network structure of RBM.

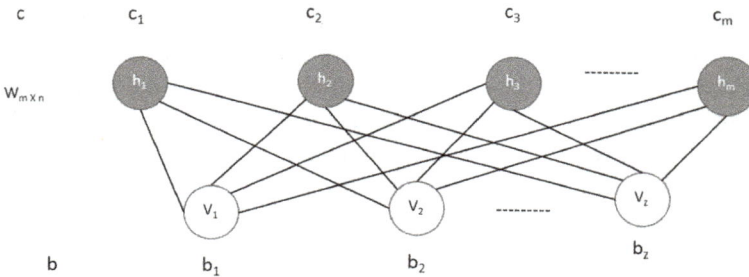

Figure 9.3: Schematic network structure of RBM.

Hm being a hidden layer and v_z being a visible layer, weights of visible and hidden layers are connected by $W_{m \times n}$, and visible and hidden layer offset arerepresented by b_z, c_m.

Stacked autoencoder: Input layer with one or more hidden layer and output layer and are composed in a feed-forward neural network without recurrence in case of the stacked autoencoder. The ability of the autoencoder to construct its output is the differentiation factor compared to the MLP. Key features can be learned from hidden layers with the space of the hidden layers being narrower compared to that of the input layer. This gives it an unsupervised learning status.

Multiple layers of sparse autoencoders compose a stacked autoencoder, where the output from one layer is in connection with the input of another layer in succession. Encoding happening in the layer can be represented as

$$a^{(l)} = f\left(z^{(l)}\right) \tag{9.9}$$

$$z^{(l=1)} = W^{(l,1)} a^{(l)} + b^{(l,1)} \tag{9.10}$$

Decode process can be represented as follows:

$$a^{(n+l)} = f\left(z^{(n+l)}\right) \tag{9.11}$$

$$z^{(n+l=1)} = W^{(n-l,1)} a^{(n+l)} + b^{(n-l,1)} \tag{9.12}$$

For a node in layer l, activation is $a^{n\ (l)}$, the input which is the weighted sum for the unit in layer l is represented by $z^{(l)} z$ which becomes input data for the first layer; weight values are $W^{(l,k)}$ and bias given as $b^{(l,k)}$

At the output of the stacked autoencoder, a neural network with training can be engaged.

CNN: Convolutional and sub-sampling layers (Lecun et al. 1998) are composed in the case of CNN's, feedforward neural network architecture. Deep architecture needs quite a bit of preprocessing which is not seen in the case of CNN. Filter-based features operate at the convolutional layer for purpose of convolution. This reduces the parameters needed in the network as the weights get shared; this leads to a better ability to generalize. Figure 9.4 depicts Convolutional neural network.

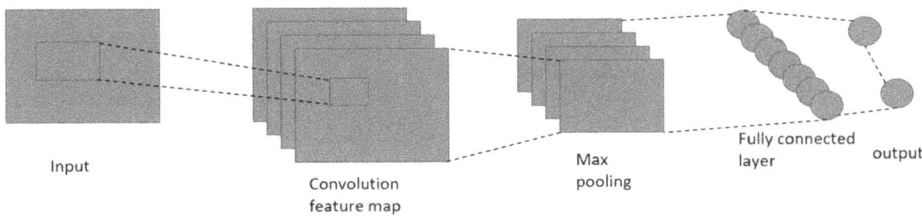

Figure 9.4: Convolutional neural network.

Ensemble methods: To enhance the accuracy of the model, ensemble models are employed which are a combination of various models (Chatterjee et al. 2014). In case of a lack of data, ensemble models help to gather better statistical representation. This being the case there is the possibility of picking hypotheses that provide higher accuracy in line with the learning algorithm. The possibility of concluding on the wrong model can be done away with the ensemble working on gathering multiple models to get the best results. Computationally, these networks work well; models like decision trees tend to conduct the best parameter search in the local space and end up choosing the local best parameter. Ensembles do well by conducting multiple local searches with a different starting point to explore a wide range of space. Taking Fourier transformation of signal processing as an example, generally representational functions cannot be depicted with a single hypothesis statement, so these provide an optimal combination of a weighted average of multiple hypothesis statements. Figure 9.5 depicts Architecture of ensemble methods.

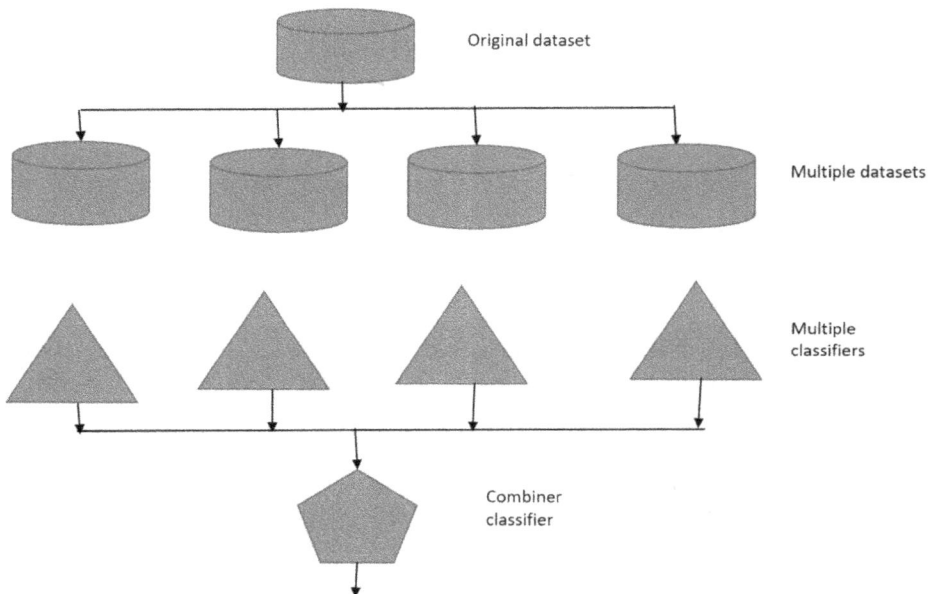

Figure 9.5: Architecture of ensemble methods.

9.2.4 General review of the section

Fault prediction finds its importance in the situations where the accidents must be prevented and the cost of operations must be optimized. Methods that have been explored in the past may not fit well in the context of real-time data processing and prediction. In these scenarios, initial state and every point of operations of the systems are to be predicted, for this online model is used for time series prediction for which SVR comes into play that will operate with combined kernels. SVR has an online version and batch processing version, and the online one works well. Since the time series data mostly involve nonlinear data, SVR cannot manage with a kernel, so using online SVR with a combination of kernels is explored. Kernels are from a local and global perspective. Local kernel focuses on learning from the neighboring states, whereas global kernel focuses on bringing out the general conditions of the fault.

In the context of regression being an optimization problem, linear regression taking a convex optimization problem and its proof is an important aspect, as this leads toward gradient descent as one of the prominent optimization algorithms that work behind many machine learning algorithms.

Kernels have been focused as the key part of the SVM; these are mathematical-based functions that sit on top of the SVM to perform a variety of the transformations as needed to the scenario to fit the data points from input space to output space. Type of data involved also influences the type of kernels chosen, like vector-based data, graphs, text, images, and so on. RBF has its prominent application as it responds with localized and finite way along the x-axis. Exploration on a wide variety of the kernels that fits the time series data especially those that can fit well with our primary focus area of predictions in the agriculture is prospective.

Looking at dynamically changing conditions of agriculture and weather, neural networks have greater potential to fair well. Also, associated data availability is also well off to leverage better. Experiments involving other neural network techniques like SVM and K-nearest neighbors have been tried. Among CNN and other neural networks, MLP has multiple connected layers where the pattern of the data is explored across the various layers, in case of CNN, convolution and max-pooling layers play a different role of compressing the patterns in to a lower form but retain the identity of the features; this facilitates them to be an automatic feature identifier. Apart from these learning rates, several layers and the size of the kernels are other key aspects.

Ensemble methods that are generally used are bagging (bootstrap aggregation) and boosting which are commonly used ones. As in the case of financial planning, the suggestion is to maintain a mix of various investment options, and the ensemble plays similar roles. Also, in scenarios where noisy data are involved, this setup assists in generating a stable performance. On the flip side, ensemble models will make model complex where the interpretability of the model is lost.

9.2.5 Ensemble deep learning methods proposed in this work

Prediction results differ when the number of epochs is changed in the case of time series and regression methods in their backpropagation methods. Combining different versions of the FNN models run with a different combination of epoch counts is also a good approach. Based on how close these predictions are to the labels, it is possible to provide the weights to these model versions to decide their influence on the final prediction. In this approach, ensembles include DBNs that are trained with different epoch combinations on top of which SVR sits that consumes the output of DBNs to produce the final output. The overall process of the algorithm is as follows:

DBN is trained with the input data matrix. About 100–2,000 backpropagation epochs are set up for backpropagation purpose; here step size of 100 is used, 20 predictions outputs are generated from y_1 to y_{20} (Xie et al. 2013). Reinitialization of DBN is done 20 times. SVR predicts the output Y using the DBNs output stored in a matrix called X_{new}, and this provides accurate forecast results.

9.2.6 A generic review of this section

In the backpropagation algorithms, stochastic also called online, batch, and mini-batch are the various versions. The outcome of the backpropagation algorithm is influenced largely by its architecture. Functions are available to change the weights of the algorithm. In the case of ensemble models', weights are, the small positive values that sum up to one, also as discussed, weight represents the prediction's confidence by each of these models. There is no specific approach to deciding weight, but the training dataset or validation dataset can be derived. Also, precaution is needed to not keep using the same dataset to arrive at the weights as that may lead to overfitting. Like we used to hold outset for validation, similarly for weight configuration, similar approach can be looked at. The more refined way would be to do a grid search to arrive at the optimal weight level. Gradient descent also would be experimented with the weight configuration.

DBNs are used in weather precipitation forecasting by representing original data features in target data space with semantic feature representation format. Though a large amount of data is available, it is challenging to get the prediction refined. Apart from the famous ARIMA model, the Markov model and gray theory-based prediction models are employed in this area. In the case of exponentially varying conditions of weather, Markov model and gray theory-based model are best suited. Figure 9.6 depicts Ensemble deep learning model proposed in the work reviewed here.

9.2.7 Outcome and comparison

In the exploration of datasets, techniques used were time series-based; datasets from electricity load demand and Glass dataset were used, and for regression, Friedman artificial domain, 2D planes and California housing datasets were used.

In the case of a blood cell regulation, Mackey–Glass equation-based time series dataset of Mackey–Glass dataset was used. This is one of the most preferred benchmark datasets in literature. 9,000 training data and 3,000 testing data were used in the experiment. Dataset breakup for train and test was done based on period, as in initial months taken for training and later months for testing. Artificial datasets generated based on the equation from these works (Breiman et al. 1984, Torgo et al. 2014) were the 2D datasets. The 1990 census data was a reference to the California Housing dataset.

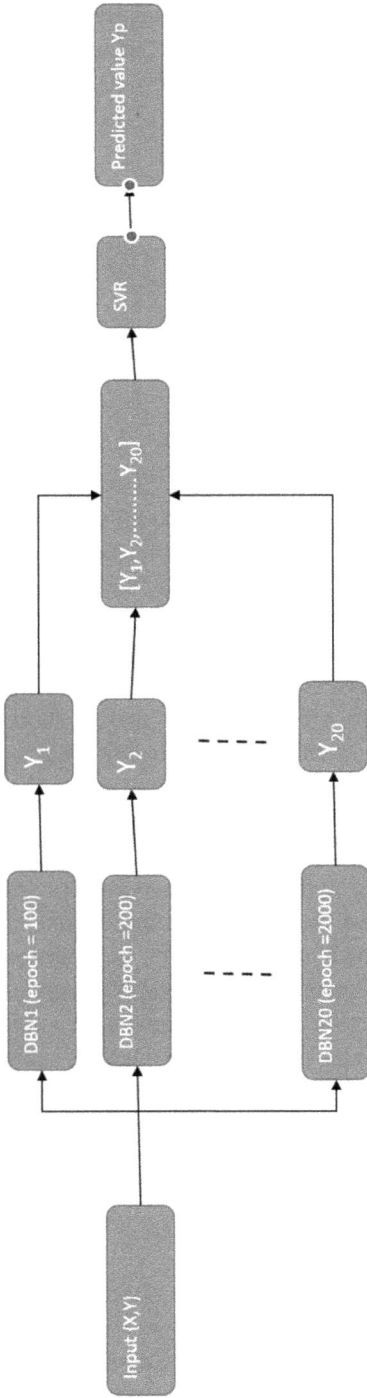

Figure 9.6: Ensemble deep learning model proposed in the work reviewed here.

The method employed: Load-based time series dataset had 24 h demand data as input, X_i and y_i being the demand data output in the next time step. Linear scaling between zero to one was done with the following equation:

$$y_i^| = (y_{max} - y_i)/(y_{max} - y_{min}) \tag{9.13}$$

SVR model simulation was done with Library for Support Vector Machine (LIBSVM) (Chang et al. 2011), and matlab deep learning tool was used for developing DBN and ensemble implementation (Palm 2012). Mean absolute percentage error (MAPE) and root mean square error (RMSE) are considered as metrics.

$$RMSE = \sqrt{1/n} \sum_{i=1}^{n} (y_i^| - y_i)^2 \tag{9.14}$$

$$MAPE = 1/n \sum_{i=1}^{n} (|y_i^| - y_i /y_i|) \tag{9.15}$$

Here, $y_i^|$ is the prediction value for y_i.

Outcomes: Grid search-based parameter selection was done with the RBF kernel in the case of SVR. C value ranged from $[2^{-4}, 2^4]$, with σ value of $[10^{-3}, 10^{-1}]$. One hidden layer model FNN was chosen with the architecture of [48, 96, 1]. [20, 20] RBM size was employed for DBN. Twenty DBNs and SVRs with the above configuration were employed. These 20 FNN models with SVR with epochs from a range of 100–2,000 were composed in the FNN model. Average forecasting error in a year was closely looked at.

Findings demonstrated that the approach used works well in the case of short-term time series data as well. The ensemble-based algorithms overtake individual algorithms. On the datasets used by the authors in this experiment beyond ANN, SVR certainly showed better performance, maybe due to the only layer used in the neural network (Romero et al. 2007). FNN is overtaken by DBN in their prediction performance, demonstrating the deep learning networks' ability. The network proposed here has done exceedingly well on training data as well as the prediction accuracy.

Network size of [10, 20, 1] is composed for Friedman datasets, and [9, 18, 1] for the California Housing dataset, with another similar parameter as earlier. RMSE computation is done with scaled data for results to be comparable. Outcomes of regression and time series data were on the same lines. Again, ensemble deep learning methods show their great impact on the results. For complicated regression problems, simple generated datasets' performance is overtaken by ensemble deep learning methods.

9.2.8 Wrapping up

Four benchmark approaches were validated, SVR, feedforward, neural network, DBN, and feed-forward neural network ensemble. In the case of regression and time series dataset ensemble, deep learning methods do well with RMSE and MASE metrics. Experiments also have shown the greater capability to work on more complex datasets.

For effective implementation of the energy policies, electricity load-demand forecasting must be effectively managed from practical perspectives. Further extension of this thought process on other deep learning models in ensemble fashion can be tried out with a variety of other datasets associated with the power industry. Prediction capabilities can be improvised with experiments on various other options of algorithms for the selection of parameters.

9.2.9 General review of the section

Back testing or hind testing is the term used for the evaluation of machine learning models in the time series data world. In time series data, it is all about the perfection with which the model learns the time step and value at that time step. K-fold cross-validation does not fit in for the time series data, as it assumes data are not dependent, but it is not true in case of time series data, where time dimension plays its role. So the temporal order of data must be accounted for. Apart from normal data breakup for train and test, walk-forward validation is the format where for every time step as the data gets added, the model gets updated. Also, data split can be done multiple times but with the expense of additional training that must be done. The selection of different split points will be the key factor here.

The industry is aligned with the thought process that data is the new oil. Understandably, the world's data is accumulated with few big names, and they have contributed to keeping it open to the world and the entry point for the world of data is very low as of now. But in case of the specific use cases and for the beginners putting these data together for experiments, it has been a big task. Apart from data scarcity, it is important to keep track of data sample complexity, computational efficiency, handling class imbalance, and so on. Synthetic data generation is a big game-changer, where the data is generated programmatically. Here the objective of the experimenter is kept in mind to put together the data that works. This also helps to take care of the security concerns that haunt the data world.

Experiments with the combination of DBN and multilayer feed-forward networks have been done. Where the DBN is initially trained for few epochs and then the weights are used to initiate a multilinear neural network, later this network is used as classifiers. Generally, DBNs are stacked RBMs, these RBMs need to be generative autoencoders. CNNs have exceeded the performance of the DBNs for image-related problems or computer vision problems. But in other areas, CNNs are exceeding.

9.3 Machine learning techniques for rainfall prediction

The meteorological department's main problem is heavy rain prediction as that has a major influence on people's lives and the economy. The root of natural disasters across the globe impacting the lives of people takes its root from here. Agriculture-oriented economy like India heavily depends on the accuracy of these predictions. The statistical analysis finds it difficult in getting through in rain prediction due to the complexity of the atmosphere behavior and its dynamics. ANN makes a mark here due to the nonlinear natured data associated. A review done in this section provides good insight into approaches and techniques used in rainfall prediction.

9.3.1 Let us begin

Rainfall prediction helps in saving life and property and is a great source of resource management. Farmers can do better planning of the crops with the information that is available historically. Meteorological scientists struggle with the time and quantity of rainfall. Across the globe, in the meteorologist's deliverables, rainfall predictions stand on top. It is a complex task begging for specialty and involves a lot of uncertain calls.

9.3.2 Theory

Numerical weather prediction model (NWP) (Drucker et al. 1997) and rainfall forecast are widely used methods. The intensity of rain, amount of rain, and rainfall frequency are the key characteristics of the rainfall data in time series. Attributes differ from point to point across the globe and across time as well. All statistical model presents some challenges, a combination of the AR and MA model as ARMA is a good time series approach. This does well for short-time rainfall prediction. It is challenging for these models to identify the nonlinear pattern in the time series (Drucker et al. 1997).

ARIMA model: Box and Jenkins (Pankratz et al. 1983) proposed a methodology as follows. Identification is the first step where the response and candidates that influence the response are figured out. Inverse autocorrelation, cross-correlation, autocorrelation, and partial autocorrelation are calculated.

In the next step, the model will estimate the parameters and identify the best-fit ones. The later forecast is done for future data and the confidence level is built for the same. Partial autocorrelation function (PACF) and autocorrelation function (ACF) are key analytical tools for time series (Hagan et al. 2002). For common data series, identification of the pattern between the data with statistical inference is the key focus of these models. ACF contributes in terms of establishing the statistical relationship between the data with a lag off course. As part of the model identification step, ACF and PACF will be identified. These parameters help in understanding parameters and what form the model will take.

ANN: Human brain-inspired ANN is a computation model where well-structured interconnected neurons operate parallelly in large numbers. Neural nets can be single or multiple layers with hidden layers housed in mid of these layers. Hidden layers house in mid of these. The single-layer case will have an input layer together with weighted neurons and the output layer called single layer feedforward (SLFF) network. Common network architecture is a multilayer feedforward network (MLFF) that will have SLFF being improvised with multiple hidden layers. Figure 9.7 depicts Box–Jenkins methodology.

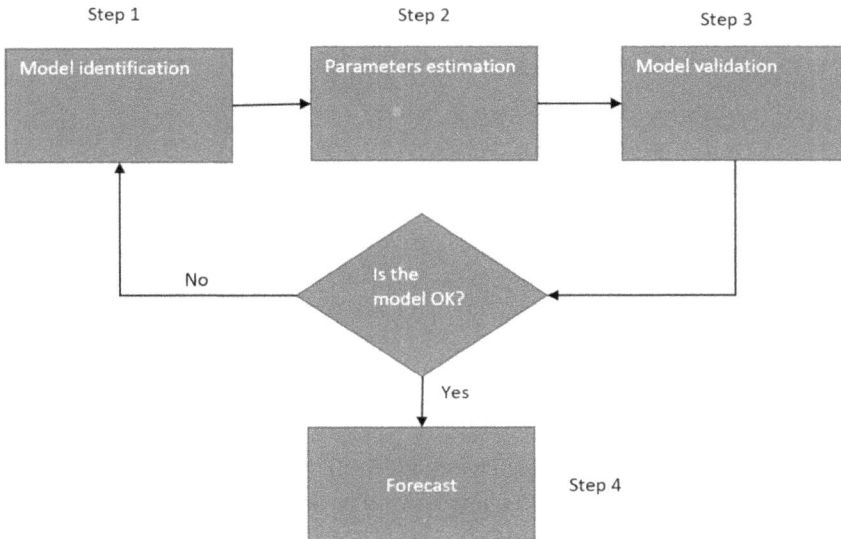

Figure 9.7: Box–Jenkins methodology.

Backpropagation neural network: This network's idea is to reduce the cost function value by sending back the learning signal in the network, where the cost is the difference of the predicted value of the network and the actual values. This network works with one input layer and output layer each and a hidden layer in between. Cost values optimization is tried until it closes toward zero.

BPNN operates with training a forward pass in the network like an FFNN, then computes the cost function and adjusts the weights and parameters with the feedback network.

ANN specializes in adjusting these weights and fine-tuning it to a scenario. One thing to take care of is to look at the training data being learned well but ending up predicting testing data badly. This hampers the network's capability to generalize, and for that sake the network needs to be watched out during the training process to analyze the error on testing data. The error threshold needs to be fixed for this study (Sivanandam et al. 2007). The basic form of network optimization for ANN would be to play around with the neurons and layers count with the output being intact. The heuristic approach stands the best method for architecture finalization of ANN. As the network architecture grows, it goes out of control in terms of pattern recognition and overfits (Haykin 1999). Figure 9.8 depicts BPNN architecture.

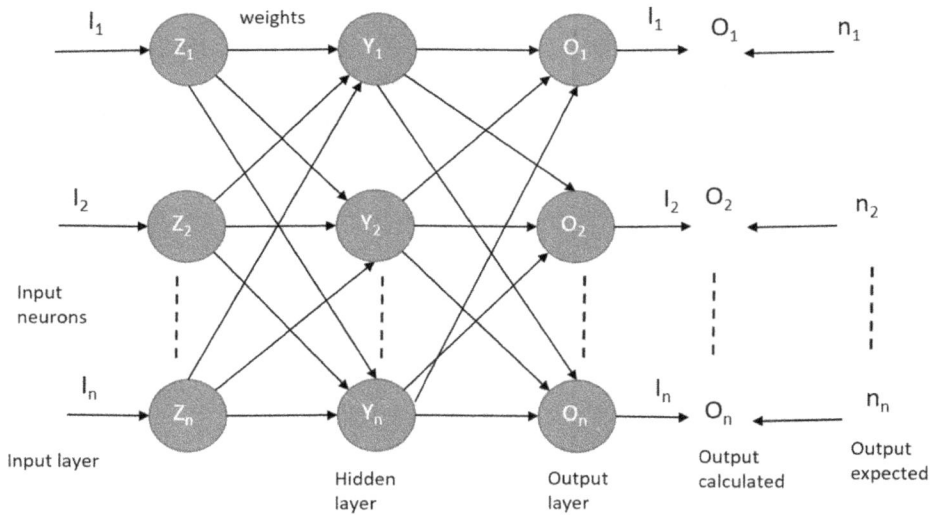

Figure 9.8: BPNN architecture.

Cascade forward backpropagation network (CFBPN): New output data is predicted with this network. Figure 9.9 depicts Cascade forward backpropagation network.

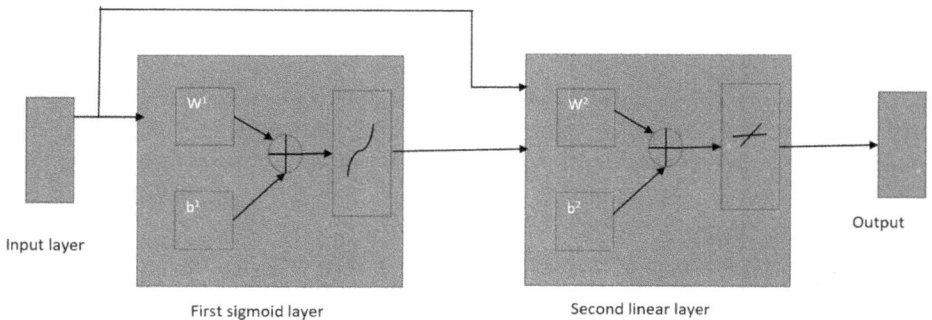

Figure 9.9: Cascade forward backpropagation network.

This network structure specializes by connecting to all previous layer and input layer as well. Input connects with all layers of the network.

Layer recurrent network (LR): Directed cycle connection between the units in the neural network in case of LR, will have an arbitrary sequence of the input that can be processed with the network's internal memory which is the unique feature in this RNN network. Every layer will receive with its input from the previous layer, the output of the network. Figure 9.10 depicts Layer-based recurrent neural network.

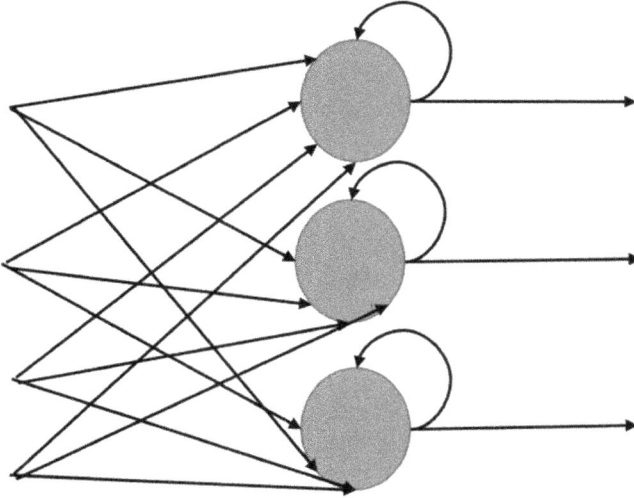

Figure 9.10: Layer-based recurrent neural network.

SVM: SVM is a feed-forward network suit for pattern classification and nonlinear regression tasks. Vapnik and team created SVM intending at supervised learning, and it takes its place due to its following features. Among neural nets, this shows good generalizability, identical solution, ability to hit optimal local minima, works well with nonvector data like graphs, and works well with a limited number of parameters. Also, its application in rainfall prediction is limited; those that are experimented with will have created some hopeful results.

Self-organizing map (SOM): SOM banks on competitive learning; itis a neural network as well. Here, there is a face-off between neurons to get activated, with the context of a neuron getting activated to get their weights updated, under the concept of "winners take all." SOM operates on unsupervised learning without human intervention. One- or two-dimension matrix is the structure. Prof. TeuvoKohonen introduced SOM as a data visualization technique to reduce the dimension with a self-organizing neural network (Haykin 1999). SOM brings out the differences (Aakash et al. 2017).

9.3.3 General review of the section

Machine learning has been leveraged by meteorologists for assessing the potential damage a storm can cause. This helps extend the understanding of the possible impact on electricity and its impact on the infrastructure. Energy suppliers are investing in these studies as they are impacted due to power outages, and they do hold data that can be leveraged on. The basic data aspect is the storm's classification into a variety of classes based on the impact that the storm makes on energy infrastructure and the effect on power outage's duration. Interestingly, the storm is parameterized by their speed, temperature, area, and similar attributes. Many such attributes are the features for machine learning to build the prediction models. Further extension of the studies to different seasons would be a good potential area to explore.

NWP models forecast for several weather-related parameters like pressure, temperature, rainfall, and wind. Lookout is on the interaction of these parameters and their influence on the day's weather. Equations working in NWP have their deficiencies as the simulation precision is not up to mark, also the contributions are from the nonavailability of the data related to mountain regions. This will impact the devising of the initial state in the forecasting. With those drawbacks also NWP stands on top of the list in weather forecasting, also it calls for an expert to interpret the results, which favors using simple forecasting techniques.

The main limitation of the time series forecasting is its inability to be deterministic. To counter this, usually, statisticians go for more large inputs and stochastic scenarios. Whether conditional variations make ARIMA take the upper hand against the ANN, has to be explored. Solar radiation concentration is also another area where ARIMA can do well. The black-box nature of ANN makes it tough for these scenarios. Some works have used Monte Carlo simulation in association with the ARIMA.

The ANN network's complexity depends on the vector product of neuron activation in each layer, which is dependent on the previous layer neuron count. Studies are focusing on arriving at the upper and lower bound for the network's complexity to hit a trade-off. There are also missing measures to assess the complexity of the network. To be fair, it will be right to claim that deeper the network is, as it will best provision to implement the complex functions compared to less deep network rather than solving complex problems. The benefit of arriving at the measures for complexity would be to draw the boundary between shallow and deep networks so that there can be a clear demarcation of both the versions' possibilities. Extension of the work on complexity measurements on to the complex neural network like network graphs will be interesting.

RNN's limitation is around the vanishing gradient and exploding gradient. Since the RNN is structured to pass in the input of RNN again and again into subsequent elements of the network, it creates a replica of the same unit repeatedly across time. So, this creates a structure of multiple ANN that has similar weight but different inputs linked by activation. With chain rule of differentiation, the gradient of every unit is multiplied across, in case of gradient being close to zero, it results in diminishing values, or in another case it may keep exploding. These situations are to be handled with a technique like a gradient clipping.

SOM is very much dependent on a good amount of data for processing. Outlier data will have a major influence on the way the network works. Having appropriate features is of utmost importance for SOM. SOM also struggles with grouping, as the groups within the map are more unique. This creates smaller clusters in turn creating multiple similar neurons. These can be handled better with the map's proper initialization, but with the final state not being clear, it makes it tougher.

9.3.4 Review of literature

Composite models have proved to be better than individual ones in the work of Goswami and Srividya (1996), with RNN and time delay neural network (TDNN) features. Indian summer monsoon rainfall assessment is done in the work of Venkatesan et al. (1997) with multilayer feedforward neural network (MLFNN), here for rainfall prediction error backpropagation (EBP) model is trained. This approach was compared with statistical models also. Summer monsoon rainfall prediction for India was conducted which was a monthly and seasonal time series based in the work of Sahai et al. (2000). In this work, the last five-year data was used from monthly and seasonal mean rainfall data. Adaptive basis function neural network (ABFNN) performed much better than Fourier analysis in the Kerala region's work in India by Philip and Joseph (Aung et al. 2012).

ANN model was used for rainfall prediction for Hyderabad in India by Somvanshi et al. (2006). This approach used the last four months' data for prediction and compared the ANN and ARIMA models. Simpler prediction with minimum and maximum temperature was used in the work of Chattopadhyay and Chattopadhyay (2007). For training of the algorithm, conjugate gradient descent and Levenberg–Marquardt algorithms were used in Chattopadhyay and Chattopadhyay's work (Chattopadhyay et al. 2008), and both performed similarly for the task. India and China's rainfall prediction work are seen in work of Wu et al. (2010), where modular artificial neural network (MANN) was employed. They were put to comparison with logistic regression, K-nearest neighbor, and ANN.

Four different focused networks were run on focused time delay neural network (FTDNN) for data on annual, bi-annual, quarterly, and monthly versions of data by Htike and Khalifa (2010). The yearly rainfall dataset gave the best performance here. K-means clustering algorithm with a decision tree was experimented by Kannan and Ghosh (2010). For the river basin with a large amount of data rainfall state prediction was done using classification and regression tree (CART). K-means clustering is used to generate historical data from multiple sites daily. Short-term rainfall is predicted in the work of Kannan (2010). In the prediction task, empirical methods are used and data for a specific month of the last five years were used. Grouping is done by clustering.

Monthly rainfall prediction with ANN was done for Chennai in India by Geetha and Selvaraj (Drucker et al. 1997). For the Patnagar region in India, rainfall and maximum and minimum temperatures with relative humidity were accounted for weekly rainfall prediction with ANN doing well over multiple linear regression. For the Australian region, time delay recurrent neural network (TDRNN) was applied by Abbot and Marohasy (2012) for month-based rainfall prediction. Average rainfall for the Udupi district in Karnataka of India was predicted by Kumar (Chattopadhyay et al.

2007), using the ANN model. This approach concluded that backpropagation algorithms lead to cascade backpropagation and layer recurrent ones.

A complex architecture of ARIMA and three ANNs were used by Nanda et al. (2013). MLP, Legendre polynomial equation (LPE), and functional-link artificial neural network (FLANN) were the ANN versions. ARIMA was ovetaken by FLANN. ANN model also employed for prediction of rainfall in the work of Naik and Pathan. In this work, modification of the backpropagation algorithm was done to make it more sophisticated than the normal one. Traditional multilinear regression was experimented for Assam's rainfall prediction in India by Pinky Saikia Dutta and Hitesh Tahbilder (Pinky et al. 2014). Parameters like wind speed, rainfall, and maximum and minimum temperatures were accounted for. Overall, this area's work can be looked at with authors involved, region, dataset time, approaches used, and the variable used.

In Wavelet-postfix-GP model, wavelet ANN are tried by Dabhi and Chaudhary in their work (Dabhi et al. 2014). Metrics that have been used for accurate measurement are relative percentage error, RMSE, correlation coefficient, mean absolute error, absolute mean difference, MAPE, Pearson correlation coefficient, high prediction accuracy, and so on. Apart from the parameters that were discussed in the section, other parameters that were considered in the approaches are precipitation, sea surface, latitude-longitude, sea surface pressure, humidity, solar radiation, evaporation, wind direction, aerosol values, climatic indices, dew point, and so on.

9.3.5 A generic review of the section

Machine learning with big data has been done for rainfall predictions. Data.gov.in is the Indian government office source of data that can be looked up for rainfall data. Overall 23 years of rainfall data is available here. To handle large amounts of data deep echo state network and echo state network can be explored; these are the effective and speedy algorithms. The fuzzy theory has been used for water outflow in the rivers.

The choice of the random forest algorithm goes well with its ability to get the regression and classification work done with the same class of algorithms. Missing values are also handled well, this can help in managing without overfitting even with many trees, categorical data also can be used for training purposes here and its ability to make the outcome interpretable will be handy. Extreme gradient boosting (XG Boost) is a great option to explore with its capabilities of limits of computational resources pushed beyond the normal limit. It employs thousands of trees under the hood, and though it is a black box model, it is effective. Its effectiveness is also contributed by its linear and solving capacity and tree learning one.

9.4 Machine learning regression model for rainfall prediction

With the growing amount of data in various fields of technology, medicine, and science, it is of utmost importance to manage them well and renovate to the benefit of humans. Understanding the data pattern quickly will be a key need for accounting complex patterns in the data and services for the information-intensive experiments. One of the approaches is to utilize the classification and clustering of the data. For the purpose of helping the industry in this work, authors have proposed two-level clustering of the large data for prediction of rainfall using SVM and SOM with ID3. Partially clustering with ID3 and hierarchical agglomerative clustering are explored. Clustering happens over two stages, first with SOM and then followed up by SVM. ID3 works well compared to the normal clustering approach.

With the extensive growth of the academic and commercialized database, manual operation of those would be a complex and time-consuming task. The condition becomes more complex when the size limit of the data hits the roof. This has called up focus on automation of data analysis and visualization in the research community. The idea is to derive the patterns from the data to get insights. SOM with its key features finds its application in speech recognition and a whole lot of applications in financial analysis.

SVM and SOM are great tools for visualization. Though background has the associated math, it is easy to implement and understand. Results are quite natural, and the scale of the algorithm is helpful. A nonprobabilistic binary linear classifier is in play with SOM and SVM built algorithm. In this work, authors are focused on weather prediction with these combos.

9.4.1 Clustering

Data mining finds its root in exploring data sources and subjecting them to analysis on a large scale. Tremendous advancements in computer hardware have put computational efficiency up and ahead in the game. SOM is based on the brain's characteristics, where a certain type of signals activates a certain part of the brain.

9.4.2 SOM

SOM specializes in bringing data from a higher dimension to a lower dimension retaining the pattern. Data surveys use SOM with its specialization in visualization. Sample data and weight vectors are key components of the SOM. Weights are representing the data concerning their location in the map coded with color.

In speech recognition, SOM contributes by developing object indicators for improving the quality of the voice. In the financial arena, SOM contributes to figuring out the enterprises' financial statements and establishing a relation between them.

9.4.3 SVM

For a series of binary classification problems, natural text needs to be classified into given categories, like sorting document by email filtering, topic, and so on. These are cases where there are multiple classes to be mapped in it.

For large datasets and multiple classification levels to be done, normal clustering methods fail and call for SOM and SVM kind of algorithm. SOM gets the data reduction to start with, then follows up with statistical modeling. For data reduction, both divisive and increasing clustering is used. In the proposed architecture in this work, key part is the hierarchical structures of the SOM. The depth of the analysis in this hierarchical setup is decided with unsupervised learning. There is a combination of SOMs and SVMs in the layers with their structure being composed in an unsupervised way. Model is suited for weather forecasting which is a hierarchical clustering situation.

9.4.4 Architecture

Clustering looks for similar elements within the cluster, and most dissimilar ones clustered away from others. As part of the clustering, the first set is to break up into two categories, first one being the predicted weather of the city that is searched with five days' information into future of temperature, humidity, and rain, followed by the second set that provides warning on cyclones, thunderstorm, and heavy rains (Naik et al. 2013). Data used covers a variety of regions to get diverse conditions. ANN learning features help the algorithms to handle noisy data as well. ANN as a data mining tool is an open area for research.

SOM: The algorithm initializes the weights with a different format, first with a random approach without being aware of the data; this may be expensive in terms of computation, as this buys enough time until the model can get some representation from the input data. Another way would be to consider random values from the data for the initialization of the weights. This gives a good appropriate starting point, by cutting down on the training dataset. This also may have drawbacks based on how good the chosen samples are.

SVM: It initiates the reference data points from the classes of the datasets. Formulation based on quadratic penalty is used to make data point being separable linearly possible in kernel space. Figuring out closest points in the kernel space needs $n2$ kernel computation, where n here stands for the number of data points.

ID3 algorithm: Decision tree is built for given examples and uses the outcome tree to classify the next set of samples. Leaf node represents the class, and non-leaf node represents the decision node. ID3 utilizes information gain to finalize how to place attributes of the decision. ID3 being supervised learners needs data. Discrete data is to be used as numerical and continuous data does not work. Outlook, temperature, wind, humidity, and rain are the categories in the dataset.

Entropy: Starting from the root node decision tree is top-down with data split into subsets based on homogeneity. ID3 utilizes entropy to compute the homogeneity. Entropy is represented as zero for completely homogeneous data and another end is the entropy of one.

$$E(S) = -(p+)^* \log 2(p+) - (p-)^* \log 2(p-)$$ (9.16)

S is the representation for set, $p+$ is several elements in the set with positive values and $p-$ for a negative one. ID3 intends at classifying the data with decision trees so that nodes of leaf become homogeneous having entropy of one.

Gain: For an attribute as the entropy reduces it results in again at every split. Decision tree construction involves picking the right attribute such that the highest information is returned with low entropy,

$$\text{Gain}(S, A) = \text{Entropy}(S) - S \left((|S_v| / |S|) \, ^*\text{Entropy}(S_v) \right)$$ (9.17)

S stands for set, A is attribute with SV as a subset of S with value v for A. $|S|$ stands for a count of elements in the set S, and $|SV|$ is the number of elements in subset SV.

9.4.5 Wrapping up

Study shows that the decision tree algorithm with ID3 does well for problems based on classification with discrete data. The outcome helps to predict the reliability of the weather for the crop. The approach is advantageous compared to other approaches in terms of prediction, cost, reliability, and computation (Usha et al. 2015).

9.5 General review of the section

Principal component analysis (PCA) is a good area of exploration for weather prediction as it comes in handy with its specialization to reduce the dimensions in the data by retaining only those dimensions or features that matter. Multivariate adaptive regression splines (MARS) is another good approach proposed by Friedman which is a local modeling technique based on the tree. This breaks the data space into several classes, with even the overlapping between them represented, and the truncated spline is fit for every region. In the case of higher dimension data, this is a great option and a good fit for nonlinear multivariate functions. MARS brings out the basis functions' contribution that help in figuring out the interaction effects and additional effects in the data.

9.6 Conclusion

In the exploration of the effective machine learning model for rainfall prediction, we looked at regression and time series approach with ensemble deep learning methods. We also explored various forecasting methods in the backdrop of machine learning and deep learning models. Specialization of RNN, DBN, RBM, and CNN was explored. Ensembling of the models and their influence on effective modeling were discussed. In the specific context of rainfall prediction, ANN and various other versions of it were unfolded. All the work is done so far by various authors, and their key features were explored.

The ensemble of SOM and SVM is a two-architecture layer to bring the capabilities of providing the visualization and efficient clustering methods helped to focus on improving the data mining capabilities.

Chapter 10
Understanding programming language structure at its full scale

10.1 Introduction

It would be essential to deep dive and understand the programming language structure and its complexity in optimizing software development processes. This kind of deeper understanding helps assess language structures' influence on the outcome produced in every phase of the software development life cycle. This exploration also will be promising to revisit some of the principles underlying software development. Getting into the language's semantics will provide an excellent chance to understand the full scale of programming language. While they are executed, the programs' information structures are essential to understand the patterns hidden in software engineering.

In the exploration, we look at abstract syntax tree (AST) representation of the source code. We draw parallels to the machine learning methods of understanding the patterns hidden in any process. Reaching an optimal combination of the influencing factors is a vital part of pattern recognition, which gets explored. While exploring the patterns is on, various issues that must be bothering the optimal software development gets discussed. In the next part, the neural attention model is explored to address some of the challenges involved in long-range inputs associated with software source code. There is a discussion on feature engineering, which is about breaking down the components related to source code and programming language.

10.2 Abstract syntax tree–based representation for source code

Software program analysis with machine learning has been the recent focus. Representing the parts of source code for detailed analysis has been challenging. Traditional methods of looking at code as natural language may miss on the source codes' semantic structure. Recent advances in this area point at AST structure-based study for software source code with neural models as the best approach. This AST makes the more extended representations for the code dependency that spans across this longer-range will be a cause of concern. In this work (Zhang et al. 2019), the author proposes a unique AST-based neural network structure represented as ASTNN. Compared to the models that is proposed thus far, the approach here concentrates on breaking up the AST into a smaller representation that gets vectorized where lexical and syntactic structures get captured. Bidirectional recurrent neural network (RNN) is involved in consuming these more miniature representations and producing vector

https://doi.org/10.1515/9783110703399-010

representations using contents' naturalness characteristics. Code clone detection and source code classification tasks get used as a reference for applying source code representation with the neural network, as these are common ways to comprehend a program. Authors have also observed that the approach here works better than state-of-art for these two tasks.

A variety of software engineering tasks have been explored for finding improvement in software development and maintenance. These include source code classification (Frantzeskou et al. 2008, Mou et al. 2016), summarization of the code (Haiduc et al. 2010, Jiang et al. 2017), detection of code clone (Kamiya et al. 2002, Sajnani et al. 2016, White et al. 2016, Wei et al. 2017), and so on. Capturing the syntactic and semantic information in the source code is a top challenge in all these efforts.

Code fragments being treated as natural language parts and modeling them as texts are the traditional way for information retrieval (IR) methods. In tasks like the classification of code authorship (Frantzeskou et al. 2008), detection of code cloning (Kamiya et al. 2002, Sajnani et al. 2016), and localization of the bug (Zhou et al. 2012), a bag of tokens or token sequence are used for the representation of programs. Source code analysis with latent Dirichlet allocation (LDA) (Blei et al. 2003) and latent semantic indexing (LSI) (Deerwester et al. 1990) are common approaches for analysis of source code (Liu et al. 2009, Lucia et al. 2012, Tairas et al. 2009). A common problem with these approaches is the considering source code as natural language texts, as confirmed by the work (Panichella et al. 2013). Text-based or token-based methods alone are not counted for source code, as there are other possible sources that hold rich information in their structures (Mou et al. 2016, Pane et al. 2001).

In comparison to the token-based methods of past, syntax-based representation makes a richer contribution to the modeling of the source code in its fullest scale (Mou et al. 2016, White et al. 2016, Wei et al. 2017). In these works, ASTs work in combination with recursive neural network (RvNN) (White et al. 2016). For capturing lexical and syntax-based information tree–long short-term memory(LSTM) (Wei et al. 2017) or tree-based convolutional neural network (CNN) (Mou et al. 2016) are used. Lexical structures are in the form of ASTs leaf nodes which are identifiers, grammar construct like WhileStatement are non-leaf nodes of ASTs. AST-based neural models work well but have their limitations. Like in the case of long text of natural language processing (NLP), these models undergo vanishing gradient, where the gradient of the training keeps losing its value. These impacts are particularly prominent in cases of large tree-structured representations. Long term context in the information will be lost in case of ASTs being tried to be completely coded and represented in a bottom-up way (White et al. 2016, Wei et al. 2017) or with the use of sliding window approach (Mou et al. 2016). Binary tree representation is done with the ASTs for their ease of visualization and representation. This takes away the syntax pattern in the code and makes it a complex structure with deep representation. Complex semantics will further lose its mark with deeper ASTs and associated transformation involved (Zhu et al. 2015). Figure 10.1 depicts Abstract syntax tree representation in case of Euclidean algorithm.

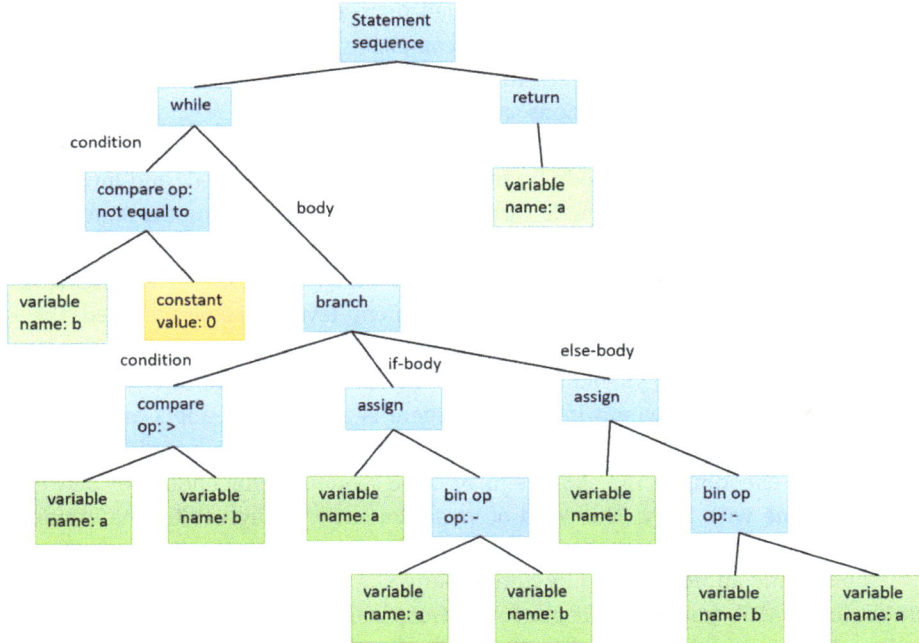

Figure 10.1: Abstract syntax tree representation in case of Euclidean algorithm (astract syntax tree for Euclidean algorithm).

The Euclidean algorithm can be represented as follows, for the case of finding the greatest common divisor for a and b (AST for Euclidean algorithm):

```
while b ≠ 0
      if a > b
a := a−b
      else
b := b−a
return a
```

For the purpose of handling the shortcomings of the AST-based models, control flow with long term and explicit solutions can be used, graph embedding technique (Ou et al. 2016) for graphs of data dependencies can be adopted for source code representation.

For the case of distance location, long-range dependencies introduced by the same function or variable are considered in the case of (Allamanis et al. 2018). For code fragments, control flow graphs are directly created in the work (Tufano et al. 2018). Control flow and data flow dependencies graph which is inter-procedure based have a dependency on intermediate compiled representation, in case of the above-mentioned works, and will not apply for incomplete fragments of the code and code which cannot be compiled. In areas with arbitrary code pieces, the code representation approach will find its limitations as above.

In the approach of this work (Zhang et al. 2019), the uniqueness of the solution comes from the fact that the code fragments need not be compiled, and large code fragment can be broken into smaller tree-based structures covering the statements, with the application of neural network based on AST called ASTNN. Statement and lexical level syntactic knowledge can be represented with vectors of the statements. In specification at the program level, AST statement–level nodes are defined, those are called as statements. Special statement node status provided for MethodDeclaration. For each of the statements like a "Try," statements that will have header and another part of the statement in its body, there is a split of headers and another part. This makes provision for the shorter sequence-based decomposition. Vectorization of the encoding of the statements and the sequential dependencies between them are done with RNN. This kind of vectorization will capture the naturalness (Hindle et al. 2012) of the code and form a neural representation.

The specific approach used in work (Zhang et al. 2019) is to use code fragment and build an AST, where smaller statement trees with a tree for a statement forming AST node are devised, with all the trees originating from the statement node. Statement vectors are also fed in with, statement-level syntax lexical information which are vectorized with recursive encoder. These statement trees are a multi-way representation. To make use of the sequence-based code, naturalness in statements vector representation for the full code part is done with bidirectional gated recurrent neural network (Bi-GRU) (Bahdanau et al. 2014, Tang et al. 2015).

In general, authors' approach in (Zhang et al. 2019) is more capable of learning syntax and semantics compared to other AST-based neural models. This approach also can be applied to general-purpose tasks like detection of code clone and classification of source code. Experiments are conducted on two such tasks of code comprehensions with public benchmark datasets and have demonstrated that these methods are effective. The source code classification approach used here has taken the accuracy from 94% to 98.2%. With two benchmark datasets, F1 scores of these approaches have enhanced the performance from 82% to 93.8% and 59.4% to 95.5% to detect code clone.

The approach stands out in its ability to capture the naturalness of the statements in code, syntactic, and semantic knowledge within the code, is brought out with neural source code. And, a demonstration of improvement seen in common code comprehension tasks with its ability to lead compared to state-of-art methods.

10.3 Further exploration of the AST

AST is a basic representation of the statements and expressions in the program represented as tokens. The compiler can generate the instructions and produce machine code with AST. A similar parallel can be drawn in case of how the machine learning represents the patterns in case of the modeled process. Let us say, in case of the simple model which is represented by $z = 2x^2 + 5y + 5$, model AST can be represented as an AST. Figure 10.2 depicts Abstract syntax tree for machine learning representation, for a case of function $z = 2x^2 + 5y + 5$.

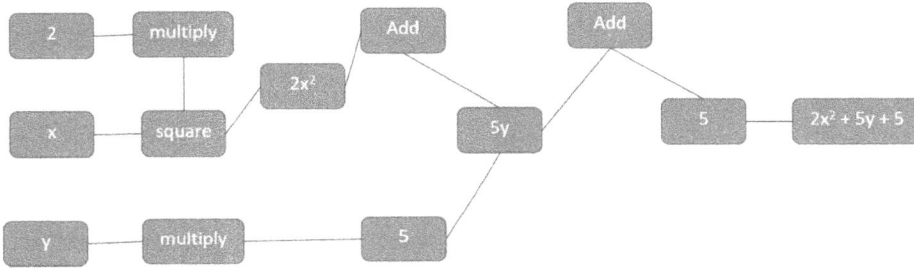

Figure 10.2: Abstract syntax tree for machine learning representation, for a case of function $z = 2x^2 + 5y + 5$.

AST uses the brain's analogy for the representation of the models to derive the numbers involved and the operations between those numbers to get the sense of the overall equation. The syntax analysis phase of the compiler makes use of the AST approach. Programs are intermittently represented with these AST until the final representation is arrived at. These will have a high level of influence on final representation. AST also facilitates getting closer to machine understanding. Intuitively, it helps to step into the code understanding any possible loopholes and provide proactive feedback. These can add quite a bit of value in areas of understanding the source code to its full potential, which will form the basis for extending smarter utilization of software engineering pipelines. ASTs features in terms of representing only semantically meaningful information unlike conventional syntax tree that cover complete-textual form for the provided input.

ASTs can also assist in understanding the software evolution cycle which can facilitate a solution for many issues faced by software engineers as part of the software development lifecycle. At a basic level, AST acts as a tool to mine the software artifacts' information. This richer understanding of the structure can facilitate one of the thought processes of fixing the bugs in software without stopping a system. In the context of dynamic program updates, it will be helpful to study the dynamics of programming like the case of understanding frequency at which the functions and variables of the program are modified. In cases of cross-project defect prediction, there is a challenge of deriving real numbers from ASTs. This transformation is needed from code to AST, further into a real number so that it can be consumed as input in the neural models and other similar approaches. The concern in the scenario is the loss of estimating the distance semantically between AST, which is important for effective training in the context of machine learning. This problem must address to make effective use of AST in source code understanding. AST can also provide a base for making complex machine learning models interpretable.

10.4 The motivation for applying AST

Now, we explore the motivation of applying AST by the authors Zhang et al. 2019. AST representation is widely used in building software engineering tools and for programming language representation. Symbols and constructs of the source code are represented as nodes in AST. These are a generic representation of the code abstractly and may not include details like delimiters and punctuations. ASTs can represent syntax and semantics of the code like method name and flow of control structures. For areas like program repairing (Weimer et al. 2009), a search of source code (Paul et al. 1994), differencing of the source code (Falleri et al. 2014), and token-based AST methods are explored. Token-based approaches have their limitation in limited syntactic structure they can read (Panichella et al. 2013).

AST inputs are being tried with tree-based neural networks (TNN). These TNNs learn bottom-up going into the computation of node embeddings, from the tree input, in a recursive way. ASTs with tree-based modeling is seen in the case of tree-based LSTM (TLSTM), tree-based convolution neural network (TBCNN), and RvNN.

Image processing and recursive structure understanding in natural language were initial areas of application for RvNN (Socher et al. 2011). Considering a parent node y_1 and children nodes (c_1, c_2) in a tree structure. Here c_1, c_2 will be an intermediate vector representation of nodes. Node y_1 will have a vector as shown below.

$$p = f\left(W^{(1)}[c_1{:}c_2] + b^{(1)}\right) \tag{10.1}$$

f represents an elementwise function, $W^{(1)}$ is a parameter matrix, $b^{(1)}$ is a term for bias.

Children are reconstructed to make sure that vector representation is of high quality, which will be a decoding layer.

$$[c_1';c_2'] = w^{(2)}P + b^{(2)} \tag{10.2}$$

Training loss is depicted as follows:

$$E(\theta) = ||C_1 - C_1'||_2^2 \tag{10.3}$$

Recursive computation and parameter optimization with the tree for root node representation of vector done in RvNN. Since ASTs demonstrate fixed-size inputs and get transformed to full binary trees in case of recursive autoencoder (RAE) used with RvNN, these are used for code clone detection (White et al. 2016).

Source code classification (Mou et al. 2016) kind of tasks is facilitated by convolution computation as part of supervised learning in the case of TBCNN. Fixed depth feature detection is done with a sliding window, across ASTs, for AST-based convolutional layers. The below equation formulates the same.

$$Y = \tanh\left(\sum_{i=1}^{1} W_{conv,i} \cdot x_i + b_{conv}\right) \tag{10.4}$$

For each sliding window x_1 to x_n represents the vectors for nodes, matrices of parameters are $W_{conv,I}$ and bias given by b_{conv}. For improvement of the localness of the representation for global information, TBCNN uses the encoding layer that is bottom-up. As the size of the convolution is fixed, ASTs will be a continuous full binary tree as treated by TBCNN even in case of more than two children representation.

For modeling tree-structured topologies, generalized LSTMs are put in place. To update the children's states across tree structures, the child sum tree version of LSTM (Tai et al. 2015) is involved. Here, current inputs are combined recursively with the children states across tree structures. Code fragments parsing into ASTs in code clone detection use cases are seen using tree LSTM (Wei et al. 2017). Full binary tree representation of the ASTs is to tackle variable count of child nodes. For the representation of the code fragments, root node vectors of the ASTs are involved with a bottom-up computation format.

Tree-based methods discussed here have limitations. Back propagation across the structures of a tree topology is involved in calculating the gradients, in training (Zhu et al. 2015). In the case of nested-like structures, deep and large AST structures are mainly due to complex programs. Thus, gradient vanishing is experienced as the computation run from lead to a root node in a bottom-up fashion and results in limitations in capturing long term dependencies (Bengio et al. 1994, Cho et al. 2003). This will result in the loss of semantic information carried by the nodes at the distance from root nodes, and leaf node identifiers are examples. Two or more child nodes of parent nodes are moved to a new tree for simplification in existing approaches of tree method views, this assumes binary trees. This leads to a change in source code semantics that it held originally, and long-term dependency problems become critical. F1 value of 57% is seen in the work (Wei et al. 2017) on the public benchmark dataset for clone detection use cases. Depth and tree size have their role to play as demonstrated by studies of NLP (Le et al. 2016, Tai et al. 2015, Zhu et al. 2015).

10.5 Activation functions and their role in machine learning architecture

We did see various activation functions wrapping the mathematical model in the case of CNN structures. It is important to study these functions' properties and identify the best fit based on the required properties of the data input space, for the architecture. Activation functions form a building block for machine learning and deep learning architectures. In the analogy of the brain, processing tons of data needs a suitable filter to segregate data based on its usefulness. Similarly, in the context of the activation function. Here, irrelevant information will be noise associated with the input data that needs to be separated.

Neural networks are constructed to mimic the brain and they have a central processing unit in them that is where the functions seat and operate. Figure 10.3 depicts Neuron representation in neural network.

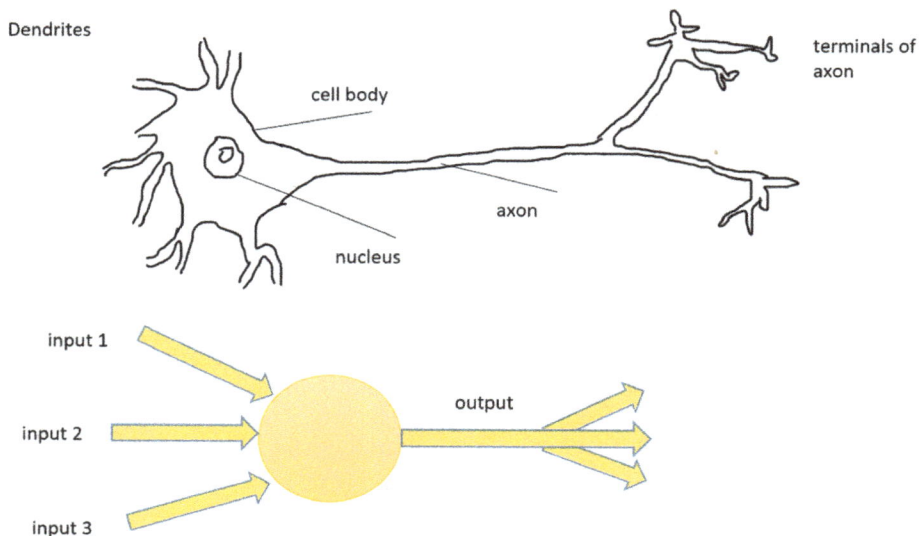

Figure 10.3: Neuron representation in neural network.

In a complex setting, a small piece of decision is taken by each of these activation functions within a neuron, and as an architecture, all these are tied together for the outcome. Apart from activation function, weights and bias play their role. Inputs are acted upon by the weight and bias, then processed input is acted upon by an activation function. The activation function provides power to the network by adding on to the nonlinear transformation capability so that the network can recognize the complex pattern in the data.

The simplest form of the activation function is a binary step function which involves the simplest decision of whether the linear transformed input is activated or not based on a certain threshold value. As its simplicity, these can be handy in cases of binary classification. Since the derivative of these function yields zero value for its gradient, there is no update of weight and bias in backpropagation,

$$f(x) = 1, x \geq\ = 0; f(x) = 0, x < 0 \tag{10.5}$$

A linear function is having a direct proportionate value with input. There would be an associated coefficient for the variable that can be constant. Even though there is an associated gradient when the derivative is taken, weight and bias get updated consistently, always leading to no learning in backpropagation, so, this may help cases of simple tasks that need interpretability.

$$f(x) = ax \tag{10.6}$$

The sigmoid activation function is the most widely used one among nonlinear activation functions. It looks like a smooth S-shaped function and it is possible to perform differentiation of the same continuously. There is a high gradient seen at a specific range of value compared to other areas, which is flatter, this indicates higher learning in these ranges only, and out of it, the learning in not significant.

$$f(x) = 1/(1 + e^{-x}) \tag{10.7}$$

At this point taking stock of these functions, it would be noteworthy to relate to AST representation of source code and further relate to various properties of these structures based on which the functions can be configured into neural network structures. Since the AST representation provides the explainable pattern in the data space, it makes it appropriate to relate it to choosing the activation function carefully, so builds an effective source code interpreting network.

The tanh function is very similar to the sigmoid function, but it is symmetrical around the data space's origin. The range of values is spread between −1 and +1, making a variable input flowing to the next layer always. tanh can probably have upper hand against sigmoid, as the function is centered around zero and gradients do not follow specific directions for their value as they feedback in backpropagation.

$$\tanh(x) = 2 * \text{sigmoid}(2x) - 1 \tag{10.8}$$

Rectified linear unit (ReLU) has its stronghold in deep learning algorithms. Differentially activating the neurons at different points of time, unlike some of the other functions that have simultaneous activation of neurons, makes ReLU stand out. Function characterizes by keeping the output to zero if input with below zero. Since limited neurons are getting activated as the input below zero is not activating neuron, computationally also this function is efficient.

$$f(x) = \max(0, x) \qquad (10.9)$$

Softmax activation function comes in to play in case of multiclass classification. It outputs the probabilities of a data point belonging to each of the classes. Softmax also plays a role to summarize the decision from various neurons in the intermittent layer of the network.

$$\sigma(z)_j = \frac{e^{z_j}}{\sum_{k=1}^{k} e^{z_k}} \qquad (10.10)$$

As discussed, looking at a scenario in hand, these activation functions need to be put in to use. Sigmoid comes into the picture in classifiers but needs to be careful about vanishing gradient issues. ReLU is the most preferred one, in the intermittent layers of the neurons.

10.6 The approach used for ASTNN

ASTNN architecture is discussed here as proposed by authors in their work in (Zhang et al. 2019). Figure 10.4 depicts AST neural network architecture.

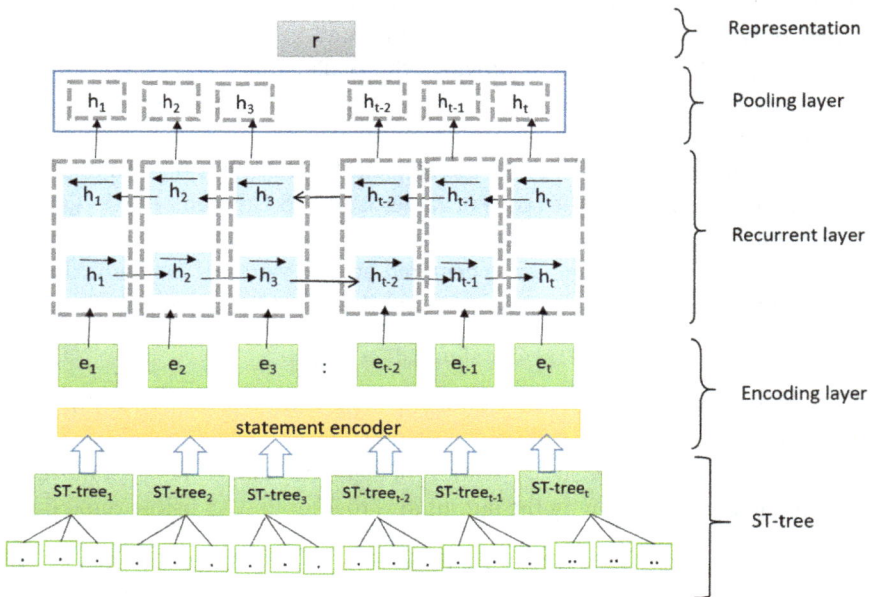

Figure 10.4: AST neural network architecture.

AST is derived by parsing the source code, statements trees sequence is derived with an algorithm that traverses and splits AST. AST nodes of the statements and statements nodes of roots tree form a statement tree. Statement encoders vectorize the statement trees. The naturalness of the statements is then handled with bi-GRU. Code fragments are represented with a single vector for pooling purposes from hidden states of bi-GRU.

Syntax analysis tools can take the first step of creating a large ASTs out of source code fragments. Preorder traversal will then assist to derive statement trees based on the granularity of the split of the ASTs. Each of the statement nodes in the AST belongs to the AST node concerning one source code statement. Nested statements are shown as separate statements, special statement node status given for MethodDeclaration. To distinguish between statement tree and binary tree, we call statement tree cases with three or more child nodes as multi-way statement trees. Multi-way statement tree sequences that are not overlapping can be created out of the one AST structure.

Traverser and constructor can make the process of dividing the AST and construction of statement trees a straightforward process. Traverser and constructor work hand in hand, as the traverser walk into AST where the depth is covered first, constructor keeps recursively adding statement tree into a sequence of statement trees. This format helps to maintain orderly update of the source code, in turn resulting in a statement tree that can be used as input for ASTNN.

The process of deciding the granularity of the split of AST is an important part. Since statement trees have the potential to carry key semantics of the source code, this work has taken up statement trees as the basic building block of the AST. Zhang et al. (2019) also has looked at the various granularities like, with brave pairs code block and AST node level. It is important to have the appropriate granularity of the ASTs so that diminishing gradient problem does not come into the picture. On the other side, if the representation is too small, it will be like tokens-based RNN which may reduce the syntactic knowledge. In this work, the authors have figured out that the granularity at the statement level is better to maintain a balance between proper syntactic information being captured with the right sizing of the statement trees.

For learning vector representation of the statements, a statement encoder that is based on RvNN is designed with a statement tree as input. Training corpus are obtained with traversing the ASTs with predefined order. Since ASTs have multiple special syntactical symbols, this is important. Symbols are vectorized in an unsupervised way by word2vec (Mikolov et al. 2013), statement encoder will then use these embeddings as a symbol that are trained as starting parameters. Symbol embeddings capture the lexical information as lexical information is associated with leaf nodes of ASTs like that of identifiers, that are part of lead modes of statement trees.

For instance, in the case of a node of MethodDeclaration that holds statement trees recursively encoder will traverse statement tree and pick new input in form of current node symbol for the purpose of computing children nodes hidden state. Figure 10.5 depicts Statement encoder representation.

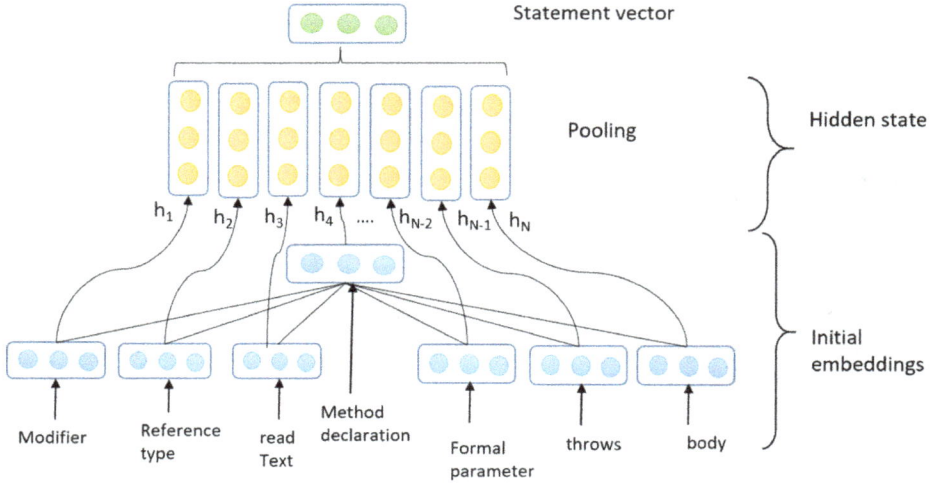

Figure 10.5: Statement encoder representation.

We may lose the original semantics if we change the statement trees to one binary tree bypassing the node into the child or its descending nodes.

For, non-leaf node is denoted by "*n*" statement tree by "*t*" and several children nodes by "*C*." Pretrained parameters of the embeddings are considered to start with.

$$W_e \in R^{|V| \times d} \tag{10.11}$$

"d" is embedding dimension for the symbol, "*V*" is the vocabulary size, node "*n*," the lexical vector can be gathered with,

$$V_n = w_e^\top x_n \tag{10.12}$$

For symbol "*n*," "x_n" is the one hot representation with embedding represented as "v_n." For the vector representation in case of node "*n*," following equation is used:

$$h = \sigma \left(\frac{W_n^\top v_n + \sum h_i + b_n}{i \in [1, C]} \right) \tag{10.13}$$

Weight matrix with encoding dimension is given by the below equation, with "b_n," "k" as bias terms for each child "*i*," hidden state is given by "h_i." The updated hidden state is given in "*h*," and activation function given by "σ," with identity function given by tanh.

$$W_n \in R^{d \times k} \tag{10.14}$$

Vectors of all nodes in the structure tree are optimized and computed in a recursive mode. For node vectors, to explore important features stacking up of all nodes, and sampling are done with max pooling. The final statement tree is gathered with the representation provided in the below equation, with "*N*" representing the statement tree nodes.

$$e_t = [\max(h_{i1}), \ldots, \max(h_{ik})], \quad i = 1, \ldots, N \qquad (10.15)$$

Information hidden in statements in the form of lexical and syntactic form is captured in statement vectors.

The batch processing algorithm will encode multiple samples to improve the training efficiency on the large datasets. Since the number of children nodes varies compared to parent nodes for a batch at the same position, batch processing is seen in multi-way statement trees. When there are two parent nodes, one with three children nodes and other with two children nodes, the vector representation for the two-parent nodes in a single batch is not possible as each of the nodes will have different child nodes. The authors here propose the design of an algorithm that can process both the samples in a dynamic way.

In this case, where the children nodes vary for each of the parent nodes, with the same position group related children nodes can be detected dynamically by the algorithm, which will make the faster computation of vector representation for the node in each group in batches, making use of the matrix operations.

The naturalness of the statement is derived with GRU which takes statement tree sequence as input. LSTM was also an option here which is experimented by authors. In GRU, average pooling or max pooling is devised for deriving key features based on their states. Every statement tree will differ in terms of the information they hold, like a case of API call in method invocation may have rich information on the functional side compared to a few other scenarios. Hence max pooling will assist in capturing the key semantics which is the default.

10.7 The criticality of vanishing gradient in the context of source code modeling

Vanishing gradient comes into picture when the more activation functions become part of the neural network and the gradients generated by forward propagation states approach zero. This masks off the real learning possible for the network. At the root, this issue crops up due to the functions ranging down the input values into specific range. Like in case of the sigmoid function, the input data space will get reduced to a data range of −1 to +1. This reduces the changes to smaller scales that may diminish. Here the gradient is in the form of a derivative of the function that is under processing. Since we are talking about the complex ecosystem of code being represented, it becomes important to keep a check on the model's architecture and influence.

Some of the solutions like a residual network, where the gradients in the backpropagation are fed back by jumping across the activation function are proposed, here it is important to relate the scenario to the domain of source code and assess the impact of this jump. As there would be scenarios that may work well even though gradient in their backpropagation skips the activation function. There may be other scenarios where this skip may result in loss of key patterns in the data. This needs closer analysis from domain experts and accordingly, the architecture must be framed.

Batch normalization method, where the range of the activation function itself is reduced to keep the gradient significant, also need to be assessed in the light of domain that we study here and based on the experiments across various scenarios we have to the device the efficient architecture.

10.8 Application of the model

ASTNN model (Zhang et al. 2019) is a general-purpose one, where many of the software comprehension tasks can benefit by representing the source code with task-specific vectors and their training.

From the perspective of understanding the program better and maintaining them well, the source code classification task will focus on fragments of code and classify them by their functionalities (Kawaguchi et al. 2006, Linares-V´asquez et al. 2014, Mou et al. 2016). The cross-entropy loss function is a widely used one and is used in this approach. For identification of two code fragments implementing the same functionality which is code cloning exercise, there is a good amount of research happening in software engineering (Tufano et al. 2018). In case, if the code fragments vectors need a distance measure between them to assess their relation semantically (Tai et al. 2015), we can take the absolute distance between those vectors. Then, the output can be treated as the sigmoid function of their input, which represents the similarity. Input here is the function (ax + b) and binary cross-entropy loss function can be used here.

ASTNN is trained with the objective of minimizing the loss, with the AdaMax (Kingma et al. 2014) approach as it provides computational effectiveness. Trained models are stored once the optimized parameters are derived from the training. Further cases that need these models to be applied, need to have their inputs structured as structured trees into these stored models. Model output will be multiple probabilities across various classes that we are classifying in to. Maximum probability values are derived to boil down to a correct class. In the case of a clone detection use case, since the value of the outcome is a single value, the comparison must be done against the acceptable threshold value to decide on the outcome.

10.9 Code cloning issue in software source code

In software, development lifecycle code cloning information is not well documented which adds the burden of maintaining the code. A key challenge is an increase in the software's maintenance cost and difficulties involved in understanding the code, which will add to the complexity across various areas. Code cloning can be structural or simple one. Simple one finds their presence localized in code snippets that are due to the repetition of the snippets. Structural code clone comes into the picture at the design of the software. This simply relates to the common pattern in the code, that can be brought out by modeling the code. This also closely relates the probabilistic modeling as the similarities of the code fragments also need probabilistic thought process. So, they go well as an approach.

To device the approach for code cloning issues, it is important to study the pattern of code clones and their similarities and characteristics features. The modeling framework can be built and refined based on the structural similarities in the data. Apart from the AST and token-based clone detection, there are metrics-based approaches as well. Here, the effort revolves around building the measurement system around code clones that can be a feeder to device the models that can identify code clones. Another key challenge that contributes to the code clone situation is, the programming language not providing the ability to abstraction capabilities due to their limitations. This problem can be helped by studying the structure of programming language well. Another key thought process is to relate the pattern of code cloning to influence the developers' external factors. Like if there is high pressure to go to market early, then code clone would possibly be one outcome. This way, the pattern of code clone would reveal a good amount of such intricacies in software development.

10.10 Experiments with ASTNN

As part of the experiments, authors have made use of two public datasets in their work (Zhang et al. 2019). Pycparser[1] and javalang[2] tools are used for extracting ASTs.

Word2vec, with the skip-gram algorithm, is used for embedding training of the symbols, with 128 embedding size. 100 hidden dimensions are used in the statement tree for bi-GRU. 15 and 5 epochs are considered for the two tasks that are experimented with and 64 mini-batch sizes. In clone detection case, the threshold was at 0.5. Training, validation, and the testing ratio of 60%, 20%, and 20% are taken for the datasets. Learning rate of 0.02 taken in with optimizer AdaMax (Kingma et al. 2014). 2.4 GHz CPU with 16 cores server are employed with Titan Xp GPU.

---

1 https://pypi.python.org/pypi/pycparser
2 https://github.com/c2nes/javalang

As part of the experiment's evaluation, authors have explored the state-of-the-art model like TBCNN (Mou et al. 2016). On top of these, other neural network–based approaches are evaluated as well. Statistical features–based support vector machines (SVMs), LSTM (Zaremba et al. 2014), program pependence graphs (PDG) embeddings (Allamanis et al. 2018, Tufano et al. 2018). Conventional IR methods are used on SVMs. Textual features are derived from LDA, N-gram, and TF-IDF. Source files are used to extract the tokens that form a corpus for training. Bi-grams are used in the N-gram model, 300 topics are targeted for LDA with 20,000 features as max.

Considering code pieces as plain text tokens in sequence, TextCNN and LSTM models were used. These models find their primary implementation in the sentence classification of NLP. The hidden state dimension is considered 100 for LSTM and 3 kernel sizes and 100 filters for TextCNN. Program features are extracted with embeddings of the statement in CNN and LSTM does the task of statement sequencing in the model LSCNN.

Control flow and data flow dependencies are used to construct program graphs in the work in (Allamanis et al. 2018) and (Tufano et al. 2018). In these works, gated graph neural network (GGNN) (Li et al. 2015) and HOPE (Ou et al. 2016) were graph embedding techniques adopted for code representation. These are part of PDG based graph embeddings. Based on the derived features, nodes of these PDGs can be represented as a numerical ID of the statements in the work of HOPE (Ou et al. 2016). These average embeddings can become the initial embedding as in GGNN. Final code representation is derived from the max-pooling layer for the nodes of PDGs. Test accuracy metrics compute the source code classification effectiveness.

Four different types of code clones generally seen are as follows (Roy et al. 2007). In the type-1, though comments and layout are different, code fragments are the same. In type-2, different identifier names and literal values are seen with identical code fragments. In the case of type-3, statement-level differences are seen which has similarity at code snippets level. Level-4 implements the same functionality but has syntactically different code snippets. For one of the datasets, clone pair similarity is defined as token-based and line-based metrics average results (Svajlenko et al. 2014). Based on the similarities range, these types are further divided into subtypes. Since the targeted programs resulted in 28 million pairs, 50,000 samples were selected to make sure that the process does not consume too much time. In one of the dataset, 6 million real clone pairs were assessed and 260,000 false ones. RAE (White et al. 2016) and CDLH (Wei et al. 2017) state that state-of-art neural models were used to compare the work here.

Supervised training format RAE + was devised using the unsupervised vectors of program obtained by the authors of RAE, an open-source tool[3]. RAE and CDLH works have already compared DECKARD (Jiang et al. 2007) and doc2vec[4], those are not experimented by authors here. DECKARD is a traditional clone detection method.

Since the code cloning problem is a classification problem to identify the clone or not, metrics like precision (P), recall (R), and F1-score are used for evaluation.

10.11 Results of the research questions

All the below research questions are based on the two tasks that authors propose to investigate.

First research question: Source code classification with the proposed approach. The test accuracy is the metrics for evaluation, samples are well balanced across all the classes in case of the code classification task. It is seen that performance is poor in the case of traditional SVM. Since the tokens used in one of the datasets is arbitrary, and the SVM method depends on token semantics or semantics features of the program for code functionality differentiation.

The capability of capturing local functional features by TextCNN and LSTM, in comparison to token-based methods, make them better performers. For instance, a sliding window of TextCNN or memory cell of LSTM will help in capturing the short scanf statement semantics. A convolutional sliding window that moves over ASTs, used by TBCNN significantly improves the accuracy. This neural network is based on full ASTs to capture tree structural features. Among the existing neural models, LSCNN does better. For recognition of the functionality from the statement, which is sequential information, accuracy will be lower than TBCNN, which is due to lack of capturing structural semantics that is rich. Out of these approaches, HOPE and GGCN that are graph-based techniques, do not perform well. One of the reasons being a node of PDGs miss on lexical knowledge being limited to representation based on the numerical ID, and here, the focus is on high abstraction based explicit dependency information (Tufano et al. 2018). Lack of syntactic information is seen in the case of GGNN with PDG that used node embeddings with tokens, though there are slight improvements.

The authors' approach demonstrates better performance over TBCNN, mainly due to RvNN performed by ASTNN with smaller statement trees compared to actual ASTs. Sliding window and binary tree transformation are not used in the proposed models, unlike other neural models. Between statement, sequential dependencies and AST statements are captured instead. Figure 10.6 depicts Illustration of binary tree transformation.

---

3 https://github.com/micheletufano/AutoenCODE
4 https://radimrehurek.com/gensim/models/doc2vec.html

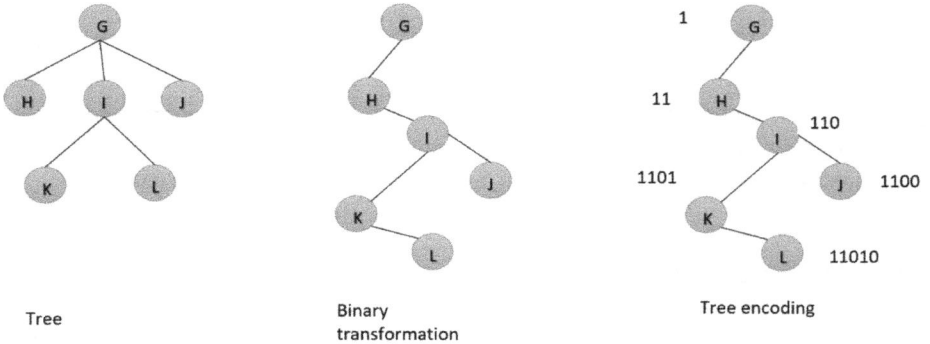

Figure 10.6: Illustration of binary tree transformation.

Research question 2 proposed by authors is the code cloning detection task and the proposed approach's performance. Here, the authors have tried the effectiveness of the approach on challenging tasks like code cloning. Data selection is carefully done to cover all types of data pairs. ASTNN model proposed in comparison to RAE + and CDLH, all perform effectively for identifying similarities between type-1 and type-2, as except for different comments and identifier names, code fragments are the same. In one of the cases of data samples, performance of RAE + was not good enough due to the lack of memorizing capacity as shown in LSTM and GRU. In the case of type-4 data, ASTNN performs better than CDLH in terms of the F1 score. There are clear indications of the ASTNNs capabilities to capture complex semantics and syntactic differences compared to CDLH, which is doing better to take care of the limitations around the naturalness of the statements sequential.

ASTNN also demonstrates its ability to learn local syntactic knowledge that provides functional similarities, and for statements, it provides naturalness in global sequential.

10.12 Binary Search Tree (BST) and its significance related to source code structure

Binary tree representation is the nodes that will have left reference, right reference, and an associated data element. The root element is central to the overall representation. Parent node will have children nodes, in case, if there are no children node, then it is called leaves. Nodes with the same parents are the siblings. Tree structures play a key role in representing structural relations within data. Insertion and searching capabilities and hierarchy representation are other advantages. Data representation becomes easier with the flexibility provided by the subtrees that can be moved around.

Traversing the nodes is another key factor and is conducted in different standard formats. Breadth-first traversal and depth-first traversal are some of the methods. Traversal is about visiting the nodes from top to bottom or left to right or in the specific order as per standard formats. Binary search trees (BST) are a specific type of binary tree that provides the library for storing by which one would do data searching, sorting, and extracting. BST also are characterized by nodes holding a key or data, keys of the left subtree are lesser than the parent node whereas the right subtree will be greater than the parent node. The insertion technique is about the tree's systematic traversing to find out the placement of the new element to make sure that the element is not already in the tree. Similarly, search and delete operations are part of data representation operations in BST.

Tree structures also facilitate the comparison of the topology of two trees. They provide the provision to search nonlinear data structures. Maintaining these trees' inappropriate structure will play a key role, else it may lead to complex search and insertion tasks. Maintaining the BST in an ideal shape has been an area of exploration for researchers. The general focus has been on optimizing the running time of the BST algorithm, there is a need for focus on what algorithm fits well for which scenario. Heights of the BST also influence their operation's performance as the heights increase the depth of the search increase.

In the previous section, exploration was around BST, from the point of their application for the source code representation. Source code structures are influenced by the programming language they use. This thought process holds its importance in the design of the tools for software development. Making the code human-readable objective adds on to the simplicity of the code. Program elements placement and commentary related to it are appropriately placed, forms a key role in program structure. Some of the structural components are indentations that separate code and comments, space in between the tokens that are adjacent to each other. Special characters representing the start of a new line.

Software tools that are based on programming language depend on the linguistic program structure. Syntax tree representation is the conventional way for representing the program, which is possible by parsing the code. Documentary structure and linguistic structures have their importance. Concerning the programmers' responsibility, there is an expectation on them to help build legible structured source code. They need to see through the lens of a graphical designer with the responsibility of making it legible and structured.

Continuing with the research questions put forward by Zhang et al. (2019). Next, they look at the influence of their design choices on the proposed model. Design choices include AST-full, AST-node, AST-block, removing pooling 1 and 2, LSTM in place of GRU, and ASTNN with long code fragments. Some of the design choices included making ASTs granular with splitting them into a different sequence of small trees that do not overlap. Considering the entire AST as one subtree which is AST-full or extracting the nodes of the ASTs as special tree AST-node. In the AST-block based splitting, ASTs are split based on compound statements or blocks that will have multiple statements within them.

Like in ASTNN, follow-up encoding after splitting and bi-directional GRU processing are on similar lines. AST-full and AST-node are overtaken by AST-block and ASTNN in their performance. The richness of syntactic information and the size of the statement tree being very well traded off makes ASTNN model perform well.

As part of pooling design, max-pooling is used on statement trees, for statement encoder in case of pooling-1 and max-pooling layer on statement sequences after recurrent layer is pooling-2, are part of ASTNN model approaches. The pooling layers' influence on the model performance is evaluated by removing each of the pooling layers and validating against only the last hidden layer. Pooling on statement sequences will provide greater performance in comparison to the statement tree-level pooling which has lesser importance. This indicates the varying importance of different code statements in the same code fragment.

As part of the LSTM design approach, GRU is used as a default option, but LSTM though showed slightly lower performance but was in par with GRU. From the efficiency of the training perspective, authors prefer GRU in their ASTNN model. Long code fragments is another design alternate where the model showed a good level of performance, these were cases where there were more than 100 statements.

In the next research question, the influence of the dynamic batching algorithm on the efficiency of the performance was looked at. With statement, encoders can supply arbitrary tree structures as input batch samples, which will increase training speed. The efficiency of the algorithm here is evaluated with different approaches like total without batching (TWB), recurrent layer batching only (PBR), and dynamic batching algorithm (DBA) where recurrent layer and encoding layers are batched. TWB computes one sample each time, one statement tree at a time is sampled in PBR though batch samples are accepted, here batching happens with padding of a sequence of statement trees on the recurrent layer. In DBA, like the PBR, statement tree's batch samples are encoded, and statement tree's sequences are dealt with. The cost of time running the train and test is associated with the experiment. Performance is not influenced by batching, but efficiency does have an influence.

PBR and DBA have a greater influence on speeding up the training in comparison to TWB.

DBA has a significantly higher performance, and more than 12 times faster in comparison to PBR and more than 20 times quicker than TWB, with a batch size of more than 64. That leads to the conclusion that the batching algorithm proposed is an efficient one as part of ASTNN. Figure 10.7 depicts Batching methods and time cost plotting.

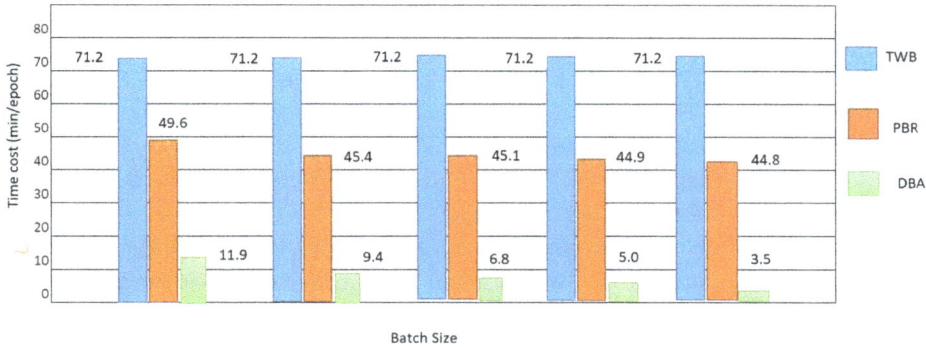

Figure 10.7: Batching methods and time cost.

10.13 Work-related to approaches

In software engineering research, effective source code representation is a key challenge. Textual token-based code representation is employed in conventional IR and other machine learning methods. In code clone detection (Kamiya et al. 2002), the regularized token sequence of code is derived from programs. Optimized inverted-index technique and token ordering are utilized in SourcererCC (Sajnani et al. 2016) which shows improvement. Deckard (Jiang et al. 2007) provides improvement with syntax structured information for clone detection task, this is besides lexical information. Source code authors and domain of the same are explored with n-gram model (Frantzeskou et al. 2008) and SVM (Linares-V´asquez et al. 2014), these are statistical machine learning models. The cohesion of the classes for the software and code fragments semantic similarities are identified with LSI in the work (Maletic et al. 2001), which uses LDA (Liu et al. 2009).

Distributed representation of the source code has been focusing areas of deep learning in the recent past (Allamanis et al. 2017). Code completion is done with RNN and n-gram models (Raychev et al. 2014). Method and class name suggestions are done with the neural context model in Allamanis et al. (2015). Deep belief networks (Wang et al. 2016) are employed for defect prediction, making use of the semantic feature from source code. Name-based bugs are explored with word2vec in DeepBugs (Pradel et al. 2018). In White et al. (2016), recursive autoencoders are preferred over ASTs that will have pretrained token embeddings for capturing syntactical information of ASTs. To learn vector representation of the code snippets, customized convolutional neural networks that work on ASTs are employed in TBCNN (Mou et al. 2016). To represent semantics in the functionality of code fragments. Tree-LSTMs are explored in CDLH (Wei et al. 2017). To detect variables, misuse, and tracking dependencies of the same functions and variables in the context of predicting variable names, GGNN is implemented in Allamanis et al. (2018).

Code functional similarity is measured with semantic matrix representation encoding code control flow and data flow in DeepSim (Zhao et al. 2018). Ensembling learning techniques (Tufano et al. 2018) can compose together different code representations such as CFGs, bytecodes, and identifiers for representing the code. The model proposed in the work reviewed here (Zhang et al. 2019) focuses on syntactic knowledge from statement level, the naturalness of the statements in a sequential way, and improving the AST methods in use.

Among emerging deep learning methods focused on software engineering, relevant API sequence are learned with natural language queries, where neural networks learn from sequence to sequence neural net, this is DeepAPI (Gu et al. 2016). Potential bug files are identified with IR techniques and deep neural networks in Lam et al. (2015). StackOverflow's related question is predicted with convolutional neural net and word embeddings in (Xu et al. 2016). Commit message automated generation is the focus of neural machine translation work in Jiang et al. (2017). Trace links generation is proposed in Guo et al. (2017), with RNN-based neural net. Semantics similarity is evaluated by putting together natural language content and source code in vector space, the joint embedding model is involved in this code search process (Gu et al. 2018). In comparison to these approaches that focus on understanding natural language text associated with source code, the author's work in Zhang et al. (2019) that is reviewed here focuses on source code representation with a neural net.

10.14 Key concerns

Some of the key concerns that may be a threat to the study's validity in Zhang et al. (2019) that we reviewed here are as follows. One of the datasets not being collected from production. But is balanced with another dataset, that has real-world code fragments. Another concern is around constructing one of the datasets, where programs related to the same problem belong to a clone pair. This causes a doubt about the validity of their clone pairing. But, this is balanced with another dataset where the clone pairs are validated manually. That balances the possible influence on the outcomes.

10.15 Summarization

Work in the (Zhang et al. 2019) was reviewed so far and the work has focused on representing the source code from point of syntactic knowledge at the statement level, the naturalness of the statements, source code fragments vector representation, and lexical knowledge, all these with ASTNN. Small sequenced statement trees are derived from large ASTs of code fragments. Multi-way statement trees are recursively encoded to gather statement vectors and gathering the vector representation for the naturalness of statement. Code clone detection and source code classification tasks are used as a reference for experimenting and evaluating ASTNN. These ASTNN largely overtakes existing approaches. Authors have provided their experiment data and code at https://github.com/zhangj1994/astnn (repository was accessible as of June 2020). The proposal is to explore deep semantics with other neural models. To validate these models with large scale datasets and various other software engineering tasks.

10.16 Neural attention model for source code summarization

Source code is associated with being of high quality when it is associated with a good quality of summaries that comes about with documentation of the code through the information posted on an online forum or documentation of the code. These summaries play a significant role but are challenging if it must be manually authored, so they are done on a smaller scale. In this work (Srinivasan et al. 2016), authors focus on the data-driven approach for extracting summaries of source code. The model proposed here CODE-NN uses attention in LSTM to generate a description for SQL queries and C# code snippets. Corpus of training data is taken from StackOverflow for CODE-NN. Experiments have shown good results for code summarization, where the approach had outperformed the benchmark approaches with end to end learning. The task of the code retrieval approach used has overtaken C# benchmarks.

Tons of source code is available online, quality of the content is influenced by code that has associated instructions and comments in the form of natural language. A code summary that is short and informative, is a good reference for tutorials and code search tasks. General-purpose language C# is the reference taken in the study, also SQL which is a database query language that is declarative. For instance, generating substring in a string that can give a high-level description to represent overall code function. One of the key challenges is the text's complexity related to the nonlocal aspects of the code. High quality of the models that can generate summaries will help interpret the code that is not commented, also helps in retrieving the code, and will assist in applying natural language programming for summarized code base that was originally large. Content selection part, which is about what must be said, realization part of how to convey the message, is the pipeline that must be built as per traditional Natural Language generation (David et al. 2010, Ehud et al. 2000, Yuk et al. 2007, Wei et al. 2011). Scaling such systems to the large domain is a challenge and needs close monitoring. CODE-NN will be an end to end neural network framework where attention mechanism is used for selecting the content and LSTMs are engaged in surface realization. With attention mechanism, one word at a time summarization is done which is done over the embeddings of the source code LSTM network generated context is used as input (Sepp et al. 1997).

Feature engineering (Gabor et al. 2010) is an algorithm for decoding which is expensive (Ioannis et al. 2013) is not needed, it would be a simple model that learns from the training data. StackOverflow, a famous help website for programming, is used as a base for the training with new code snippets that has short descriptions. This is an open data available on the public website. There is quite a bit of noise associated with terms of content being a non-parsable, grammatic error, and lacks content as a part, but holistically could provide a strong signal for learning. To make a meaningful evaluation of the data, human-constructed clean data is added on.

Code retrieval and summarization are targeted. METEOR and BLEU-4 are the metrics chosen for assessing the summarization, this is also inclusive of the human study of the naturalness of the code. Authors also see their approach showing performance as better than most other techniques and has its uniqueness in source code summary gathering from the online generated data. Programming related questions of C# benchmark where code retrieval is tried out with CODE-NN, results show greater performance when compared to its benchmark, for metric Mean Reciprocal Rank (MRR).

10.17 Related approaches

As a general task, code snippets of source code are used to generate Natural Language (NL) summary, by CODE-NN. Also, there was attempt to generate source code with questions in NL inversely.

Though the focus will be around a high-level summary of the source code, abstracting the code description at various other levels has been an exploration of some of the work. Class-level comments are predicted by learning from topic models and n-gram from the java projects, which are open source character-saving metrics used for testing the same for already existing (Dana et al. 2013). Methods and class names are suggested by the models created in Miltiadis et al. (2015) that embed the outcome in continuous space of higher dimension. Summaries of Java are generated in (Giriprasad et al. 2010) where the pipeline is built, relevant content is chosen and then template-based summaries are extracted. Summarizing the bugs in software and cross-cutting concerns in source code is identified as part of program comprehension (Sonia et al. 2010). Authors here highlight that their work, which is the primary work that we are reviewing in this section (Srinivasan et al. 2016), constructing the new sentences from the code snippets with learning technique, stands unique. This aspect provides a good base for our exploration of understanding the structure of software source code.

Formal meaning representation has its commonalities with the source code summarization task. Utilizing an inverted semantic parser, sentences are generated in the system with lambda calculus in work (Yuk et al. 2007). Database records are used for text generation with a learning algorithm, with an assumption that the formal meaning of the representation is arrived at with pairing data (Gabor et al. 2010, Hongyuan et al. 2016, Ioannis et al. 2013). About these works, authors here use web data available easily and learn from them.

SQL database systems that can talk to users, was a need identified in work (Alkis et al. 2009), from the database community. Similarly, in LOGOS (Georgia et al. 2010), an interactive system for translating SQL queries to NL template with database schemas was explored. Rules are used to generate NL from SPARQL queries, where, for each query dependency, trees are generated, also effort toward making the outcome natural (Axel-Cyrille et al. 2013). Here, there is a need for manual intervention and quite a bit of effort involved as they are not based on learning.

Here authors propose attention-based LSTM, RNN, where source code and NL are jointly modeled. Some of the tasks, where RNN finds its popularity are machine translation-based text generation, description of video and images, summarization of the sentence, Chinese poetry generation, and so on. (Andrej et al. 2015, Alexander et al. 2015, Illya et al. 2011, Jacob et al. 2015, Subhashini et al. 2015, Xingxing et al. 2014). Closest work (Tsung-Hsien et al. 2015) has a two-stage approach, to generate spoken dialog–based text with LSTM decoder where speech acts can be logically represented. Stackoverflow data and web search logs are utilized for C# code snippets extraction with a model that takes NL question input and vice versa. Code structure and language are represented in bidirectional with multiplicative and additive models with language and code being scored, where retrieval works well but generation task struggle. Search logs are not used in the neural generation model which learns to score the code in retrieval task and demonstrate high accuracy, this is characteristic of the proposed model.

NLP and systems research community have focused study on the language being translated into the code. With the proposed model being utilized for SQL and C#, this will yield a domain agnostic method.

10.18 Data management

Each post of the StackOverflow that has been used as data may have multiple tags associated with them. Segregation of data based on tags is done for C#, SQL, database, and oracle tags for SQL. The data's general structure is the short title for the post, the detailed question, with few responses with one of it marked right. It also confirmed that the post's questions and responses are domain-specific and verbose, with quite a bit of information that is not relevant. Also, it is observed that the code parts in the responses that are not accepted are mostly tangential to the question or irrelevant. The key part of the data that was picked was the title and the code snippet from the accepted response and the same was tagged and added on to the corpus.

As part of the data cleaning, the semi-supervised classifier will assist in irrelevant titles that do not connect with the code snippets. For the purpose sample annotation was created, for every language, and then used in the algorithm with bootstrapping.

The rest of the corpus is run as an unsupervised classifier and led to 73% accuracy in the process. Here, the test set was manually labeled for both languages. After the finalization of the corpus 80%, 10%, and 10% split of data is done for training, validation, and testing, respectively. Figure 10.8 depicts Machine learning workflow.

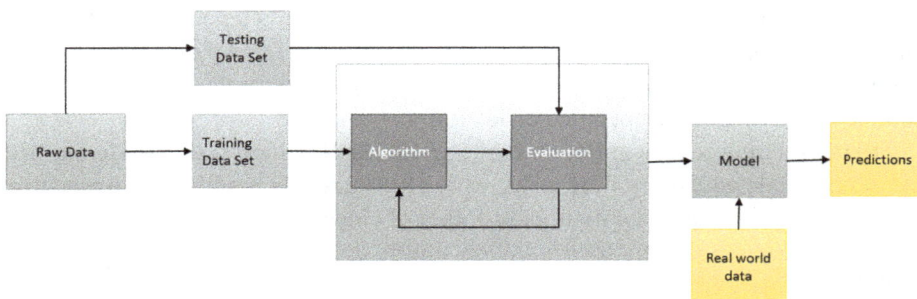

Figure 10.8: Machine learning workflow.

Parsing the data is the next option, StackOverflow being an informal setup where code snippets only partially relate to answers. For instance, in the case of SQL, only 12% of the content is free of syntactic errors when running with SQL queries parser. Literals are replaced with tokens to make them free of being content-specific. Similarly, in the case of SQL tables and column names, numerical placeholders are provided keeping note of retention of the dependencies, if any.

When looked at data statistics, the size of the structure's code and complexity make this a challenging task. To give an instance of complexity, in the C# corpus, 40% of the code snippets contain three and more functions and statements and loops and conditionals make a 20%. One-third of SQL queries include multiple tables, subqueries, columns, and functions.

Concerning the corpus's actual title, the summary is generated by the model, here n-gram based metrics are utilized. Titles will be short and code snippets can be explained in multiple ways, where the overlapping content is less. To make up for the limited data, more hand-labeled data are plugged into the study. In this manual process, annotators were shown only code snippets to come up with code summaries after a few examples of the code summary was reviewed by them. Here, expectation was for annotators to come up with a question that they may have asked the help website, if they, in turn, would want to see a code snippet of that type in their response. So, the focus was to come up with the key aspect that the code snippet was trying to convey. Here, half of the test set went into the tuning of the model and the rest of them for evaluation.

10.19 Feature engineering in source code

If we look at the programming mechanics, it involves transforming patterns, models, and design into source code. Understanding this is key for the discussion of features in source code. Feature engineering is about understanding the key data associated with the process that can be leveraged to build a prediction system as we do in case of machine learning. It is also the thought process of transforming a subjective system into an objective one to make an intelligent assessment of the system that would help in building the future process.

Here, we have a context of understanding the software source code in its full scale, which would contribute to building a smart software development lifecycle and efficient one. These features will capture the variabilities associated with the process from which they come from. Understanding the program is in terms of its structural programming. Every component will store its state in the variable, other components are referred based on calculations done with variable values. Data exchange also is done with variables across the components. Figure 10.9 depicts Software source code construct.

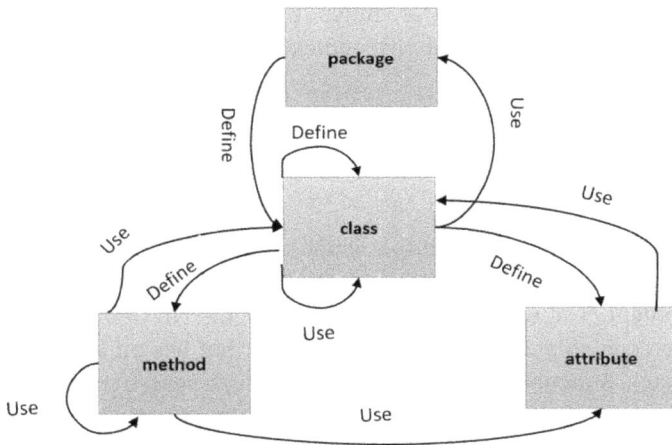

Figure 10.9: Software source code construct.

Feature derivation also has its role to play in the reusability of the code. One of the important considerations is the domain experts' inputs in terms of devising these feature engineering pipeline for the source code. Based on the specific area of interest, there is a need to device the feature engineering focus. Interacting features is also important to be assessed, as there will be potential features that can together contribute better rather than individually. Identifying such combinations will strengthen the feature engineering process. Feature representation is another thought process where the expert inputs become key to device a way of representing the features, which will help bring out the information effectively. Another area worth exploring is combining the features derived from multiple sources. This can help tackle the problem of data deficit and put together the integrated knowledge from multiple connected systems.

As part of the feature engineering, there is a greater dependency on how the source code data collection and cleaning pipelines are established. Publicly available data and their content relevance is a big challenge. Unavailability of the data owned by companies would be a concern. Amid all these, like any predictive system building process, cleaning the data to make it appropriate to the prediction pipeline is key. Some of the basic data cleaning aspects like managing the missing information, scaling and normalizing the information, managing some of the special information like dates, encoding character-based data, and data outliers are important aspects. Beyond these, domain knowledge in software development must be put together to ensure comprehensive handling of the data before it is modeled.

Long term dependencies management within source code is a characteristic feature that should be accounted while we build the features. The feature also needs to provide the generalizable capability for models. Since NLP closely associates with the source code modeling, representation approaches like encoder-decoder will fit in well. Here feature engineering will play a prominent role. The idea would be to represent the input data with encoder and derive contextual representation to decode and derive output representation.

10.20 CODE-NN model

Moving back to the work (Srinivasan et al. 2016), which is primary work being reviewed. Authors employ an end to end system that specializes in surface realization and selection of the content. The conditional distribution of the NL (natural language) summary on the code snippet with a neural network that has an attention mechanism is the approach. Summarizing one word at a time with source code snippet having an attention mechanism in LSTM. With N being the summaries' vocabulary for $n = n_1, \ldots \ldots, n_l$ representing the 1-hot vector for a sequence of summaries. With code snippets $c_1, \ldots, c_k$, title $n = n_1, \ldots, n_l$ is generated, Distribution representation for the code snippet, t_i, is computed with attention cell with LSTM hidden state h_i. To feedback, the next LSTM cell, h_i, and t_i combination are used and the next work n_i is generated. This process continues until the required number of words is generated or the sequence moves to the end. Softmax operations are involved in the end. Figure 10.10 depicts CODE-NN model.

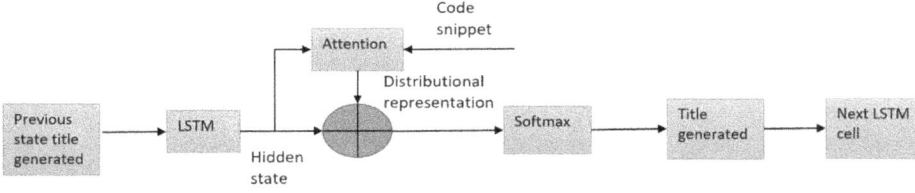

Figure 10.10: CODE-NN model.

Scoring function *s* is computed for the probability of *n*, which will be the product of the conditional probability for the next set of word probabilities,

$$s(c, n) = \prod_{i=1}^{l} p(n_i | n_1, \ldots, n_{i-1}) \tag{10.16}$$

with,

$$p(n_i | n_1, \ldots, n_{i-1}) \propto W \tanh(W_1 h_i + W_2 t_i) \tag{10.17}$$

Here, $W \in \mathbb{R}^{|N| \times H}$ and $W1, W2 \in \mathbb{R}^{H \times H}$

For the summaries, embedding dimensionality is provided by *H*. For the source code, t_i will be the attention model contribution. For the current time step, hidden state of the LSTM is represented by h_i, previous words form input for this, with m_{i-1} cell state of LSTM and LSTM hidden state h_{i-1}.

$$m_i; h_i = f(n_{i-1}E, m_{i-1}, h_{i-1}; \theta) \tag{10.18}$$

Here, $E \in \mathbb{R}^{|N| \times H}$ will provide word embeddings summary matrix. Function *f* here has been computed with the reference of the LSTM cell architecture as in (Zaremba et al. 2014).

Attention is represented as follows. The weighted sum of the embeddings is computed for code snippets by the global attention model as in (Thang et al. 2015), this is based on the current LSTM state. Set of one-hot vectors c1, ...,$C_K \in \{0,1\}^{|C|}$ for tokenized source code. For all the tokens in code snippet vocabulary, C is considered. The computation of the attention model is given as follows:

$$t_i = \sum_{j=1}^{k} \alpha_{i,j} \cdot c_j F \tag{10.19}$$

Here, $F \in \mathbb{R}^{|N| \times H}$ is an embedding matrix for token and each value of $\alpha_{i,j}$ will be proportionate for dot product between token embedding c_j and corresponding current internal LSTM hidden state h_i.

$$\alpha_{i,j} = \frac{exp(h_i^T c_j F)}{\sum_{j=1}^{k} exp(h_i^T c_j F)} \tag{10.20}$$

Training is performed with supervised learning backpropagation, where parameters are learned from embedding matrices to derive the LSTM cell's parameters that computes the function. To reduce negative log-likelihood in the training set, minibatch stochastic gradient descent with multiple epochs is employed. Dropout layers avoid overfitting and assisted with the softmax layer at the end.

Implementation is as follows, UNK token used for all tokens that appear less than 3 times, START, and END tokens are included in the case of the training sequence. Based on the performance on the validation set, hyperparameters are finalized. Minibatch size of 100 is considered, 400 dimensions considered for hidden states of LSTM, summary embeddings and token embeddings. Model parameters are all initialized with values between −0.35 and 0.35. A starting learning rate of 0.5 is reduced by 0.8 after every 60 epochs, if validation set accuracy goes down. 0.01 is the leaning rate considered for termination. The parameter gradient is proposed to be capped at 5 and drop out of 0.5 considered.

GPU training conducted with the Torch framework. METEOR metric (Satanjeev et al. 2005) is used for every epoch in the development set to finalize the best model. In the case of decoding, 10 beam size preferred and the length is chosen to be maxed at 20 words.

10.21 Qualitative analysis and conclusion

Content selection is of high quality in CODE-NN in which informative tokens in code snippets with key summary words. Most of the cases show meaningful output for simple code snippets but with longer ones, there is the degradation of the performance. Redundancy is a major source of error, with missing content and generation of extraneous phrases. This came out in manual error analysis done with random samples selected from each language's development training set. In the case of low-frequency tokens, the model output would go out of context, concerning inputs. These are cases with a frequency of less than 7% and in these cases, CODE-NN would use generic phrases to realize these.

Authors here presented CODE-NN, an end-to-end neural attention model with LSTM to generate summaries from SQL and C# code from the noise involved in online source of data. Models have outperformed the baseline models competing in this space and achieved a state-of-the-art performance for metrics like METEOR, BLEU (Kishore et al. 2002), and human valuation. Most appropriate code snippets are derived from the corpus to answer programming questions that have done better than earlier baselines. Authors see a potential extension of the work toward better models that can capture the input structure and use these systems to apply in other areas such as source code documentation automation.

10.22 Overall conclusion

AST and neural attention model covers a lot of ground in the objective of under-standing the software development life cycle landscape. This provides a good chance to study the software programming language on a full scale. As we explore various characteristics of the software source code and programming language, we develop deeper insight in to applying data science into the area and solving problems. Also, this understanding helps to make use of the knowledge gained in modeling across the various domains. So, the investment needed for building the system from scratch can be done away with. Building an efficient data pipeline for data modeling also is explored and experimented with. Natural language–related components of the source code also play a prominent role in the gamut of exploration. With a wider area of source code subjected to experiments, there is a greater possibility of making space for a software developer to focus on the creative part of the work rather than worrying about the basics. This will contribute to building smarter systems.

References

M. Abadi et al. 2016. TensorFlow: Large-Scale Machine Learning on Heterogeneous Distributed Systems. http://arxiv.org/abs/1603.04467

J. Abbot, J. Marohasy. 2012. Application of artificial neural networks to rainfall forecasting in Queensland, Australia. Adv. Atmos. Sci., 29(4), 717–730.

Abstract syntax tree for Euclidean algorithm. https://commons.wikimedia.org/wiki/File:Abstract_syntax_tree_for_Euclidean_algorithm.svg. Accessed on October 4, 2020

K. Akkarajitsakul, E. Hossain, D. Niyato, D. I. Kim. 2011. Game theoretic approaches for multiple access in wireless networks: A survey. IEEE Commun. Surveys Tutorials., 13(3), 372–395.

A. Alipour, A. Hindle, E. Stroulia. 2013. A contextual approach towards more accurate duplicate bug report detection. 10th Working Conference on Mining Software Repositories (MSR), San Francisco, CA. 183–192. Doi: 10.1109/MSR.2013.6624026.

H. A. Al-Jamimi, L. Ghouti. 2011. Efficient prediction of software fault proneness modules using support vector machines and probabilistic neural networks. In 5th Malaysian Conference in Software Engineering (MySEC2011). 251–255.

M. Allamani, E. T. Barr, P. Devanbu. 2018. A Survey of Machine Learning for Big Code and Naturalness. https://arxiv.org/pdf/1709.06182.pdf. Accessed October 3, 2020.

M. Allamanis, E. T. Barr, C. Bird, C. Sutton. 2015. Suggesting accurate method and class names. In Proceedings of the 2015 10th Joint Meeting on Foundations of Software Engineering. ACM, 2015a. https://homepages.inf.ed.ac.uk/csutton/publications/accurate-method-and-class.pdf. Accessed October 3, 2020.

M. Allamanis, E. T. Barr, C. Bird, C. Sutton. 2015. Suggesting accurate method and class names. In Proceedings of the 2015 10th Joint Meetingon Foundations of Software Engineering. ACM. 38–49.

M. Allamanis, E. T. Barr, C. Bird, C. Sutton. 2015a. Suggesting accurate method and class names. In Proceedings of the 2015 10th Joint Meeting on Foundations of Software Engineering. 38–49.

M. Allamanis, E. T. Barr, P. Devanbu, C. Sutton. 2017. A survey of machine learning for big code and naturalness. arXiv preprint arXiv:1709.06182.

M. Allamanis, M. Brockschmidt, M. Khademi. 2018. Learning to represent programs with graphs. In International Conference on Learning Representations, 2018. https://openreview.net/forum?id=BJOFETxR-. Accessed on October 5, 2020

M. Allamanis, D. Devanbu, C. Sutton. A Survey of Machine Learning for Big Code and Naturalness. 2018. arxiv.org. Accessed Sep 2020

M. Allamanis, H. Peng, C. Sutton. 2016. A convolutional attention network for extreme summarization of source code. https://arxiv.org/abs/1602.03001. Accessed October 3, 2020.

M. Allamanis, C. Sutton. 2013. Mining source code repositories at massive scale using language modeling. In Proceedings of the 10th Working Conference on Mining Software Repositories (MSR '13).

M. Allamanis, C. Sutton. 2014. Mining idioms from source code. In Proceedings of the 22nd ACM SIGSOFT International Symposium on Foundations of Software Engineering, FSE 2014. 472–483. Doi: 10.1145/2635868.2635901. https://arxiv.org/abs/1404.0417. Accessed October 3, 2020.

M. Almukaynizi, E. Nunes, K. Dharaiya, M. Senguttuvan, J. Shakarian, P. Shakarian. 2019. Patch Before Exploited: An Approach to Identify Targeted Software Vulnerabilities. In: AI in Cybersecurity, L. F. Sikos, Ed., Cham, Springer International Publishing, 2019, 81–113.

U. Alon, M. Zilberstein, O. Levy, E. Yahav. 2019. Code2vec: Learning distributed representations of code. Proc. ACM Program. Lang, 3, (POPL), 40. 1–40, 29. https://arxiv.org/abs/1803.09473. Accessed October 3, 2020.

https://doi.org/10.1515/9783110703399-011

D. G. Altman, J. M. Bland. 1994. Statistics notes: diagnostic tests 1: sensitivity and specificity. Br. Med. J., 308, 1552. https://www.bmj.com/content/308/6943/1552. Accessed October 2, 2020.

Android – OpenCV library. https://opencv.org/platforms/android/. Accessed October 3, 2020.

Android Profiler for Android Studio.https://developer.android.com/studio/profile/android-profiler. Accessed October 3, 2020.

Android Studio. https://developer.android.com/studio/. Accessed October 3, 2020.

G. Angeli, P. Liang, D. Klein. 2010. A simple domain-independent probabilistic approach to generation. In Proceedings of the 2010 Conferenceon Empirical Methods in Natural Language Processing. 502–512.

Ansible Inc. 2014. WhitePaper: Ansible In Depth. https://www.ansible.com/hubfs/pdfs/Ansible-InDepth-WhitePaper.pdf. Accessed October 3, 2020.

G. Antoniol, K. Ayari, M. DiPenta, F. Khomh, Y.-G. Gu´eh´eneuc. 2008. Is it a bug or an enhancement? a text-based approach to classify change requests. In Proceedings of the 2008 conference of the center for advanced studies on collaborative research: meeting of minds. ACM.23-23. https://doi.org/10.1145/1463788.1463819.

G. Antoniol, G. Canfora, G. Casazza, A. De Lucia, E. Merlo. 2002. Recovering traceability links between code and documentation. IEEE Trans. Softw. Eng., 28–10, 970–983. Doi: 10.1109/TSE.2002.1041053.

Approximate Bayesian computation. https://en.wikipedia.org/wiki/Approximate_Bayesian_computation

Arduino. https://www.arduino.cc/. Accessed October 3, 2020.

C. Arora, M. Sabetzadeh, L. C. Briand, F. Zimmer, R. Gnaga. 2013. Automatic checking of conformance to requirement boilerplates via textchunking: An industrial case study. In 2013 ACM / IEEE International Symposium on Empirical Software Engineering and Measurement, Baltimore, Maryland, USA, October 10–11, 2013. 35–44. Doi: 10.1109/ESEM.2013.13.

H. U. Asuncion, A. U. Asuncion, R. N. Taylor. 2010. Software traceability with topic modeling. 2010 ACM/IEEE 32nd International Conference on Software Engineering, Cape Town. 95–104. Doi: 10.1145/1806799.1806817.

T. Atwood. 2009. RF channel characterization for cognitive radio using support vector machines. Ph.D. dissertation, University of New Mexico.

Z. Aung, M. Toukhy, J. R. Williams, A. Sanchez, S. Herrero. 2012. Towards accurate electricity load forecasting in smart grids. In The Fourth International Conference on Advances in Databases, Knowledge, and Data Applications (DBKDA2012). 51–57.

Autoencoder. https://en.wikipedia.org/wiki/Autoencoder. Accessed October 3, 2020.

N. N. Axel-Cyrille, L. B¨uhmann, C. Unger, J. Lehmann, D. Gerber. 2013. Sorry, I don't speak sparql: Translating sparqlqueries into natural language. In Proceedings of the 22nd International Conference on World Wide Web. 977–988.

D. Bahdanau, K. Cho, Y. Bengio. 2014. Neural machine translation by jointly learning to align and translate. arXiv Preprint, arXiv:1409.0473. https://arxiv.org/abs/1409.0473. Accessed October 3, 2020.

D. Bahdanau, K. Cho, Y. Bengio. 2014. Neural machine translation by jointly learning to align and translate. arXiv preprint arXiv:1409.0473.

N. Baldo, B. Tamma, B. Manojt, R. Rao, M. Zorzi. 2009. A neural network based cognitive controller for dynamic channel selection. In IEEE International Conference on Communications (ICC '09). 1–5.

N. Baldo, M. Zorzi. 2008. Learning and adaptation in cognitive radios using neural networks. In 5th IEEE Consumer Communications and Networking Conference (CCNC '08). 998–1003

S. Banerjee, A. Lavie. 2005. Meteor: An automatic metric for mt evaluation with improved correlation with human judgments. Proceedings of the acl workshop on intrinsic and extrinsic evaluation measures for machine translation and/or summarization, 29, 65–72.

J. Baxter, P. L. Bartlett. 2001. Infinite-horizon policy-gradient estimation. J. Artif. Intell. Res., 15, 319–350, 2001.

Bayes' theorem. https://en.wikipedia.org/wiki/Bayes%27_theorem. Accessed October 2, 2020

Bayesian inference. https://en.wikipedia.org/wiki/Bayesian_inference

A. Begel, Y. P. Khoo, T. Zimmermann. 2010. Codebook: discovering and exploiting relationships in software repositories. 2010 ACM/IEEE 32nd International Conference on Software Engineering, Cape Town. 125–134. Doi: 10.1145/1806799.1806821.

Y. Bengio, R. Ducharme, P. Vincent, C. Jauvin. 2003. A neural probabilistic language model. J. Mach. Learn. Res., 3, 1137–1155.

Y. Bengio, P. Simard, P. Frasconi. 1994. Learning long-term dependencies with gradient descent is difficult. IEEE Trans. Neural Netw., 5(2), 157–166.

A. Bennaceur, R. Haehnle, K. Meinke, (Editors). 2016. Machine Learning for Dynamic Software Analysis: Potentials and Limits. Springer, Switzerland AG.

J. O. Benson, J. J. Prevost, P. Rad, 2016. Survey of automated software deployment for computational and engineering research. 2016 Annual IEEE Systems Conference (SysCon), Orlando, FL. 1–6. Doi: 10.1109/SYSCON.2016.7490666.

P. Bhattacharya, M. Iliofotou, I. Neamtiu, M. Faloutsos. 2012. Graph-based analysis and prediction for software evolution. Published in: 2012 34th International Conference on Software Engineering (ICSE). 419–429.

P. Bhattacharya, I. Neamtiu. 2010. Fine-grained incremental learning and multi-feature tossing graphs to improve bug triaging. 2010 IEEE International Conference on Software Maintenance, Timisoara. 1–10. Doi: 10.1109/ICSM.2010.5609736.

P. Bhattacharya, I. Neamtiu. 2011. Assessing programming language impact on development and maintenance: a study on c and c++. 2011 33rd International Conference on Software Engineering (ICSE), Honolulu, HI. 171–180. Doi: 10.1145/1985793.1985817.

A. Bhoopchand, T. Rockt¨aschel, E. Barr, S. Riedel. 2016. Learning Python Code Suggestion with a Sparse Pointer Network. 2016. https://arxiv.org/abs/1611.08307. Accessed October 3, 2020.

C. Bird, T. Menzies, T. Zimmermann, (Editors). 2015. The Art and Science of Analyzing Software Data. Elsevier, Amsterdam, Netherlands.

T. F. Bissyandé, F. Thung, S. Wang, D. Lo, L. Jiang. 2013. Réveillère. Empirical evaluation of bug linking. 17th European Conference on Software Maintenance and Reengineering, Genova. 89–98. Doi: 10.1109/CSMR.2013.19.

M. Bkassiny, S. K. Jayaweera, K. A. Avery. 2011. Distributed reinforcement learning based MAC protocols for autonomous cognitive secondary users. In 20th Annual Wireless and Optical Communications Conference (WOCC '11), Newark, NJ. 1–6.

M. Bkassiny, S. K. Jayaweera, Y. Li. 2012. Multidimensional Dirichlet process-based non-parametric signal classification for autonomous selflearningcognitive radios. IEEE Trans. Wireless Commun.

M. Bkassiny, S. K. Jayaweera, Y. Li. 2013. Multidimensional Dirichlet Process-Based Non-Parametric Signal Classification for Autonomous Self-Learning Cognitive Radios. IEEE Trans. Wireless Commun., 12(11), 5413–5423. Doi: 10.1109/TWC.2013.092013.120688.

M. Bkassiny, S. K. Jayaweera, Y. Li, K. A. Avery. 2012. Wideband spectrum sensing and non-parametric signal classification for autonomousself-learning cognitive radios. IEEE Trans. Wireless Commun., 11–7, 2596–2605.

M. Bkassiny, Y. Li, S. K. Jayaweera. 2013. A survey on machine-learning techniques in cognitive radios. In IEEE Commun. Surv. Tutorials, 15(3), 1136–1159. Doi: 10.1109/SURV.2012.100412.00017.

D. M. Blei, A. Y. Ng, M. I. Jordan. 2003. Latent Dirichlet allocation. J. Mach. Learn. Res, 3, 993–1022.

P. Bojanowski, E. Grave, A. Joulin, T. Mikolov. 2016. Enriching word vectors with subword information. arXiv Preprint, arXiv:1607.04606.

N. K. Bose, P. Liang. 1998. Neural Network Fundamentals with Graphs. Algorithms, and Applications. Tata McGraw-Hill, New Delhi.

B. E. Boser, I. M. Guyon, V. N. Vapnik. 1992. A training algorithm for optimal margin classifiers. In Proceedings of the fifth annual workshop on Computational Learning Theory, ser. COLT '92. New York, NY, USA: ACM, 1992. 144–152. http://doi.acm.org/10.1145/130385.130401. Accessed October 3, 2020.

G. E. Box, G. M. Jenkins, G. C. Reinsel. 2013. Time Series Analysis: Forecasting and Control. John Wiley & Sons, Hoboken, New Jersey, United States.

M. Bozorgi, L. K. Saul, S. Savage, G. M. Voelker. 2010. Beyond heuristics: learning to classify vulnerabilities and predict exploits. In 16th ACM SIGKDD International Conference on Knowledge Discovery and Data Mining, Washington, DC, USA, 2010. 105–114. https://doi.org/10.1145/1835804.1835821

P. Brebner, W. Emmerich. Deployment of Infrastructure and Services in the Open Grid Services Architecture (OGSA). In: Component Deployment. CD 2005. Lecture Notes in Computer Science, A. Dearle, S. Eisenbach, Eds., Vol. 3798, Springer, Berlin, Heidelberg, 3798. Doi: https://doi.org/10.1007/11590712_15.

L. Breiman. 1996. Bagging Predictors. In: Machine Learning, Kluwer Academic Publishers, 24, 3: 123–140.

L. Breiman. 1996. Bias, variance, and arcing classifiers. http://citeseerx.ist.psu.edu/viewdoc/summary?doi=10.1.1.30.8572. Accessed on October 4, 2020

L. Breiman, J. H. Freidman, R. A. Olshen, C. J. Stone. 1984. Classification and Regression Trees. Wadsworth.

S. Brin, L. Page. 1998. The anatomy of a large-scale hypertextual web search engine. Comput. Netw. ISDN Syst., 30(1–7), 107–117.

M. Bruch, E. Bodden, M. Monperrus, M. Mezini. 2010. Ide 2.0: collective intelligence in software development. In Proceedings of the FSE/SDP workshop on Future of software engineering research. ACM 53–58. https://hal.archives-ouvertes.fr/hal-01575346/document. Accessed October 3, 2020.

M. Bruch, E. Bodden, M. Monperrus, M. Mezini. 2010. IDE 2.0: collective intelligence in software development. In Proceedings of the FSE/SDP workshop on Future of software engineering research. ACM, 2010. 53–58. https://hal.archives-ouvertes.fr/hal-01575346/document. Accessed October 3, 2020.

M. Bruch, M. Monperrus, M. Mezini.2009. Learning from examples to improve code completion systems. In Proceedings, ACM SIGSOFT ESEC/FSE. https://hal.archives-ouvertes.fr/hal-01575348/file/Learning-from-Examples-to-Improve-Code-Completion-Systems.pdf. Accessed October 3, 2020.

E. Bruneton, T. Coupaye, J. B. Stefani. 2004. The Fractal Component Model. http://fractal.objectweb.org/specification/index.html. Accessed October 3, 2020.

F. Buschmann, K. Henney, D. C. Schmidt. Pattern-Oriented Software Architecture Volume 5: On Patterns and Pattern Languages. 2007. Wiley, Hoboken, NJ.

E. Busseti, I. Osband, S. Wong. 2012. Deep learning for time series modeling. Technical report, Stanford University, Tech. Rep.

D. Cabric. 2008. Addressing feasibility of cognitive radios. IEEE Signal Proc. Mag., 25(6), 85–93.

Y. Cao, J. Xu, T. Y. Liu, T. Y. Liu, H. Li, Y. Huang, H. W. Hon. 2006. Adapting ranking SVM to document retrieval. SIGIR, 06, 186–193.

P. Carbonnelle. Pypl popularity of programming language. http://pypl.github.io/PYPL.html. Accessed October 3, 2020.

C. Catal, A. Akbulut, E. Ekenoglu, M. Alemdaroglu. 2017. Development of a software vulnerability prediction web service based on artificial neural networks. Conference: Pacific-Asia Conference on Knowledge Discovery and Data Mining. Doi: 10.1007/978-3-319-67274-8_6

C. C. Chang, C.-J. Lin. 2011. Libsvm: a library for support vector machines. TIST, 2(3), 27–27.

S. Chatterjee, A. Dash, S. Bandopadhyay. 2014. Ensemble support vector machine algorithm for reliability estimation of a mining machine. Quality and Reliability Engineering International.

S. Chattopadhyay, G. Chattopadhyay. 2008. Comparative study among different neural net learning algorithms applied to rainfall time series. Meteorol. Appl., 15(2), 273–280.

S. Chattopadhyay, M. Chattopadhyay. 2007. A soft computing technique in rainfall forecasting. Int. Conf. on IT, HIT. 19–21.

Chef Software, Inc. 2015. https://www.chef.io/customers/. Accessed October 3, 2020.

D. L. Chen, J. Kim, J. Raymond. 2010. Training a multilingual sportscaster: Using perceptual context to learn language. J. Artif. Intell. Res., 397–435.

X. Chen, J. Grundy. 2011. Improving automated documentation to code traceability by combining retrieval techniques. 26th IEEE/ACM International Conference on Automated Software Engineering (ASE 2011), Lawrence, KS, 2011. 223–232. Doi: 10.1109/ASE.2011.6100057.

X. Chen, L. Xu, Z. Liu, M. Sun, H. Luan. 2015. Joint learning of character and word embeddings. In Proc. Int. Joint Conf. Artificial Intelligence. 1236–1242.

K. Cho, B. Van Merriënboer, D. Bahdanau, Y. Bengio. 2014. On the properties of neural machine translation: Encoder–decoder approaches. 103–103. https://arxiv.org/abs/1409.1259. Accessed October 3, 2020.

K. Cho, B. Van Merriënboer, C. Gulcehre, D. Bahdanau, F. Bougares, H. Schwenk, Y. Bengio. 2014. Learning phrase representations using RNN encoder-decoder for statistical machine translation. arXiv Preprint, arXiv: 1406.1078. https://arxiv.org/abs/1406.1078. Accessed October 3, 2020.

S.-Y. Cho, Z. Chi, W.-C. Siu, A. C. Tsoi. 2003. An improved algorithm for learning long-term dependency problems in adaptive processing ofdata structures. IEEE Trans. Neural Netw., 14(4), 781–793.

L. Chu, R. Qiu, H. Liu, Z. Ling, T. Zhang, J. Wang. 2017. Individual Recognition in Schizophrenia using Deep Learning Methods with Random Forest and Voting Classifiers: Insights from Resting-State EEG Streams. http://arxiv.org/abs/1707.03467. Accessed October 3, 2020.

C. Clancy, J. Hecker, E. Stuntebeck, T. O'Shea. 2007. Applications of machine learning to cognitive radio networks. IEEE Wireless Commun., 14(4), 47–52.

C. Claus, C. Boutilier. 1998. The dynamics of reinforcement learning in cooperative multiagent systems. In Proceedings of the Fifteenth National Conference on Artificial Intelligence, Madison, WI. 746–752

J. Cleland-Huang, A. Czauderna, M. Gibiec, J. Emenecker. 2010. A machine learning approach for tracing regulatory codes to product specific requirements. In Proceedings of the 32nd ACM/IEEE International Conference on Software Engineering, ICSE, Cape Town, South Africa, 1–8 May 2010. 1:155–164. Doi: 10.1145/1806799.1806825.

J. Cleland-Huang, R. Settimi, C. Duan, X. Zou. 2005. Utilizing supporting evidence to improve dynamic requirements traceability. 13th IEEE International Conference on Requirements Engineering (RE'05), Paris, 2005. 135–144. Doi: 10.1109/RE.2005.78.

J. Cleland-Huang, R. Settimi, X. Zou, P. Solc. 2006. The detection and classification of non-functional requirements with application to early aspects. 14th IEEE International Requirements Engineering Conference (RE'06), Minneapolis/St. Paul, MN. 39–48. Doi: 10.1109/RE.2006.65.

J. Cleland-Huang, R. Settimi, X. Zou, P. Solc. 2007. Automated classification of non-functional requirements. Requir. Eng., 12(2), 103–120. Doi: https://doi.org/10.1007/s00766-007-0045-1.

R. Collobert, J. Weston. 2008. A unified architecture for natural language processing: Deep neural networks with multitask learning. In Proc. 25th Int. Conf. Machine Learning. 160–167.

R. Collobert, J. Weston, L. Bottou, M. Karlen, K. Kavukcuoglu, P. Kuksa. 2011. Natural language processing (almost) from scratch. J. Mach. Learn. Res, 12, 2493–2537.

M. L. Core https://developer.apple.com/documentation/coreml. Accessed October 3, 2020.

Coremltools. https://apple.github.io/coremltools. Accessed October 3, 2020.

L. F. Cort´es-coy, M. Linares-v´asquez, J. Aponte, D. Poshyvanyk. 2014. On automatically generating commit messages via summarization of source code changes. In IEEE 14th International Working Conference on Source Code Analysis and Manipulation (SCAM). IEEE. 275–284.

C. Cortes, V. Vapnik. 1995. Support-vector networks. Mach Learn, 20(3), 273–297.

T. Costlow. 2003. Cognitive radios will adapt to users. IEEE Intell Syst., 18(3), 7. Doi: 10.1109/MIS.2003.1200720.

G. D. Croon, M. F. V. Dartel, E. O. Postma. 2005. Evolutionary learning outperforms reinforcement learning on non-Markovian tasks. In 8th European Conference on Artificial Life Workshop on Memory and Learning Mechanisms in Autonomous Robots, Canterbury, Kent, UK.

Curse of dimensionality. https://en.wikipedia.org/wiki/Curse_of_dimensionality. Accessed October 3, 2020.

V. K. Dabhi, S. Chaudhary. 2014. Hybrid Wavelet-Postfix-GP model for rainfall prediction of Anand region of India. Adv. Artif. Intell, 1–11.

B. Dagenais, M. P. Robillard. 2012. Recovering traceability links between an API and its learning resources. In Proceedings of the 34th International Conference on Software Engineering 47–57.

F. Dandurand, T. Shultz. 2009. Connectionist models of reinforcement, imitation, and instruction in learning to solve complex problems. IEEE Trans. Auton. Ment. Dev., 1(2), 110–121.

Y. N. Dauphin, A. Fan, M. Auli, D. Grangier. 2016. Language modeling with gated convolutional networks. arXiv Preprint, arXiv:1612.08083. https://arxiv.org/abs/1612.08083.Accessed October 3, 2020.

A. De Lucia, M. Di Penta, R. Oliveto, A. Panichella, S. Panichella. 2012. Using IR methods for labeling source code artifacts: Is it worthwhile? In Program Comprehension (ICPC), 2012 IEEE 20th International Conference on. IEEE, 2012. 193–202.

A. Dearle. 2007. Software deployment, past, present and future. Future of Software Engineering (FOSE '07), Minneapolis, MN. 269–284. Doi: 10.1109/FOSE.2007.20.

A. Dearle, G. Kirby, A. McCarthy, J. Diaz Y Carballo. 2004. A Flexible and Secure Deployment Framework for Distributed Applications. In Lecture Notes in Computer Science 3083, (Eds.), In Proceedings of the 2nd International Working Conference on Component Deployment (CD 2004), Edinburgh, Scotland, W. Emmerich, A. L. Wolf Ed.: Springer. 219–233.

Debian – the universal operating system. https://www.debian.org/.

deep-learning-model-convertor: The convertor/conversion of deep learning models for different deep learning.

S. Deerwester, S. T. Dumais, G. W. Furnas, T. K. Landauer, R. Harshman. 1990. Indexing by latent semantic analysis. J. Am. Soc. Inf. Sci., 41(6), 391–391.

L. Deng et al., 2013. Recent advances in deep learning for speech research at Microsoft. Proc. IEEE Int. Conf. Acoust. Speech Signal Process. 8604–8608. https://www.microsoft.com/en-us/research/publication/recent-advances-in-deep-learning-for-speech-research-at-microsoft/. Accessed October 3, 2020.

M. Denil, A. Demiraj, N. Kalchbrenner, P. Blunsom, N. deFreitas. 2014. Modelling, visualising and summarising documents with a single convolutional neural network. 26th Int. Conf. Computational Linguistics. 1601–1612.

J. Devlin, H. Cheng, H. Fang, S. Gupta, L. Deng, H. Xiaodong, Z. Geoffrey, M. Margaret. 2015. Language models for imagecaptioning: The quirks and what works. In Proceedings of the 53rd

Annual Meeting of the Association for Computational Linguistics and the 7th InternationalJoint Conference on Natural Language Processing (Volume 2: Short Papers). 100–105.

R. Di Cosmo, A. Eiche, J. Mauro, G. Zavattaro, S. Zacchiroli. 2015. Automatic Deployment of Software Components in the Cloud with the Aeolus Blender. https://www.researchgate.net/publication/281159486_Automatic_Deployment_of_Software_Components_in_the_Cloud_with_the_Aeolus_Blender. Accessed October 3, 2020.

X. Dong Gu, H. Zhang, D. Zhang, S. Kim. 2016. Deep API learning. In Proceedings of the International Symposium on Foundations of Software Engineering (FSE).

L. Dong, F. Wei, M. Zhou, K. Xu. 2015. Question answering over freebase with multi-column convolutional neural networks. In Proceedings of the Annual Meeting Association Computational Linguistics. 260–269.

X. Dong, Y. Li, C. Wu, Y. Cai. 2010. A learner based on neural network for cognitive radio. In 12th IEEE International Conference on Communication Technology (ICCT '10), Nanjing, China. 893 –896.

N. Dragan, M. L. Collard, J. I. Maletic. Reverse engineering method stereotypes. 2006. 22nd IEEE International Conference on Software Maintenance, Philadelphia, PA. 24–34. Doi: 10.1109/ICSM.2006.54.

H. Drucker, C. J. Burges, L. Kaufman, A. Smola, V. Vapnik. 1997. Support vector regression machines. Adv. Neural Inf. Process. Syst., 9, 155–161.

C. Duan, J. Cleland-Huang. 2007. Clustering support for automated tracing. ASE, 2007, 244–253. Doi: https://doi.org/10.1145/1321631.1321668.

L. Duan, I. W. Tsang, D. Xu, M. Stephen J. 2009. Domain transfer svm for video concept detection. 1375–1381.

S. T. Dumais. 2004. Latent semantic analysis. Annu. Rev. Inf. Sci. Tech, 38(1), 188–230.

P. Dutta, H. Tahbilder. 2014. Prediction of rainfall using data mining technique over Assam. IJCSE, 5, 2.

M. Edkrantz, S. Truvé, A. Said. 2015. Predicting vulnerability exploits in the wild. In 2015 IEEE 2nd International Conference on Cyber Security and Cloud Computing. 513–514. Doi: 10.1109/CSCloud.2015.56.

J. L. Elman. 1990. Finding structure in time. Cogn. Sci., 14(2), 179–211.

J. L. Elman. 1991. Distributed representations, simple recurrent networks, and grammatical structure. Mach. Learn, 7(2–3), 195–225.

W. Eric Wong, R. Gao, Y. Li, R. Abreu, F. Wotawa. 2016. A Survey on Software Fault Localization. ACM, New York, USA.

M. D. Escobar. 1994. Estimating normal means with a Dirichlet process prior. J. Am. Stat. Assoc., 89(425), 268–277. http://www.jstor.org/stable/2291223. Accessed October 3, 2020.

W. E. EWong. 2015. OpenSource Automation in Cloud Computing. In Proceedings of the 4th International Conference on Computer Engineering and Networks, Springer International Publishing. 805–812

J.-R. Falleri, F. Morandat, X. Blanc, M. Martinez, M. Monperrus. 2014. Fine-grained and accurate source code differencing. In Proceedings of the 29th ACM/IEEE international conference on Automated software engineering. ACM. 313–324.

M. Faloutsos, P. Faloutsos, C. Faloutsos. 1999. On Power-Law Relationships of the Internet topology. SIGCOMM'99. https://doi.org/10.1145/316194.316229

J. Fern´andez-ramil, D. Izquierdo-Cortazar, T. Mens. 2009. What does it take to develop a million lines of open source code? OSS 2009: Open Source Ecosystems: Diverse Communities Interacting. 170–184.

Fortify. https://www.fortify.com/ (2011)

frameworks/softwares. https://github.com/ysh329/deep-learning-modelconvertor. Accessed October 3, 2020.

C. Franks, Z. Tu, P. Devanbu, V. Hellendoorn. 2015. CACHECA: A cache language model based code suggestion tool. 2015 IEEE/ACM 37th IEEE International Conference on Software Engineering, Florence. 705–708. Doi: 10.1109/ICSE.2015.228.

G. Frantzeskou, S. MacDonell, E. Stamatatos, S. Gritzalis. 2008. Examining the significance of high-level programming features in source codeauthor classification. J. Syst. Software, 81(3), 447–460.

A. Galindo-Serrano, L. Giupponi. 2010. Distributed Q-learning for aggregated interference control in cognitive radio networks. IEEE Trans. Veh. Technol., 59(4), 1823–1834.

F. Galton. 1989. Kinship and correlation (reprinted 1989). Stat. Sci., 4(2), 80–86.

E. Gamma, R. Helm, R. Johnson, J. M. Vlissides. 1994. Design patterns: Elements of reusable object-oriented software: addison wesley. Gradle Build Dependencies for Android Studio. https://developer.android.com/studio/build/dependencies. Accessed October 3, 2020.

G. Geetha, R. S. Selvaraj. 2011. Prediction of monthly rainfall in Chennai using Back Propagation Neural Network model. Int. J. Eng. Sci. Technol., 3(1), 211213.

G. Geetha, R. S. Selvaraj. 2011. Prediction of monthly rainfall in Chennai using Back Propagation Neural Network model. Int. J. Eng. Sci. Technol., 3(1), 211, 213.

P. Geurts, D. Ernst, L. Wehenkel. 2006. Extremely randomized trees. Mach Learn, 63(1), 3–42.

A. Gittens, D. Achlioptas, M. W. Mahoney. 2017. Skip-gram-zipf + uniform = vector additivity. In Proceedings of the 55th Annual Meeting Association Computational Linguistics.1:69–76.

L. Giupponi, A. Galindo-Serrano, P. Blasco, M. Dohler. 2010. Docitive networks: an emerging paradigm for dynamic spectrum management [Dynamic Spectrum Management]. IEEE Wireless Commun., 17(4), 47–54. Doi: 10.1109/MWC.2010.5547921.

A. M. Glenberg, D. A. Robertson. 2000. Symbol grounding and meaning: A comparison of high-dimensional and embodied theories of meaning. J. Memory Lang., 43(3), 379–401.

X. Glorot, A. Bordes, Y. Bengio. 2011. Domain adaptation for large-scale sentiment classification: A deep learning approach. In Proceedings of the 28th International Conference on Machine Learning. 513–520.

Y. Goldberg. 2016. A primer on neural network models for natural language processing. J. Artif. Intell. Res., 57, 345–420.

S. Gong, W. Liu, W. Yuan, W. Cheng, S. Wang. 2009. Thresholdlearning in local spectrum sensing of cognitive radio. In IEEE 69th Vehicular Technology Conference (VTC Sp. '09), Barcelona, Spain. 1–6.

Google Cloud Platform. 2015. Compute Engine Management with Puppet, Chef, Salt, and Ansible. https://cloud.google.com/developers/articles/google-computeengine-management-puppet-chef-salt-ansible/. Accessed February 5, 2015.

A. Gordon, T. Henzinger, A. Nori, S. Rajamani. 2014. Probabilistic programming. FOSE 2014: Future of Software Engineering Proceedings. ACM. New York, USA.

O. C. Z. Gotel, C. W. Finkelstein. 1994. An analysis of the requirements traceability problem. Proceedings of IEEE International Conference on Requirements Engineering, Colorado Springs, CO, USA. 94–101. Doi: 10.1109/ICRE.1994.292398.

A. Graves. 2013. Generating sequences with recurrent neural networks. arXiv Preprint, arXiv:1308.0850. https://arxiv.org/abs/1308.0850. Accessed October 3, 2020.

A. Graves, M. Liwicki, S. Fern´andez, R. Bertolami, H. Bunke, J. Schmidhuber. 2009. A novel connectionist system for unconstrained handwriting recognition. IEEE Trans. Pattern Anal. Mach. Intell., 31(5), 855–868.

A. Graves, G. Wayne, I. Danihelka. 2014. Neural Turing Machines. https://arxiv.org/abs/1410.5401. Accessed October 3, 2020.

X. Gu, H. Zhang, S. Kim. 2018. Deep code search. In Proceedings of the 2018 40th International Conference on Software Engineering (ICSE 2018). ACM.

X. Gu, H. Zhang, D. Zhang, S. Kim. 2016. Deep API learning.In Proceedings of the 2016 24th ACM SIGSOFT International Symposium on Foundations of Software Engineering. ACM. 631–642.

J. Guo, J. Cheng, J. Cleland-Huang. 2017. Semantically enhanced software traceability using deep learning techniques. In Proceedings of the 39th International Conference on Software Engineering. IEEE Press. 3–14.

J. Guo, J. Cheng, J. Cleland-Huang.2017. Semantically enhanced software traceability using deep learning techniques. 2017 IEEE/ACM 39th International Conference on Software Engineering (ICSE), Buenos Aires. 3–14. Doi: 10.1109/ICSE.2017.9.

R. Gupta, A. Kanade, S. Shevade. 2018. Deep Reinforcement Learning for Programming Language Correction. https://arxiv.org/abs/1801.10467. Accessed October 3, 2020.1631–1642.

M. Haddad, S. Elayoubi, E. Altman, Z. Altman. 2011. A hybrid approach for radio resource management in heterogeneous cognitive networks. IEEE J. Sel. Areas Commun., 29(4), 831–842.

M. T. Hagan, H. B. Demuth, M. Beale. 2002. Neural Network Design. Thomson Asia Pte. Ltd, Singapore.

S. Haiduc, J. Aponte, A. Marcus. 2010. Supporting program comprehension with source code summarization. In Proceedings of the 32nd ACM/IEEE International Conference on Software Engineering. 2. ACM, 223–226.

S. Haiduc, J. Aponte, A. Marcus. 2010. Supporting program comprehension with source code summarization. In Proceedings of the 32nd ACM/IEEE International Conference on Software Engineering. 2, 223–226.

B. Hamdaoui, P. Venkatraman, M. Guizani. 2009. Opportunistic exploitation of bandwidth resources through reinforcement learning. In IEEE Global Telecommunications Conference (GLOBECOM '09), Honolulu. 1–6.

S. Han, H. Mao, W. J. Dally. 2015. Deep Compression: Compressing Deep Neural Networks with Pruning, Trained Quantization, and Huffman Coding. http://arxiv.org/abs/1510.00149. Accessed October 3, 2020.

Z. Han, R. Fan, H. Jiang. 2009. Replacement of spectrum sensing in cognitive radio. IEEE Trans. Wireless Commun., 8(6), 2819–2826.

Z. Han, X. Li, Z. Xing, H. Liu, Z. Feng. 2017. Learning to Predict Severity of Software Vulnerability Using Only Vulnerability Description. In 2017 IEEE International Conference on Software Maintenance and Evolution (ICSME). 125–136. Doi: 10.1109/ICSME.2017.52.

Z. Han, C. Pandana, K. Liu. 2007. Distributive opportunistic spectrum access for cognitive radio using correlated equilibrium and no-regretlearning. In IEEE Wireless Commun. and Networking Conference (WCNC '07), Hong Kong, China. 11–15

Z. Han, R. Zheng, H. Poor. 2011. Repeated auctions with Bayesian nonparametric learning for spectrum access in cognitive radio networks. IEEE Trans. Wireless Commun., 10(3), 890–900.

S. HanD, R. Wallace, R. C. Miller. 2009. Code Completion from Abbreviated Input. 2009 IEEE/ACM International Conference on Automated Software Engineering, Auckland. 332–343. Doi: 10.1109/ASE.2009.64.

S. Haykin. 1999. Neural Networks, a Comprehensive Foundation. Macmillan College Publishing Co, New York.

S. Haykin. 2005. Cognitive radio: brain-empowered wireless communications. IEEE J. Sel. Areas Commun., 23(2), 201–220.

S. S. Haykin. 1999. Neural networks: A Comprehensive Foundation. 2nd ed. Prentice Hall.

A. He, K. K. Bae, T. Newman, J. Gaeddert, K. Kim, R. Menon, L. Morales-Tirado, J. Neel, Y. Zhao, J. Reed, W. Tranter. 2010. A survey of artificial intelligence for cognitive radios. IEEE Trans. Veh. Technol., 59(4), 1578–1592.

K. He, X. Zhang, S. Ren, J. Sun. 2015. Deep Residual Learning for Image Recognition. http://arxiv.org/abs/1512.03385. Accessed October 3, 2020.

M. A. Hearst, J. O. Pedersen. 1996. Reexamining the cluster hypothesis: Scatter/Gather on retrieval results. In SIGIR. 76–84. https://doi.org/10.1145/243199.243216

A. Herbelot, M. Baroni. 2017. High-risk learning: Acquiring new word vectors from tiny data. arXiv Preprint, arXiv:1707.06556. https://arxiv.org/abs/1707.06556. Accessed October 3, 2020.

A. Hindle, E. T. Barr, Z. Su, M. Gabel, P. Devanbu. 2012. On the naturalness of software. In Proceedings of the 34th International Conference on Software Engineering, ICSE '12.837–847. http://dl.acm.org/citation.cfm?id=2337223.2337322. Accessed October 3, 2020.

A. Hindle, E. T. Barr, Z. Su, M. Gabel, P. Devanbu. 2012. On the naturalness of software. In Software Engineering (ICSE), 2012 34th International Conference on. IEEE, 837–847.

G. Hinton, S. Osindero, Y.-W. Teh. 2006. A fast learning algorithm for deep belief nets. Neural Comput., 18(7), 1527–1554.

P. Hitzler, A. Seda. Mathematical Aspects of Logic Programming Semantics. 2017. CRC Press, Boca Raton, FL.

S. Hochreiter, J. Schmidhuber. 1997. Long short-term memory. Neural Comput., 9(8), 1735–1780.

W. Hong, Y. Dong, L. Chen, S. Wei, 2012. Seasonal support vector regression with chaotic genetic algorithm in electric load forecasting. 2012 Sixth International Conference on Genetic and Evolutionary Computing, Kitakushu. 124–127. Doi: 10.1109/ICGEC.2012.128.

A. Hovsepyan, R. Scandariato, W. Joosen. 2012. Software vulnerability prediction using text analysis techniques. MetriSec '12: Proceedings of the 4th international workshop on Security measurements and metrics. 7–10. https://doi.org/10.1145/2372225.2372230.

A. G. Howard et al. 2017. MobileNets: Efficient Convolutional Neural Networks for Mobile Vision Applications. http://arxiv.org/abs/1704.04861. Accessed October 3, 2020.

K. K. Htike, O. O. Khalifa. 2010. Rainfall forecasting models using Focused Time-Delay Neural Networks. Comput. and Commun. Eng. (ICCCE), Int. Conf. on IEEE.

H. Hu, J. Song, Y. Wang. 2008. Signal classification based on spectral correlation analysis and SVM in cognitive radio. In Advanced Information Networking and Applications, 2008. AINA 2008. 22nd International Conference on, Mar. 2008. 883–887

J. Hu, J. Lu, Y. Tan. 2015. Deep transfer metric learning. 2015 IEEE Conference on Computer Vision and Pattern Recognition (CVPR), Boston, MA. 325–333. Doi: 10.1109/CVPR.2015.7298629.

C.C. Huang, F.-Y. Lin, F. Y.-S. Lin, Y. S. Sun. 2013. A novel approach to evaluate software vulnerability prioritization. J. Syst. Software, 86, 2822–2840. Doi: 10.1016/j.jss.2013.06.040.

G. Huang, Z. Liu, L. Van Dermaaten, K. Q. Weinberger. 2016. Densely Connected Convolutional Networks. http://arxiv.org/abs/1608.06993. Accessed October 3, 2020.

Y. Huang, H. Jiang, H. Hu, Y. Yao. 2009. Design of learning engine based on support vector machine in cognitive radio. In InternationalConference on Computational Intelligence and Software Engineering (CiSE '09), Wuhan, China. 1–4.

B. HuaZ, M. Li, F. ZhaoJ, Z. ZouY, B. Xie, C. Li. 2014. Code function mining tool based on topic modeling technology. Comput. Sci., 41(9), 52–59, in Chinese.

Y. Hussain, Z. Huang, S. Wang, Y. Zhou. 2019. Codegru: Context-aware deep learning with gated recurrent unit for source code modeling. https://arxiv.org/abs/1903.00884. Accessed October 3, 2020.

Y. Hussaina, Z. Huanga, Y. Zhoua, S. Wanga. 2019. Deep Transfer Learning for Source Code Modeling. https://arxiv.org/abs/1910.05493. Accessed October 3, 2020.

M. Iliofotou, B. Gallagher, T. Eliassi-Rad, G. Xie, M. Faloutsos. 2010. Profiling-by-association: A resilient traffic profiling solution for the internet backbone. In CoNEXT'10. https://doi.org/10.1145/1921168.1921171

Instruments for Xcode IDE. https://help.apple.com/instruments/mac/current/. Accessed October 3, 2020.

"Inventory," Ansible, Inc. http://docs.ansible.com/intro_inventory.html. Accessed October 3, 2020.

S. Iyer, I. Konstas, A. Cheung, L. Zettlemoyer. 2016. Summarizing source code using a neural attention model. Proceedings of the 54th Annual Meeting of the Association for Computational Linguistics (Volume 1: Long Papers). 2073–2083.

S. Iyer, I. Konstas, A. Cheung, L. Zettlemoyer. 2016. Summarizing source code using a neural attention model. Proceedings of the 54th Annual Meeting of the Association for Computational Linguistics (Volume 1: Long Papers). 2073–2083

S. Iyer, I. Konstas, A. Cheung, L. Zettlemoyer. 2016. Summarizing source code using a neural attention model. In Proceedings of the Annual Meeting of the Association for Computational Linguistics (ACL).

F. Jacob, R. Tairas. 2010. Code template inference using language models. In Proceedings of the 48th Annual SoutheastRegional Conference. https://www.researchgate.net/publication/220995646_Code_template_inference_using_language_models. Accessed October 3, 2020.

S. Jayaweera, Y. Li, M. Bkassiny, C. Christodoulou, K. Avery. 2011. Radiobots: The autonomous, self-learning future cognitive radios. In International Symposium on Intelligent Signal Processing and Communications Systems (ISPACS '11), Chiangmai, Thailand.1–5.

S. K. Jayaweera, C. G. Christodoulou. 2011. Radiobots: Architecture, algorithms and realtime reconfigurable antenna designs for autonomous, self-learning future cognitive radios. University of New Mexico, Technical Report EECE-TR-11-0001. http://repository.unm.edu/handle/1928/12306. Accessed October 3, 2020.

S. Jha, U. Phuyal, M. Rashid, V. Bhargava. 2011. Design of OMCMAC: An opportunistic multi-channel MAC with QoS provisioning for distributed cognitive radio networks. IEEE Trans. Wireless Commun., 10(10), 3414–3425.

Y. Jia et al. 2014. Caffe: Convolutional Architecture for Fast Feature Embedding. http://arxiv.org/abs/1408.5093. Accessed October 3, 2020.

S. JialinPan, J. TKwok, Q. Yang. 2008. Transfer learning via dimensionality reduction. Proceedings of the Twenty-Third AAAI Conference on Artificial Intelligence.8, 677–682.

L. Jiang, G. Misherghi, Z. Su, S. Glondu. 2007. DECKARD: Scalable and accurate tree-based detection of code clones. In Proceedings of the 29th International Conference on Software Engineering, ser. ICSE '07. Washington, DC, USA: IEEE Computer Society. 96–105. http://dx.doi.org/10.1109/ICSE.2007.30

S. Jiang, A. Armaly, C. McMillan. 2017. Automatically generating commit messages from diffs using neural machine translation. In Proceedings of the 32nd IEEE/ACM International Conference on Automated Software Engineering. IEEE Press. 135–146.

T. Jiang, D. Grace, P. Mitchell. 2011. Efficient exploration in reinforcement learning-based cognitive radio spectrum sharing. IET Commun., 5(10), 1309–1317.

P. Johnson, R. Lagerstrom, M. Ekstedt, U. Franke. 2018. Can the common vulnerability scoring system be trusted? A Bayesian analysis. IEEE Trans. Dependable Secure Comput., 15, 1–1. Doi: 10.1109/TDSC.2016.2644614.

R. Johnson, T. Zhang. 2015. Semi-supervised convolutional neural networks for text categorization via region embedding. In Proceedings of the Advances Neural Information Processing Systems. 919–927.

G. E. Kaiser, P. H. Feiler, S. S. Popovich. 1988. Intelligent assistance for software development and maintenance. IEEE Softw., 5(3), 40–49. Doi: 10.1109/52.2023.

N. Kalchbrenner, E. Grefenstette, P. Blunsom. 2014. A convolutional neural network for modelling sentences. Proc. 52nd Annu. Meet. Ass. Comput. Ling., 1, 655–665.

T. Kamiya, S. Kusumoto, K. Inoue. 2002. CCFinder: a multilinguistic token-based code clone detection system for large scale source code. IEEE Trans. Softw. Eng., 28(7), 654–670.

M. Kannan, S. Prabhakaran, P. Ramachandran. 2010. Rainfall forecasting using data mining technique. Int. J. Eng. Technol., 2(6), 397–401.

S. Kannan, S. Ghosh. 2010. Prediction of daily rainfall state in a river basin using statistical downscaling from GCM output. Springer-Verlag, July- 2010.

S. Karaivanov, V. Raychev, M. Vechev. 2014. Phrase-based statistical translation of programming languages. In International Symposium on New Ideas, New Paradigms, and Reflections on Programming & Software.

A. Karpathy, L. Fei-Fei 2015. Deep visual semantic alignments for generating image descriptions. In IEEE Conference on Computer Vision andPattern Recognition, CVPR 2015. 3128–3137.

M. B. Kassiny, S. K. Jayaweera, Y. Li, K. A. Avery. 2012. Blind cyclostationary feature detection based spectrum sensing for autonomous self-learning cognitive radios. 2012 IEEE International Conference on Communications (ICC), Ottawa, ON. 1507–1511. Doi: 10.1109/ICC.2012.6363649.

S. Kawaguchi, P. K. Garg, M. Matsushita, K. Inoue. 2006. Mudablue: An automatic categorization system for open source repositories. J. Syst. Software, 79(7), 939–953.

X. Kelvin, B. Jimmy, K. Ryan, C. Kyunghyun, C. Aaron, S. Ruslan, Z. Richar, Y. Bengio. 2015. Show, attend and tell: Neural image caption generationwith visual attention. In ICML, 2015. https://arxiv.org/abs/1502.03044. Accessed October 3, 2020.

R. S. Kenett, F. Ruggeri, F. W. Faltin, (Editors). 2018. Analytic Methods in Systems and Software Testing. Wiley, Hoboken, NJ.

Keras. https://keras.io/. Accessed October 3, 2020.

Y. Kim. 2014. Convolutional neural networks for sentence classification. arXiv Preprint, arXiv:1408.5882. https://arxiv.org/abs/1408.5882. Accessed October 3, 2020.

J. King. 1976. Symbolic execution and program testing. Commun ACM, 19, 385–394. Doi: https://doi.org/10.1145/360248.360252.

D. P. Kingma, J. Ba. 2014. Adam: A method for stochastic optimization.arXiv. preprint, arXiv:1412.6980.

I. Konstas, M. Lapata. 2013. A global model for concept-to-text generation. J. Artif. Intell. Res., 48(1), 305–346.

J. Koskinen. 2015. Software maintenance costs. https://ocw.unican.es/pluginfile.php/1408/course/section/1805/SMCOSTS.pdf. Accessed October 3, 2020.

G. Koutrika, A. Simitsis, E. Yannis. 2010. Explaining structured queries in natural language. In Data Engineering (ICDE), 2010 IEEE 26th International Conference. 333–344.

A. Krizhevsky, I. Sutskever, G. E. Hinton. 2012. ImageNet classification with deep convolutional neural networks. In NIPS, 2012. https://papers.nips.cc/paper/4824-imagenet-classification-with-deep-convolutional-neural-networks.pdf. Accessed October 3, 2020.

A. Krizhevsky, I. Sutskever, G. E. Hinton. 2012. ImageNet classification with deep convolutional neural networks. Adv. Neural Inf. Process. Syst., 1–9.

T. Kuremoto, S. Kimura, K. Kobayashi, M. Obayashi. 2012. Time Series Forecasting using Restricted Boltzmann Machine. In: Emerging Intelligent Computing Technology and Applications, Springer, 17–22.

I. Labutov, H. Lipson. 2013. Re-embedding words.In Proceedings of the Annual Meeting Association Computational Linguistics. 489–493.

A. N. Lam, A. T. Nguyen, H. A. Nguyen, T. N. Nguyen. 2015. Combining deep learning with information retrieval to localize buggy files for bug reports (n). In Automated Software Engineering (ASE), 2015 30th IEEE/ACM International Conference on. IEEE. 476–481.

G. Lample, M. Ballesteros, S. Subramanian, K. Kawakami, C. Dyer. 2016. Neural architectures for named entity recognition. arXiv Preprint, arXiv:1603.01360. https://arxiv.org/abs/1603.01360. Accessed October 3, 2020.

N. D. Lane et al. 2016. DeepX: A software accelerator for low-power deep learning inference on mobile devices. 2016 15th ACM/IEEE International Conference on Information Processing in Sensor Networks (IPSN), Vienna. 1–12. Doi: 10.1109/IPSN.2016.7460664.

N. D. Lane, S. Bhattacharya, A. Mathur, P. Georgiev, C. Forlivesi, F. Kawsar. 2017. Squeezing deep learning into mobile and embedded devices. IEEE Pervasive Comput., 16(3), 82–88.

B. Latifa, Z. Gao, S. Liu. 2012. No-regret learning for simultaneous power control and channel allocation in cognitive radio networks. In Computing, Communications and Applications Conference (Com- ComAp '12), Hong Kong, China. 267–271

E. L. Lawler, J. K. Lenstra, A. H. G. R. Kan, D. B. Shmoys. 1985. The Traveling Salesman Problem: A Guided Tour of Combinatorial Optimization. John Wiley & Sons Ltd, Hoboken, New Jersey, United States.

Y. Le Cun et al. 1989. Handwritten digit recognition: applications of neural network chips and automatic learning. IEEE Commun. Mag., 27(11), 41–46.

P. Le, W. H. Zuidema. 2016. Quantifying the vanishing gradient and long distance dependency problem in recursive neural networks and recursiveLSTMs. In Proceedings of the 1st Workshop on Representation Learning for NLP, Berlin, Germany. 8793–8793.

T. B. Le, M. Linares-Vasquez, D. Lo, D. Poshyvanyk. 2015. RCLinker: Automated linking of issue reports and commits leveraging rich contextual information. IEEE 23rd International Conference on Program Comprehension, Florence. 36–47. Doi: 10.1109/ICPC.2015.13.

Y. Lecun, Y. Bengio, G. Hinton. 2015. Deep learning. Nature, 521(7553), 436–444.

Y. Lecun, L. Bottou, Y. Bengio, P. Haffner, 1998. Gradient-based learning applied to document recognition, In Proceedings of the IEEE. 86, 11, 2278–2324. Doi: 10.1109/5.726791.

Y. LeCun, L. Bottou, Y. Bengio, P. Haffner. 1998. Gradient-based learning applied to document recognition. Proc. IEEE, 86(11), 2278–2324.

K. B. Letaief, W. Zhang. 2009. Cooperative communications for cognitive radio networks. Proc. IEEE, 97(5), 878–893. Doi: 10.1109/JPROC.2009.2015716.

A. Leuski. 2001. Evaluating document clustering for interactive information retrieval. In CIKM. 223–232. Doi: 10.1145/502585.502592

Y. Li, D. Tarlow, M. Brockschmidt, R. Zemel. 2015. Gated graph sequence neural networks. arXiv preprint arXiv:1511.05493.

B. Liblit, M. Naik, A. Zheng, A. Aiken, M. Jordan. 2005. Scalable statistical bug isolation. In Proceedings of ACM SIGPLAN Conference on Programming Language Design and Implementations. 15–26. Chicago, USA.

Z. Lin, B. Xie, Y. Zou et al. 2017. Intelligent development environment and software knowledge graph. J. Com. Sci. Technol., 32(2), 242–249. Doi: 10.1007/s11390-017-1718-y.

M. Linares-v´asquez, C. McMillan, D. Poshyvanyk, M. Grechanik. 2014. On using machine learning to automatically classify software applicationsinto domain categories. Empir. Softw. Eng., 19(3), 582–618.

W. Ling, E. Grefenstette, K. M. Hermann, T. Kocisky, A. Senior, F. Wang, P. Blunsom. 2016. Latent predictor networks for code generation. https://arxiv.org/abs/1603.06744. Accessed October 3, 2020.

K. LinY, Y. LiuZ, S. SunM, Y. Liu, X. Zhu. 2015. Learning entity and relation embeddings for knowledge graph completion. In Proceedings of the 29th AAAI Conference on Artificial Intelligence. 2181–2187.

Q. Liu, Y. Li, H. Duan, Y. Liu, Z. G. Qin. 2016. Knowledge graph construction techniques. J. Com. Res. Dev., 53(3), 582–600.

Y. Liu, D. Poshyvanyk, R. Ferenc, T. Gyim´othy, N. Chrisochoides. 2009. Modeling class cohesion as mixtures of latent topics. In Software Maintenance, 2009. ICSM 2009. IEEE International Conference on. IEEE, 2009. 233–242.

P. Louridas, D. Spinellis, V. Vlachos. 2008. Power laws in software. ACM TOSEM, 18(1), 1–26.

H. Lu, B. Cukic, M. Culp. 2012. Software defect prediction using semi supervised learning with dimension reduction. In Automated Software Engineering (ASE), 2012 Proceedings of the 27th IEEE/ACM International Conference on. IEEE. 314–317. Doi: 10.1145/2351676.2351734.

H. Lu, E. Kocaguneli, B. Cukic. 2014. Defect Prediction between Software Versions with Active Learning and Dimensionality Reduction. 2014 IEEE 25th International Symposium on Software Reliability Engineering, Naples. 312–322. Doi: 10.1109/ISSRE.2014.35.

W. Lu, H. TouNg. 2011. A probabilistic forest-to-string model for language generation from typed lambda calculus expressions. In Proceedings of the 2011 Conference on Empirical Methods in Natural Language Processing. 1611–1622.

L. Lucy, J. Gauthier. 2014. Are distributional representations ready for the real world? Evaluating word vectors for grounded perceptual meaning.arXiv Preprint, arXiv:1705.11168. https://arxiv.org/abs/1705.11168. Accessed October 3, 2020.

T. Luong, H. Pham, C. D. Manning. 2015. Effective approaches to attention-based neural machine translation. In Proceedings of the2015 Conference on Empirical Methods in Natural Language Processing. 1412–1421.

P. Mahadevan, C. Hubble, D. Krioukov, B. Huffaker, A. Vahdat. 2007. Orbis: rescaling degree correlations to generate annotated internet topologies. In ACM SIGCOMM. https://doi.org/10.1145/1282427.1282417.

J. I. Maletic, A. Marcus. 2001. Supporting program comprehension using semantic and structural information. In Proceedings of the 23rdInternational Conference on Software Engineering. IEEE Computer Society. 103–112.

A. Marcus, A. Sergeyev, V. Rajlich, J. I. Maletic. 2004. An information retrieval approach to concept location in source code. In Proceedings of the 11th Working Conference on Reverse Engineering. 214–223.

A. Mariakakis et al. 2017. PupilScreen: Using smartphones to assess traumatic brain injury. In Proceedings of the ACM Interactive, Mobile, Wearable Ubiquitous Technol. 1:3:1–27.

A. Mariakakis, M. A. Banks, L. Phillipi, L. Yu, J. Taylor, S. N. Patel. 2017. BiliScreen: Scleral Jaundice detection with a smartphone. Proc. ACM Interactive, Mobile, Wearable Ubiquitous Technol. 1(2), 1–26.

Mark Stamp. Introduction to Machine Learning with Applications in Information Security. 2017. CRC Press.Boca Raton, FL.

Markov decision process. https://en.wikipedia.org/wiki/Markov_decision_process. Accessed October 2, 2020

K. McCallumA. 2002. MALLET: A machine learning for language toolkit. http://mallet.cs.umass.edu. Accessed October 3, 2020.

M. McPherson, L. Smith-Lovin, J. M. Cook. 2001. Birds of a feather: Homophily in social networks. Annu. Rev. Sociol., 27, 1.

J.-W. Meent, B. Paige, H. Yang, F. Wood. 2018. An Introduction to Probabilistic Programming. https://arxiv.org/abs/1809.10756.

H. Mei, M. Bansal, M. R. Walter. 2016. What to talk about and how? selective generation using lstms with coarse-to-fine alignment. In Proceedings of the 2016 Conference of the North American Chapter of the Association for Computational Linguistics: Human Language Technologies.

T. Menzies, L. Williams, T. Zimmermann, (Editors). 2016. Perspectives on Data Science for Software Engineering. Elsevier, Amsterdam, Netherlands.

Microsoft Cognitive Toolkit. https://www.microsoft.com/en-us/cognitive-toolkit. Accessed
    October 3, 2020.
R. Mikkilineni, A. Comparini, G. Morana. 2012. The Turing O-Machine and the DIME Network
    Architecture: Injecting the Architectural Resiliency into Distributed Computing. In Proceedings
    of the Turing-100 – The Alan Turing Centenary, Manchester, UK.
T. Mikolov, K. Chen, G. Corrado, J. Dean. 2013. Efficient estimation of word representations in
    vector space.arXiv Preprint, arXiv:1301.3781. https://arxiv.org/abs/1301.3781. Accessed
    October 3, 2020.
T. Mikolov, M. Karafiát, L. Burget, J. Cernocký, S. Khudanpur. 2010. Recurrent neural network-based
    language model. In Proc. Interspeech. vol., 2, 3.
T. Mikolov, I. Sutskever, K. Chen, G. S. Corrado, J. Dean. 2013. Distributed representations of words
    and phrases and their compositionality. In Proceedings of the Advances Neural Information
    Processing Systems. 3111–3119.
T. Mikolov, I. Sutskever, K. Chen, G. S. Corrado, J. Dean. 2013. Distributed representations of words
    and phrases and their compositionality. In Advances in neural information processing systems.
    3111–3119.
T. H. Minh Le, B. Sabir, M. AliBabar. 2019. Automated software vulnerability assessment with
    concept drift. 2019 IEEE/ACM 16th International Conference on Mining Software Repositories
    (MSR), Montreal, QC, Canada, 2019. 371–382. Doi: 10.1109/MSR.2019.00063.
J. Mitola. 2000. Cognitive radio: An integrated agent architecture for software defined radio. Ph.D.
    dissertation, Royal Institute of Technology (KTH), Stockholm, Sweden.
J. Mitola. 2009. Cognitive Radio architecture evolution. In Proceedings of the IEEE. 97(4), 626–641.
    Doi: 10.1109/JPROC.2009.2013012.
J. Mitola, G. Q. Maguire. 1999. Cognitive radio: making software radios more personal. In IEEE
    Personal Communications. 6(4), 13–18. Doi: 10.1109/98.788210.
MITRE, Common Weakness Enumeration. https://cwe.mitre.org/data/index.html.
MNIST handwritten digit database. http://yann.lecun.com/exdb/mnist/. Accessed October 3,
    2020.
Mobile Operating System Market Share. 2018. http://gs.statcounter.com/os-market-share/mobile/
    worldwide. Accessed October 3, 2020.
M. Mooty, A. Faulring, J. Stylos, B. A. Myers. 2010. Calcite: Completing code completion for
    constructors using crowds. 2010 IEEE Symposium on Visual Languages and Human-Centric
    Computing, Leganes. 15–22. Doi: 10.1109/VLHCC.2010.12.
L. Mou, G. Li, L. Zhang, T. Wang, Z. Jin. 2016. Convolutional neural networks over tree structures for
    programming language processing. In AAAI. 2(3), 4-4.
D. Movshovitz-Attias, W. W. Cohen. 2013. Natural language models for predicting programming
    comments. In Proceedings of the Annual Meeting of the Association for Computational
    Linguistics (ACL).
D. Movshovitz-Attias, W. W. Cohen. 2013. Natural Language Models for Predicting Programming
    Comments. http://www.cs.cmu.edu/~dmovshov/papers/dma_acl2013.pdf. Accessed
    October 3, 2020.
D. Movshovitz-Attias, W. W. Cohen. 2013. Natural language models for predicting programming
    comments. In Proceedings of the 51st AnnualMeeting of the Association for Computational
    Linguistics. 35–40.
A. Mukherjee, B. Liu. 2012. Aspect extraction through semi-supervised modeling. In Proc. 50th
    Annu. Meeting Association Computational Linguistics. 339–348.
V. Murali, S. Chaudhuri, C. Jermaine. 2017b. Finding Likely Errors with Bayesian Specifications.
    arXiv preprint arXiv:1703.01370 (2017).

S. Murtaza, W. Khreich, A. Hamou-Lhadj, A. B. Bener. 2016. Mining trends and patterns of software vulnerabilities. J. Syst. Software, 117, 218–228. Doi: https://doi.org/10.1016/j.jss.2016.02.048.

C. R. Myers. 2003. Software systems as complex networks: Structure, function, and evolvability of software collaboration graphs. Phys. Rev. E, 68(4), 046116.

N. Nagappan, T. Ball. 2005. Use of relative code churn measures to predict system defect density. Proceedings 27th International Conference on Software Engineering, 2005. ICSE 2005, Saint Louis, MO, USA. 284–292. Doi: 10.1109/ICSE.2005.1553571.

A. R. Naik, S. K. Pathan. 2013. Indian monsoon rainfall Classification and Prediction using Robust Back Propagation Artificial Neural Network. Int. J. Emerging Technol Adv. Eng., 3(11), 99–101.

S. K. Nanda, D. P. Tripathy, S. K. Nayak, S. Mohapatra. 2013. Prediction of rainfall in India using Artificial Neural Network (ANN) models. Int. J. of Intell. Syst. Applicat., 5(12), 1–22.

S. Neuhaus, T. Zimmermann, C. Holler, A. Zeller. 2007. Predicting vulnerable software components. In: Proceedings of the 14th ACM Conference on Computer and Communications Security. ACM. 529–540.

M. E. J. Newman. 2002. Assortative mixing in networks. Phys. Rev. Lett. 89, 20. https://www.win.tue.nl/~wstomv/edu/2ip30/references/criteria_for_modularization.pdf. Accessed October 3, 2020.

A. T. Nguyen, T. T. Nguyen, H. A. Nguyen, T. N. Nguyen. 2012. Multilayered approach for recovering links between bug reports and fixes. In Proceedings of the ACM SIGSOFT 20th International Symposium on the Foundations of Software Engineering. ACM63-63. https://doi.org/10.1145/2393596.2393671

A. T. Nguyen, T. T. Nguyen, T. N. Nguyen. 2015. Divide-and-conquer approach for multi-phase statisticalmigration for source code. In Proceedings of the International Conference onAutomated Software Engineering (ASE).

H. NguyenV, M. S. TranL. 2010. Predicting vulnerable software components with dependency graphs. Proceedings of the 6th International Workshop on Security Measurements and Metrics. Bolzano, IT.

NIST Report. http://www.abeacha.com/NIST_press_release_bugs_cost.htm

NIST. Juliet test suite v1.3, 2017. https://samate.nist.gov/SRD/testsuite.php.

N. Niu, A. Mahmoud. 2012. Enhancing candidate link generation for requirements tracing: The cluster hypothesis revisited. 20th IEEE International Requirements Engineering Conference (RE), Chicago, IL, 2012. 81–90. Doi: 10.1109/RE.2012.6345842.

NMedvidovic. 1996. ADLs and dynamic architecture changes.Proc. Joint Proceedings of the Second International Software Architecture Workshop (ISAW-2) and International Workshop on Multiple Perspectives in Software Development (Viewpoints '96), San Francisco, California. 24–27.

E. Nunes, A. Diab, A. Gunn, E. Marin, V. Mishra, V. Paliath, et al. 2016. Darknet and Deepnet mining for proactive cybersecurity threat intelligence. In 2016 IEEE Conference on Intelligence and Security Informatics (ISI). 7–12. Doi: 10.1109/ISI.2016.7745435.

Object Management Group. 2003. Specification for Deployment and Configuration of Component-based Distributed Applications. OMG. https://www.omg.org/spec/DEPL/4.0/PDF. Accessed October 3, 2020.

Object Management Group. 2005. Unified Modelling Language: Superstructure version 2.0. https://www.omg.org/spec/UML/2.0/About-UML/.Accessed October 3, 2020.

Occam's razor. https://en.wikipedia.org/wiki/Occam%27s_razor. Accessed October 2, 2020

V. Oriol, F. Meire, J. Navdeep. 2015. Pointer networks. In NIPS, 2015.https://arxiv.org/abs/1506.03134. Accessed October 3, 2020.

M. Ou, P. Cui, J. Pei, Z. Zhang, W. Zhu. 2016. Asymmetric transitivity preserving graph embedding. In Proceedings of the 22nd ACM SIGKDD International Conference on Knowledge Discovery and

Data Mining, ser. KDD '16. New York, NY, USA: ACM, 2016. 1105–1114. http://doi.acm.org/10.1145/2939672.2939751. Accessed on October 4, 2020

S. P. Goswami. 1996. A novel Neural Network design for long range prediction of rainfall pattern. Current Sci. (Bangalore), 70(6), 447–457.

S. Palkar, J. J. Thomas, A. Shanbhag, D. Narayanan, H. Pirk, M. Schwarzkopf, S. Amarasinghe, M. Zaharia. 2017. Weld: A common runtime for high performance data analytics. In Proceedings of the Conference on Innovative Data Systems Research (CIDR), Chaminade, CA, USA.

R. B. Palm. 2012. Prediction as a candidate for learning deep hierarchical models of data.Master's thesis.

J. F. Pane, B. A. Myers et al. 2001. Studying the language and structure in non-programmers' solutions to programming problems. Int. J. Hum. Comput. Stud., 54(2), 237–264.

Y. Pang, X. Xue, A. S. Namin. 2015. Predicting vulnerable software components through n-gram analysis anstatistical feature selection," In 2015 IEEE 14th Int. Conf. Machine Learning and Applications (ICMLA), 2015.

A. Panichella, B. Dit, R. Oliveto, M. DiPenta, D. Poshynanyk, A. De Lucia. 2013. How to effectively use topic models for software engineering tasks? an approach based on genetic algorithms. In Software Engineering (ICSE), 2013 35th International Conference on. IEEE, 2013. 522–531.

A. Panichella, R. Oliveto, A. D. Lucia. 2014. Cross-project defect prediction models: L'union fait la force. In 2014 Software Evolution Week – IEEE Conference on Software Maintenance, Reengineering, and Reverse Engineering, CSMR-WCRE. Antwerp, Belgium, February 3–6, 2014. 164–173. Doi: 10.1109/CSMR-WCRE.2014.6747166.

A. Pankratz. 1983. Forecasting With Univariate Box-Jenkins Models Concepts and Cases. John Wiley Sons, Inc, New York.

K. Papineni, S. Roukos, T. Ward, W.-J. Zhu. 2002. Bleu: a method for automatic evaluation of machine translation. In Proceedings of the 40th annual meeting on association for computational linguistics. 311–318.

S. Parekh, N. Gandhi, J. Hellerstein, D. Tilbury, T. Jayram, J. Bigus. 2001. Using control theory to achieve service level objectives in performance management. 2001 IEEE/IFIP International Symposium on Integrated Network Management Proceedings. Integrated Network Management VII. Integrated Management Strategies for the New Millennium (Cat. No.01EX470), Seattle, WA, USA. 841–854. Doi: 10.1109/INM.2001.918084.

A. Parmar, K. Mistree, M. Sompura. Machine learning techniques for rainfall prediction: A review 2017. 2017 International Conference on Innovations in information Embedded and Communication Systems (ICIIECS).

D. L. Parnas. 1972. On the criteria to be used in decomposing systems into modules. Commun. ACM., 15, 1053–1058.

D. Patel, A. Hindle, J. Nelson, A. Eddie, A. Santos, J. C. Campbell. 2018. Syntax and sensibility: Using language models to detect and correct syntax errors. 3D Digital Imaging and Modeling, International Conference. 311–322.

S. Paul, A. Prakash. 1994. A framework for source code search using program patterns. IEEE Trans. Softw. Eng., 20(6), 463–475.

R. Paulus, C. Xiong, R. Socher. 2017. A deep reinforced model for abstractive summarization.arXiv Preprint, arXiv:1705.04304. https://arxiv.org/abs/1705.04304. Accessed October 3, 2020.

H. Peng, E. Cambria, X. Zou. 2017. Radical-based hierarchical embeddings for Chinese sentiment analysis at sentence level. In Proc. Int. Florida Artificial Intelligence Research Society Conf., 2017. 347–352.

J. Pennington, R. Socher, C. D. Manning. 2014. Glove: Global vectors for word representation. In Proc. Conf. Empirical Methods Natural Language Processing. 14, 1532–1543.

M. Petrova, P. Ma Andho Andnen, A. Osuna. 2010. Multi-class classification of analog and digital signals in cognitive radios using support vector machines.In 7th International Symposium on Wireless Communication Systems (ISWCS '10). 986–990

N. S. Philip, K. B. Joseph. 2003. A Neural Network tool for analyzing trends in rainfall. Comput. Geosci., 29(2), 215–223.

Y. Pinter, R. Guthrie, J. Eisenstein. 2017. Mimicking word embeddings using subword rnns.arXiv Preprint, arXiv:1707.06961. https://arxiv.org/abs/1707.06961. Accessed October 3, 2020.

S. Poria, E. Cambria, D. Hazarika, P. Vij. 2016. A deeper look into sarcastic tweets using deep convolutional neural networks. In Proc. Int. Conf. Computational Linguistics. 1601–1612.

M. Pradel, K. Sen. 2018. Deep Bugs: A learning approach to name-based bug detection. Proc. ACM Program. Lang., vol. 2, no. OOPSLA.2:147:1–147: 25. http://doi.acm.org/10.1145/3276517. Accessed on October 5, 2020

M. L. Puterman. 1994. Markov Decision Processes: Discrete Stochastic Dynamic Programming. John Wiley and Sons, Hoboken, New Jersey, United States.

PyTorch. https://pytorch.org/. Accessed October 3, 2020.

X. Qiu, L. Zhang, Y. Ren, P. N. Suganthan, G. Amaratunga. 2014. Ensemble deep learning for regression and time series forecasting. 2014 IEEE Symposium on Computational Intelligence in Ensemble Learning (CIEL), Orlando, FL1–6. Doi: 10.1109/CIEL.2014.7015739.

P. Rad, R. Boppana, P. Lama, G. Berman, M. Jamshidi. 2015. Low-latency software defined network for high performance clouds. In Proceedings of the 8th IEEE System of Systems. 805–812. Doi: 10.1109/SYSOSE.2015.7151909.

A. Rajasekharan. 2014. What is representation learning in deep learning? Quora. https://www.quora.com/What-is-representation-learning-in-deep-learning. Accessed September 26, 2020.

S. M. Rajasooriya, C. P. Tsokos, P. K. Kaluarachchi. 2017. Cyber security: Nonlinear Stochastic Models for predicting the exploitability. J. Inf. Secur., 8, 125–140. Doi: http://dx.doi.org/10.4236/jis.2017.82009.

M. M. Ramon, T. Atwood, S. Barbin, C. G. Christodoulou. 2009. Signal classification with an SVM-FFT approach for feature extraction in cognitive radio. In SBMO/IEEE MTT-S International Microwave and Optoelectronics Conference (IMOC '09), Belem, Brazil. 286–289

R. Ranca, M. Allamanis, M. Lapata, C. Sutton, J. Fowkes, P. Chanthirasegaran. 2017. Autofolding for source code summarization. IEEE Trans. Softw. Eng., 43(12), 1095–1109.

P. Raspberry. https://www.raspberrypi.org/. Accessed October 3, 2020.

B. Ray, V. Hellendoorn, S. Godhane, Z. Tu, A. Bacchelli, P. Devanbu. 2016. On the naturalness of buggy code. In Proceedings of the International Conference on Software Engineering (ICSE).

V. Raychev, M. Vechev, E. Yahav. 2014. Code completion with statistical language models. Proceedings of the 35th ACM SIGPLAN Conference on Programming Language Design and Implementation – PLDI '14. 419–428.

V. Raychev, M. Vechev, E. Yahav. 2014. Code completion with statistical language models. ACM SIGPLAN Not., 49(6), 419–428.

V. Raychev, M. Vechev, E. Yahav. 2014. Code completion with statistical language models. Acm Sigplan Not., 49(6), 419–428.

Recurrent neural network. https://en.wikipedia.org/wiki/Recurrent-neural-network. Accessed on October 4, 2020

E. Reiter, R. Dale. 2000. Building Natural Language Generation Systems. Cambridge University Press, New York. Doi: https://doi.org/10.1017/CBO9780511519857.

S. Ren, K. He, R. Girshick, J. Sun. 2017. Faster R-CNN: Towards Real-Time Object Detection with Region Proposal Networks. IEEE Trans. Pattern Anal. Mach. Intell., 39(6), 1137–1149. Doi: 10.1109/TPAMI.2016.2577031.

M. Renieris, S. P. Reiss. 2013. Fault Localization with Nearest Neighbor Queries. In Proceedings of International Conference on Automated Software Engineering. 30–39. Montreal, Canada.

M. Robillard, R. Walker, T. Zimmermann. 2010. Recommendation Systems for Software Engineering. IEEE Softw., 27(4), 80–86. Doi: 10.1109/MS.2009.161.

E. Romero, D. Toppo. 2007. Comparing support vector machines and feedforward neural networks with similar hidden-layer weights. IEEE Trans. Neural Netw., 18(3), 959–963.

Y. Rouman, J. K. Nwankpa, Y. F. Roumani. 2015. Time series modeling of vulnerabilities. Computers & Security (2015). Doi: 10.1016/j.cose.2015.03.003.

C. K. Roy, J. R. Cordy. 2007. A survey on software clone detection research. Queens Sch. Comput. TR, 541(115), 64–68.

A. M. Rush, S. Chopra, J. Weston. 2015. A neural attention model for abstractive sentence summarization.arXiv Preprint, arXiv:1509.00685. https://arxiv.org/abs/1509.00685. Accessed October 3, 2020.

A. M. Rush, S. Chopra, J. Weston. 2015. A neural attention model for abstractive sentence summarization. In Proceedings of the 2015Conference on Empirical Methods in Natural Language Processing. 379–389.

M. RushAlexander, C. Sumit, W. Jason. 2015. A neural attention model for abstractive sentence summarization. In EMNLP, 2015. https://arxiv.org/abs/1509.00685. Accessed October 3, 2020.

O. Russakovsky et al. 2014. ImageNet Large Scale Visual Recognition Challenge. https://arxiv.org/abs/1409.0575v3. Accessed October 3, 2020.

R. Russell, L. Kim, L. Hamilton, T. Lazovich, J. Harer, O. Ozdemir, P. Ellingwood, M. McConley. 2018. Automated vulnerability detection in source code using deep representation learning. 17th IEEE International Conference on Machine Learning and Applications. 757–762. Doi: 10.1109/ICMLA.2018.00120.

C. Sabottke, O. Suciu, T. Dumitras, 2015. Vulnerability disclosure in the age of social media: Exploiting twitter for predicting real-world exploits. In USENIX Security Symposium. 1041–1056.

A. K. Sahai, M. K. Soman, V. Satyan. 2000. All India summer monsoon rainfall prediction using an Artificial Neural Network. Clim. Dyn., 16(4), 291–302.

H. Sajnani, V. Saini, J. Svajlenko, C. K. Roy, C. V. Lopes. 2016. SourcererCC: Scaling code clone detection to big-code. In SoftwareEngineering (ICSE), 2016 IEEE/ACM 38th International Conference on. IEEE, 2016. 1157–1168.

SaltStack – Github. 2015. RAET (Reliable Asynchronous Event Transport) Protocol. https://github.com/saltstack/raet/blob/master/README.md. Accessed October 3, 2020.

SaltStack, Reactor System. http://docs.saltstack.com/en/latest/topics/reactor/index.html. Accessed October 3, 2020.

SamirMittal. 2017. Cognitive computing architectures for Machine (Deep) learning at scale. Presented at the IS4SI 2017 Summit DIGITALISATION FOR A SUSTAINABLE SOCIETY, Gothenburg, Sweden, 12–16 June 2017. Doi: 10.3390/IS4SI-2017-04025

A. Sampaio, N. Loughran, A. Rashid, P. Rayson. 2005. Mining Aspects in Requirements. Workshop on Early Aspects (held with AOSD 2005). https://doi.org/10.1016/j.scico.2010.04.013

M. Schuhmacher, S. P. Ponzetto. 2014. Knowledge-based graph document modeling. In Proc. the 7th ACM International Conference on Web Search and Data Mining. 543–552.

S. Schumate, 2004. "Implications of Virtualization", Technical Report 2004. www.dell.com/downloads/global/power/ps4q04-20040152-Shumate.pdf. Accessed October 3, 2020.

D. Sculley, G. Hold, D. Golovin, E. Davydov, H. Phillips, D. Ebner, V. Chaudhary, M. Young, J. F. Crespo, D. Dennison. 2015. Hidden technical debt in machine learning systems. In Proceedings of the 28th International Conference on Neural Information Processing Systems, Montreal, QC, Canada, 7–12 December 2015. 2503–2511

A. Sehgal, N. Kehtarnavaz. 2018. A convolutional neural network smartphone app for real-time voice activity detection. IEEE Access, 6, 9017–9026.

A. Sehgal, N. Kehtarnavaz. 2019. Guidelines and Benchmarks for Deployment of Deep Learning Models on Smartphones as Real-Time Apps. https://arxiv.org/abs/1901.02144.Accessed October 3, 2020.

A. Sethi, A. Sankaran, N. Panwar, S. Khare, S. Mani. 2017. DLPaper2Code: Auto-generation of Code from Deep Learning Research Papers. https://arxiv.org/abs/1711.03543. Accessed October 3, 2020.

L. Shar, L. Briand, H. Tan. 2014. Web application vulnerability prediction using hybrid program analysis and machine learning. IEEE Trans. Dependable Secure Comput., 12, 688–707. Doi: 10.1109/TDSC.2014.2373377.

M. Shepperd, D. Bowes, T. Hall. 2014. Researcher bias: The use of machine learning in software defect prediction. IEEE Trans. Softw. Eng., 40(6), 603–616.

Y. Shin, A. Meneely, L. Williams, J. A. Osborne. 2011. Evaluating complexity, code churn, and developer activity metrics as indicators of software vulnerabilities. IEEE Trans. Software Eng., 37(6), 772–787. Doi: 10.1109/TSE.2010.81.

Y. Shin, A. Meneely, L. Williams, J. A. Osborne. 2011. Evaluating complexity, code churn, and developer activity metrics as indicators of software vulnerabilities. IEEE Trans. Softw. Eng., 37(6), 772–787. Doi: 10.1109/TSE.2010.81.

A. Simitsis, Y. E. Ioannidis. 2009. Dbmss should talk back too. In CIDR 2009, Fourth Biennial Conference on Innovative Data Systems Research, Online Proceedings.

S. N. Sivanandam, S. N. Deepa. 2007. Principles of soft computing. Wiley India (P) Ltd, New Delhi.

R. Socher, C. C. Lin, C. Manning, A. Y. Ng. 2011. Parsing natural scenes and natural language with recursive neural networks. In Proceedings of the 28th international conference on machine learning (ICML-11). 129–136.

R. Socher, A. Perelygin, J. Wu, J. Chuang, C. D. Manning, A. Ng, C. Potts. 2013. Recursive deep models for semantic compositionality over a sentiment treebank. In Proc. Conf. Empirical Methods Natural Language Processing. 1631–1642.

V. K. Somvanshi, O. P. Pandey, P. K. Agrawal, N. V. Kalanker1, M. RaviPrakash, R. Chand. 2006. Modeling and prediction of rainfall using Artificial Neural Network and ARIMA techniques. J. Ind. Geophys. Union, 10(2), 141–151.

D. Soni. 2018. Introduction to Bayesian networks. Towards data science. https://towardsdatascience.com/introduction-to-bayesian-networks-81031eeed94e. Accessed September 26, 2020.

G. Soundararajan, C. Amza, A. Goel. 2006. Database replication policies for dynamic content applications. In Proc. EuroSys. https://www.eecg.utoronto.ca/~ashvin/publications/eurosys-2006.pdf. Accessed October 3, 2020.

J. C. Sousa, H. M. Jorge, L. P. Neves. 2014. Short-term load forecasting based on support vector regression and load profiling. Int. J. Energy Res., 38(3), 350–362.

G. Sridhara, E. Hill, D. Muppaneni, L. Pollock, K. Vijay-Shanker. 2010. Towards automatically generating summary comments for java methods. In Proceedings of the IEEE/ACM international conference on Automated software engineering. 43–52.

Statista. 2014–2020. Number of smartphone users worldwide 2014–2020. https://www.statista.com/statistics/330695/number-ofsmartphone-users-worldwide/. Accessed October 3, 2020.

Z. Sun, G. Bradford, J. Laneman. 2011. Sequence detection algorithms for PHY-layer sensing in dynamic spectrum access networks. IEEE J. Sel. Topics Signal Process, 5(1), 97–109.

M. Sundermeyer, H. Ney, R. Schlüter. 2015. From feedforward to recurrent LSTM neural networks for language modeling. IEEE Trans. Audio, Speech, Lang. Process, 23(3), 517–529.

I. Sutskever, J. Martens, G. EHinton. 2011. Generating text with recurrent neural networks. In Proceedings of the 28th InternationalConference on Machine Learning (ICML-11). 1017–1024.

R. Sutton, D. Mcallester, S. Singh, Y. Mansour. 2001. Policy gradient methods for reinforcement learning with function approximation. In Proceedings of the 12th conference on Advances in Neural Information Processing Systems (NIPS '99). Denver, CO: MIT Press. 1057–1063

R. S. Sutton, A. G. Barto. 1998. Reinforcement Learning: An Introduction. Cambridge, MA, MIT Press.

J. Svajlenko, J. F. Islam, I. Keivanloo, C. K. Roy, M. M. Mia. 2014. Towards a big data curated benchmark of inter-project code clones. InSoftware Maintenance and Evolution (ICSME), 2014 IEEE International Conference on. IEEE. 476–480.

C. Szyperski. 2003. Component technology – what, where, and how? 25th International Conference on Software Engineering, 2003. Proceedings., Portland, OR, USA. 684–693. Doi: 10.1109/ICSE.2003.1201255.

K. S. Tai, R. Socher, C. D. Manning. 2015. Improved semantic representations from tree-structured long short-term memory networks.arXiv Preprint, arXiv:1503.00075. https://arxiv.org/abs/1503.00075. Accessed October 3, 2020.

K. S. Tai, R. Socher, C. D. Manning. 2015. Improved semantic representations from tree-structured long short-term memory networks. In Proceedings of the 53rd Annual Meeting of the Association for Computational Linguistics and the 7th International Joint Conferenceon Natural Language Processing (Volume 1: Long Papers).1, 1556–1566.

R. Tairas, J. Gray. 2009. An information retrieval process to aid in the analysis of code clones. Empir. Softw. Eng., 14(1), 33–56.

D. Tang, B. Qin, T. Liu. 2015. Document modeling with gated recurrent neural network for sentiment classification. In Proceedings of the Conference on Empirical Methods Natural Language Processing.1422–1432.

D. Tang, B. Qin, T. Liu. 2015. Document modeling with gated recurrent neural network for sentiment classification. In Proceedings of the 2015 conference on empirical methods in natural language processing. 1422–1432.

D. Tang, F. Wei, N. Yang, M. Zhou, T. Liu, B. Qin. 2014. Learning sentiment-specific word embedding for twitter sentiment classification. In Proceedings of the Annual Meeting Association Computational Linguistics. 1555–1565.

M. Tang, M. Alazab, Y. Luo. 2016. Exploiting vulnerability disclosures: Statistical framework and case study. In 2016 Cybersecurity and Cyberforensics Conference (CCC). 117–122. Doi: 10.1109/CCC.2016.10.

Y. W. Teh, M. I. Jordan, M. J. Beal, D. M. Blei. 2006. Hierarchical Dirichlet processes. J. Am. Stat. Assoc., 101(476), 1566–1581.

tf-coreml: TensorFlow to CoreML Converter. https://github.com/tf-coreml/tf-coreml. Accessed October 3, 2020.

B.-T. Thouraya, B.-B. Lydia, R. Stuart H, (Editors). 2019. Theory and Application of Reuse, Integration, and Data Science. Springer, Switzerland AG.

L. Torgo. 2014. Regression datasets. https://www.dcc.fc.up.pt/~ltorgo/Regression/DataSets.html. Accessed on October 4, 2020

K. M. Tran, A. Bisazza, C. Monz. 2016. Recurrent memory networks for language modeling. In NAACL HLT 2016, The 2016 Conference of the North American Chapter of the Association for Computational Linguistics: Human Language Technologies, San Diego California, USA, June 12–17. 321–331. https://arxiv.org/abs/1601.01272. Accessed October 3, 2020.

F. Trautsch, S. Herbold, P. Makedonski, J. Grabowski. 2016. Addressing problems with external validity of repository mining studies through a smart data platform. In Proc. 13th International Conference on Mining Software Repositories. 97–108.

M. Tufano, C. Watson, G. Bavota, M. Di Penta, M. White, D. Poshyvanyk. 2018. Deep learning similarities from different representations of source code. 2018 IEEE/ACM 15th International Conference on Mining Software Repositories (MSR), Gothenburg, 2018. 542–553.

V. Tumuluru, P. Wang, D. Niyato. 2010. A neural network-based spectrum prediction scheme for cognitive radio. In IEEE International Conference on Communications (ICC '10). 1–5.

S. Upadhyay, K. Chang, M. Taddy, A. Kalai, J. Zou. 2017. Beyond bilingual: Multi-sense word embeddings using multilingual context.arXiv Preprint, arXiv: 1706.08160. https://arxiv.org/abs/1706.08160. Accessed October 3, 2020.

B. Urgaonkar, P. Shenoy, A. Chandra, P. Goyal. 2005. Dynamic provisioning of multi-tier internet applications. Second International Conference on Autonomic Computing (ICAC'05), Seattle, WA. 217–228. Doi: 10.1109/ICAC.2005.27.

R. Usha Rani, T. K. R. Krishna Rao, R. Kiran Kumarreddy. 2015. An Efficient machine learning regression model for rainfall prediction. Int. J. Comput. Appl. Technol., Doi: 10.5120/20292-2681.

S. Valverde, R. V. Sol´e. 2007. Hierarchical small worlds in software architecture. https://arxiv.org/abs/cond-mat/0307278. Accessed October 3, 2020.

M. Van Der Schaar, F. Fu. 2009. Spectrum access games and strategic learning in cognitive radio networks for delay-critical applications. Proc. IEEE, 97(4), 720–740.

V. N. Vapnik. 1995. The Nature of Statistical Learning Theory. NewYork, Springer-Verlag.

V. N. Vapnik. 1998. Statistical Learning Theory. New York, Wiley.

R. Vasa, J.-G. Schneider, O. Nierstrasz. 2007. The inevitable stability of software change. In ICSM, 2007. 4–13.

C. Venkatesan, S. D. Raskar, S. S. Tambe, B. D. Kulkarni, R. N. Keshavamurty. 1997. Prediction of all India summer monsoon rainfall using Error-Back-Propagation Neural Networks. Meteorology and Atmospheric Physics. 225–240.

S. Venugopalan, H. Xu, J. Donahue, M. Rohrbach, R. J. Mooney, K. Saenko. 2015. Translating videos to natural languageusing deep recurrent neural networks. In Proceedings of the 2015 Conference of the North American Chapter of the Association for ComputationalLinguistics: Human Language Technologies. 1494–1504.

O. Vinyals, M. Fortunato, N. Jaitly. 2015. Pointer networks. In Proceedings of the Advances Neural Information Processing Systems. 2692–2700.

O. Vinyals, A. Toshev, S. Bengio, D. Erhan. 2015. Show and tell: A neural image caption generator. Proc. IEEE Conf. Comput. Vision Pattern Recognit., 3156–3164.

Vision Solutions. 2014. Assessing the Financial Impact of Downtime: Understand the factors that contribute to the cost of downtime and accurately calculate its total cost in your organization. http://courseware.cutm.ac.in/wp-content/uploads/2020/06/Assessing-the-Financial-Impact-of-Downtime-UK.pdf. Accessed October 3, 2020.

Wadic. https://wadic.net/software-development-methodologies-evolution/. accessed Sep 2020

Y. WahWong, J. Raymond. 2007. Generation by inverting a semantic parser that uses statistical machine translation. In Proceedings of the 2007 Conference of the North American Chapter of the Association for Computational Linguistics: Human Language Technologies. 172–179.

A. Waibel, T. Hanazawa, G. Hinton, K. Shikano, K. J. Lang. 1989. Phoneme recognition using time-delay neural networks. IEEE Trans. Acoust., Speech, Signal Process, 37(3), 328–339.

B. Wang, K. Ray Liu, T. Clancy. 2010. Evolutionary cooperative spectrum sensing game: how to collaborate? IEEE Trans. Commun., 58(3), 890–900.

L. J. Wang, L. Fang, L. Y. Wang, G. Li, B. Xie, Q. YangF. 2011. API Example: An effective web search based usage example recommendation system for Java APIs. In Proceedings of the 26th IEEE/ACM International Conference on Automated Software Engineering. 592–595.

S. Wang, T. Liu, L. Tan. 2016. Automatically learning semantic features for defect prediction. In Proceedings of the 38th International Conferenceon Software Engineering. ACM. 297–308.

S. Wang, T. Liu, L. Tan. 2016b. Automatically learning semantic features for defect prediction. In Proceedings of the International Conference on Software Engineering (ICSE).

W. Wang. 2009. Spectrum sensing for cognitive radio. 2009 Third International Symposium on Intelligent Information Technology Application Workshops, Nanchang. 410–412. Doi: 10.1109/IITAW.2009.49.

X. Wang, G. Lai, C. Liu. 2009. Recovering relationships between documentation and source code based on the characteristics of software engineering.Electronic Notes in Theoretical Computer. Science, 243, 121–137.

V. M. Ware. 2006. Building the Virtualized Enterprise with VMware Infrastructure, Technical Report 2006. https://www.vmware.com/pdf/vmware_infrastructure_wp.pdf.Accessed October 3, 2020.

-H.-H. Wei, M. Li. 2017. Supervised deep features for software functional clone detection by exploiting lexical and syntactical information insource code. In Proceedings of the 26th International Joint Conference on Artificial Intelligence. AAAI Press, 2017. 3034–3040.

W. Weimer, T. Nguyen, C. Le Goues, S. Forrest. 2009. Automatically finding patches using genetic programming. In Proceedings of the 31st International Conference on Software Engineering. IEEE Computer Society. 364–374.

M. Weiser. 1979. Program slicing: formal, psychological, and practical investigations of an automatic program abstraction method. Ph.D. thesis, University of Michigan, Ann Arbor, US.

Weld: A Common Runtime for High Performance Data Analytics. 2017. ShoumikPalkar, James J. Thomas, AnilShanbhag, DeepakNarayanan, HolgerPirk, MalteSchwarzkopf, SamanAmarasinghe, MateiZaharia. https://dawn.cs.stanford.edu/pubs/weld-cidr2017.pdf. Accessed on October 9, 2020.

T.-H. Wen, M. Gasic, N. Mrk˘si˘c, P. H. Su, D. Vandyke, S. Young. 2015. Semantically conditioned lstm-based natural languagegeneration for spoken dialogue systems. In Proceedings of the 2015 Conference on Empirical Methods in Natural Language Processing. 1711–1721.

M. White, M. Tufano, C. Vendome, D. Poshyvanyk. 2016. Deep learning code fragments for code clone detection. In Proceedings of the 31st IEEE/ACM International Conference on Automated Software Engineering. ACM, 2016. 87–98.

M. White, C. Vendome, M. Linares-Vasquez, D. Poshyvanyk. 2015. Toward Deep Learning Software Repositories. 2015 IEEE/ACM 12thWorking Conference on Mining Software Repositories. 334–345.

M. Wieloch, S. Amornborvornwong, J. Cleland-Huang. 2013. Trace-by-classification: A machine learning approach to generate trace links for frequently occurring software artifacts. 7th International Workshop on Traceability in Emerging Forms of Software Engineering (TEFSE), San Francisco, CA. 110–114. Doi: 10.1109/TEFSE.2013.6620165.

Y. W. Wong, R. J. Mooney. 2007. Generation by inverting a semantic parser that uses statistical machine translation. In Proceedings of the2007 Conference of the North American Chapter of the Association for Computational Linguistics: Human Language Technologies. 172–179.

C. L. Wu, K. W. Chau, C. Fan. 2010. Prediction of rainfall time series using Modular Artificial Neural Networks coupled with datapreprocessing techniques. J. Hydrol., 389(1), 146–167.

R. Wu, H. Zhang, S. Kim, S.-C. Cheung. 2011. Relink: recovering links between bugs and changes. In Proceedings of the 19th ACM SIGSOFT symposium and the 13th European conference on Foundations of software engineering. ACM. 15–25. https://doi.org/10.1145/2025113.2025120

Xcode. https://developer.apple.com/xcode. Accessed October 3, 2020.

J. Xie, B. Xu, Z. Chuang. 2013. Horizontal and vertical ensemble with deep representation for classification.arXiv preprint arXiv:1306.2759.

Y. Xing, R. Chandramouli. 2008. Human behavior inspired cognitive radio network design. IEEE Commun. Mag., 46(12), 122–127.

B. Xu, D. Ye, Z. Xing, X. Xia, G. Chen, S. Li. 2016. Predicting semantically linkable knowledge in developer online forums via convolutional neural network. In Proceedings of the 31st IEEE/ACM International Conference on Automated Software Engineering. ACM. 51–62.

K. Xu, J. Ba, R. Kiros, K. Cho, A. Courville, R. Salakhudinov, R. Zemel, Y. Bengio. 2015. Show, attend and tell: Neural image caption generation with visual attention. In Proc. Int. Conf. Machine Learning. 2048–2057.

A. Yau, P. Komisarczuk, P. D. Teal. 2010. Applications of reinforcement learning to cognitive radio networks. In IEEE International Conference on Communications Workshops (ICC), 2010, Cape Town, South Africa.1–6.

K.-L. A. Yau, P. Komisarczuk, P. D. Teal. 2010. Applications of reinforcement learning to cognitive radio networks. In IEEE International Conference on Communications Workshops (ICC), 2010, Cape Town, South Africa.1–6.

X. Ye, R. Bunescu, C. Liu. 2014. Learning to rank relevant files for bug reports using domain knowledge. In Proc. 22nd ACM SIGSOFT International Symposium on Foundations of Software Engineering. 689–699.

X. Ye, H. Shen, X. Ma, R. Bunescu, C. Liu. 2016. From word embeddings to document similarities for improved information retrieval in software engineering. 2016 IEEE/ACM 38th International Conference on Software Engineering (ICSE), Austin, TX. 404–415. Doi: 10.1145/2884781.2884862.

J. Yi, F. Maghoul. 2009. Query clustering using click-through graph. In WWW. 1055–1056. https://doi.org/10.1145/1526709.1526853

W. Yih, X. He, C. Meek. 2014. Semantic parsing for single-relation question answering. In Proc. Annu. Meeting Association Computational Linguistics. 643–648.

C. Yilmaz, A. Paradkar, C. Williams. 2008. Time will tell: fault localization using time spectra. In Proceedings of International Conference on Software Engineering. 81–90. Leipzig, Germany.

J. Yosinski, J. Clune, Y. Bengio, H. Lipson. 2014. How transferable are featuring in deep neural networks? http://arxiv.org/abs/1411.1792. Accessed October 3, 2020.

T. Young, D. Hazarika, S. Poria, E. Cambria. 2018. Recent Trends in Deep Learning Based Natural Language Processing [Review Article]. In IEEE Commun. Intell. Mag., 13(3), 55–75. Doi: 10.1109/MCI.2018.2840738.

T. Yucek, H. Arslan. 2009. A survey of spectrum sensing algorithms for cognitive radio applications. In IEEE Commun. Surv. Tutorials, 11(1), 116–130. Doi: 10.1109/SURV.2009.090109.K.-L.

W. Zaremba, I. Sutskever. 2014. Learning to execute.arXiv preprint arXiv:1410.4615.

W. Zaremba, I. Sutskever. 2014. Learning to execute. CoRR,abs/1410.4615. https://arxiv.org/abs/1410.4615. Accessed on October 6, 2020.

C. S. Zegedy, V. Vanhoucke, S. Ioffe, J. Shlens, Z. Wojna. 2015. Rethinking the Inception Architecture for Computer Vision. http://arxiv.org/abs/1512.00567. Accessed October 3, 2020.

H. Zhang, L. Gong, S. Versteeg. 2013. Predicting bug-fixing time: an empirical study of commercial software projects. In 35th InternationalConference on Software Engineering, ICSE '13, San Francisco, CA, USA, May 18–26, 2013. 1042–1051. Doi: 10.1109/ICSE.2013.6606654.

J. Zhang, X. Wang, H. Zhang, H. Sun, K. Wang, X. Liu. 2019. A novel neural source code representation based on abstract syntax tree. 2019 IEEE/ACM 41st International Conference on Software Engineering (ICSE), Montreal, QC, Canada. 783–794. Doi: 10.1109/ICSE.2019.00086.

X. Zhang, M. Lapata. 2014. Chinese poetry generation with recurrent neural networks. In Proceedings of the 2014 Conference onEmpirical Methods in Natural Language Processing (EMNLP). 670–680.

G. Zhao, J. Huang. 2018. DeepSim: Deep learning code functional similarity. In Proceedings of the 2018 26th ACM Joint Meeting on European Software Engineering Conference and Symposium on the Foundations of Software Engineering, ser. ESEC/FSE 2018. New York, NY, USA: ACM. 141–151. http://doi.acm.org/10.1145/3236024.3236068. Accessed on October 5, 2020

A. Zheng, M. Jordan, B. Liblit, A. Aiken. 2003. Statistical Debugging of Sampled Programs. ACM, New York, USA.

A. Zheng, M. Jordan, B. Liblit, M. Naik, A. Aiken. 2006. Statistical Debugging: Simultaneous Identification of Multiple Bugs. ICML '06: Proceedings of the 23rd international conference on Machine learning. ACM. New York, USA.

X. Zheng, H. Chen, T. Xu. 2013. Deep learning for Chinese word segmentation and pos tagging. In Proc. Conf. Empirical Methods Natural Language Processing. 647–657.

J. Zhou, H. Zhang, D. Lo. 2012. Where should the bugs be fixed? more accurate information retrieval-based bug localization based on bugreports. In 2012 34th International Conference on Software Engineering (ICSE), June 2012. 14–24.

X. Zhu, P. Sobhani, H. Guo. 2015. Long short-term memory over recursive structures. In Proceedings of the 32nd International Conference on International Conference on Machine Learning. 37. 1604–1612. http://dl.acm.org/citation.cfm?id=3045118.3045289. Accessed on October 4, 2020

Z. H. Zhu, T. Basar. 2010. No-regret learning in collaborative spectrum sensing with malicious nodes. In IEEE International Conference on Communications (ICC '10), Cape Town, South Africa. 1–6.

Index

www.ingramcontent.com/pod-product-compliance
Lightning Source LLC
Chambersburg PA
CBHW080906220326
41598CB00034B/5488